T0156096

UNITEXT for Physics

Series editors

Michele Cini, Roma, Italy
Attilio Ferrari, Torino, Italy
Stefano Forte, Milano, Italy
Guido Montagna, Pavia, Italy
Oreste Nicrosini, Pavia, Italy
Luca Peliti, Napoli, Italy
Alberto Rotondi, Pavia, Italy

More information about this series at http://www.springer.com/series/13351

Luigi E. Picasso

Lectures in Quantum Mechanics

A Two-Term Course

 Springer

Luigi E. Picasso
Dipartimento di Fisica
Pisa
Italy

In collaboration with Emilio d'Emilio

ISSN 2198-7882   ISSN 2198-7890 (electronic)
UNITEXT for Physics
ISBN 978-3-319-37386-7   ISBN 978-3-319-22632-3 (eBook)
DOI 10.1007/978-3-319-22632-3

Springer Cham Heidelberg New York Dordrecht London
© Springer International Publishing Switzerland 2016
Softcover re-print of the Hardcover 1st edition 2016
This work is subject to copyright. All rights are reserved by the Publisher, whether the whole or part
of the material is concerned, specifically the rights of translation, reprinting, reuse of illustrations,
recitation, broadcasting, reproduction on microfilms or in any other physical way, and transmission
or information storage and retrieval, electronic adaptation, computer software, or by similar or dissimilar
methodology now known or hereafter developed.
The use of general descriptive names, registered names, trademarks, service marks, etc. in this
publication does not imply, even in the absence of a specific statement, that such names are exempt from
the relevant protective laws and regulations and therefore free for general use.
The publisher, the authors and the editors are safe to assume that the advice and information in this
book are believed to be true and accurate at the date of publication. Neither the publisher nor the
authors or the editors give a warranty, express or implied, with respect to the material contained herein or
for any errors or omissions that may have been made.

Printed on acid-free paper

Springer International Publishing AG Switzerland is part of Springer Science+Business Media
(www.springer.com)

Preface

This book is intended for a course on nonrelativistic quantum mechanics for physics students: it is the English version (due to my friend and colleague Emilio d'Emilio) of a book based on the lectures I gave to the students of the University of Pisa. The encouraging and long-lasting acceptance of the Italian version led d'Emilio to consider the opportunity of the English version. He eventually convinced me to undertake the task.

In the present version there are a few additions with respect to the original Italian edition: for instance, a chapter introductive to the theory of collisions and a section on Bell's inequalities. It certainly covers more than the program of a two-term course, thus giving the lecturer some possibility of choice.

Although oversized for a two-term course, the book skips some important topics: for instance, the WKB approximation and the Stern–Gerlach experiment. The last subject is presented in the form of a problem in the book written with E. d'Emilio: "Problems in Quantum Mechanics (with solutions)" (Springer 2011) that is to be considered as a complementary tool for the learning of the subject: for quantum mechanics (much as for many other subjects, but maybe a little more) a step-by-step painstaking verification of the understanding of the new ideas is advisable and, for this purpose, problem solving is – in my opinion – a particularly effective mean.

There are many excellent books more complete than the present one and I do not intend to compete with those: when first writing this book (the Italian version) the objective I had in mind was to help students in their first approach to this beautiful, although not easy, subject: I did not underestimate the difficulties a student may encounter, after two years profitably spent learning classical physics, in accepting that a particle (a photon, a neutron, an electron ...) does not follow a well definite path but two, or even more, at the same time (the which-way problem).

For this reason, after a reasonable space dedicated to the issues that determined the crisis of classical physics and the Einstein–Bohr–de Broglie ideas that were seminal for the birth of quantum mechanics, the principles of quantum mechanics are presented in an inductive way through real experiments (as those of neutron interferometry), and their abstract formulation is given a 'psychological preparation'. After that, the student is taken step by step from the very birth of quantum mechanics to the more affordable applications of the theory, to end up with the still alive and more intriguing aspects connected with the paradoxes of quantum mechanics and a mention to the Bell inequalities and Aspect's experiments.

The collaboration of Emilio d'Emilio I took advantage of in writing this book must be acknowledged: not only he translated the Italian version, but from the very beginning of my task he assisted me day after day with advices and suggestions, outcome of his everyday experience with students undertaking the study of this subject. It is an understatement to assert that this book would had never been written without his collaboration.

Neither would had been written if, long time ago, I did not benefit of such great masters as Franco Bassani, Elio Fabri and Luigi A. Radicati.

Pisa, June 2015

Contents

Chapter 1

The Crisis of Classical Physics

1.1 Atomic Models

Let us shortly summarize the development of our knowledge about the structure of the atom, taking as starting point the notions that were firmly acquired by the end of the 19-th century.

- The existence of the atoms was introduced as a hypothesis suitable to explain two fundamental laws in chemistry: the law of definite proportions and that of multiple proportions.
- As a consequence, to any single chemical element there corresponds a single type of atom.
- Electrolysis, thermionic effect, photoelectric effect, electric conduction in gases and in metals, the fact that atoms can absorb and emit radiation, namely electromagnetic waves, etc. – all these facts hint at the necessity of admitting that atoms, that per se are neutral, contain both positive and negative charges.
- The negative charges contained in atoms are corpuscles all equal to one another called *electrons*. This is proven by the fact that measurements of mass and charge performed on the negative charges extracted by different methods (thermionic effect, photoelectric effect etc.) and from the most diverse atoms, always provide the same result.

The ratio e/m_e (charge/mass of the electron) was measured by J.J. Thomson (electric and magnetic field spectrometer); later R.A. Millikan measured e. We report here the numerical values of m_e and e:

$$m_e = 0.9 \times 10^{-27}\, \text{g} \simeq 10^{-27}\, \text{g}$$

$$e = 1.6 \times 10^{-19}\, \text{C} = 4.8 \times 10^{-10}\, \text{esu (electrostatic units)}$$

(by e we shall always denote the absolute value of the electron charge).

We recall here the definition of an energy unit that is extremely significant in atomic physics, the electronvolt (eV): $1\,\text{eV}$ equals the kinetic energy an

© Springer International Publishing Switzerland 2016
L.E. Picasso, *Lectures in Quantum Mechanics*, UNITEXT for Physics,
DOI 10.1007/978-3-319-22632-3_1

electron gains when it goes from a point to another point, the difference in electric potential between the two points being 1 volt. As a consequence:

$$1\,\text{eV} = 1.6 \times 10^{-19}\,\text{C} \times 1\,\text{V} = 1.6 \times 10^{-19}\,\text{J} = 1.6 \times 10^{-12}\,\text{erg}\,.$$

Since a gram-atom of any element consists of $N_A = 6.02 \times 10^{23}$ atoms (N_A is Avogadro's number), one can extract the mass of a single atom. For example, for hydrogen H, whose atomic weight is $A = 1$, one obtains:

$$M_H = N_A^{-1}\,\text{g} = 1.7 \times 10^{-24}\,\text{g}\,, \qquad M_H = 1836\,m_e\,.$$

This teaches us that the contribution of the electrons to the mass of an atom is negligible. We can also estimate the size of an atom: let us take a solid of some monatomic substance, e.g. gold, Au. We know that its atomic weight is $A = 197$, which amounts to say that in 197 g of Au there are N_A atoms; gold density is about 19 g/cm^3, so a gram-atom occupies 10 cm^3. It follows that the volume per atom is about $10/N_A \simeq 17 \times 10^{-24}$ cm^3. Since we are dealing with a solid substance, the atoms are very close to each other so that this can be identified, to a good approximation, with the volume of one atom. Therefore the linear size of an atom is about 10^{-8} cm. This quantity, of common use in atomic physics, takes its name after A.J. Ångström:

$$1\,\text{Å} = 10^{-8}\,\text{cm}\,.$$

Let us summarize: any atom is made out of a positive charge that carries practically all the mass ($10^{-24} \div 10^{-23}$ g): this charge is balanced by a certain number of electrons, each of them endowed with charge -4.8×10^{-10} esu and mass 0.9×10^{-27} g. The linear atomic size is about 1 Å.
One has now to establish in which way the electrons and the positive charge are arranged within the atom – in other words one has to build up a 'model'.

The first somewhat successful model was proposed by J.J. Thomson in the beginning of the 20-th century and monopolized the attention of the community of physicists for about ten years. Thomson suggested that the atom consisted of a sphere of uniformly distributed positive charge (whose linear size is about 1 Å) within which are the electrons – taken as pointlike corpuscles – in a such a number as to make the whole atom neutral (the number of the electrons in an atom is called *atomic number* and is denoted by Z). When the system – i.e. the atom – is in equilibrium, all the electrons should occupy a position where the attractive force toward the center is balanced by the repulsive forces among them.

Thomson was able to explain several phenomena by means of this model. For example, the absorption and emission of electromagnetic radiation was attributed to the motion of the electrons around their equilibrium positions. Likewise other phenomena were explained, at least qualitatively, by this model: it had been shown that the conditions for electrostatic equilibrium imply that the electrons arrange themselves at the vertices of regular concentric polygons, each of them with no more than eight vertices – a fact that suggested the possibility of explaining the regularity of the periodic table of the elements.

Let us consider for example the hydrogen atom ($Z = 1$). The potential due to the distribution of positive charge is (a stands for the radius of the sphere within which the charge is distributed):

$$\varphi(r) = \frac{3}{2}\frac{e}{a} - \frac{1}{2}\frac{e\,r^2}{a^3} . \tag{1.1}$$

The equilibrium position of the electron is in the center of the sphere and the ionization energy of the atom – i.e. the energy that is necessary to take the electron at infinity with vanishing kinetic energy – is

$$E_{\mathrm{i}} = -e\left(\varphi(\infty) - \varphi(0)\right) = \frac{3}{2}\frac{e^2}{a} . \tag{1.2}$$

Since it is known from experiments that $E_{\mathrm{i}} = 13.6\,\mathrm{eV}$, it follows that $a \simeq 1.6 \times 10^{-8}\,\mathrm{cm}$. The motion of the electron within the charge distribution is harmonic with frequency:

$$\nu = \frac{1}{2\pi}\sqrt{\frac{e^2}{m_{\mathrm{e}}\,a^3}} = 1.2 \times 10^{15}\,\mathrm{s}^{-1} \tag{1.3}$$

therefore the atom should emit electromagnetic radiation with a wavelength $\lambda = c/\nu = 2400\,\text{Å}$.

Another atomic model that, as the previous one, provided an explanation for quite a lot of experimental facts, was due to E. Rutherford (indeed Rutherford was not the first to propose the model that now brings his name). Rutherford suggested that the positive charge was concentrated in a 'kernel' or *nucleus* whose size is much smaller than that of the atom. The nucleus should carry a charge equal in magnitude, but opposite in sign, to the sum of the charges of all the electrons in the atom. It should also carry a mass 'practically' equal to the mass of the atom. According to this model, the electrons 'orbit' around the nucleus at an average distance of about $10^{-8}\,\mathrm{cm}$.

The atom is therefore looked at as a small planetary system – the substantial difference being that the forces among electrons are repulsive whereas those among planets are attractive. In addition the repulsive electrostatic forces are of the same order of magnitude as the electrostatic attraction by the nucleus, while in the planetary system the attraction exerted by the Sun is considerably stronger than the attractions among planets.

Let us again consider the hydrogen atom, but now according to the Rutherford model. Let us assume that the electron goes a circular orbit of radius a. In this case the (kinetic+potential) energy is

$$E = \frac{1}{2}m_{\mathrm{e}}\,v^2 - \frac{e^2}{a} = -\frac{1}{2}\frac{e^2}{a} \tag{1.4}$$

whence:

$$E_I = -E = 13.6\,\mathrm{eV} \qquad \Rightarrow \qquad a = 0.53\,\text{Å} . \qquad (1.5)$$

The revolution frequency of the electron is

$$\nu = \frac{1}{2\pi} \sqrt{\frac{e^2}{m_e\,a^3}} = 16.6 \times 10^{15}\,\mathrm{s}^{-1} \qquad \Rightarrow \qquad \lambda = 455\,\text{Å} . \qquad (1.6)$$

In the days of Thomson and Rutherford the question was: which of the two models is 'right'? It is important to remark that the word 'right' has – in physics – no absolute signification: one can only ask which of the two models is more suitable to explain the experimental data. This amounts to say that, if one wants to choose between the two models, "Nature must be questioned" by means of an experiment whose result can be understood in terms of only one of the two models.

One of the merits of Rutherford was to understand how crucial was, in this respect, the role of the experiments that H. Geiger and E. Marsden had been performing already since 1909. Geiger and Marsden sent α particles produced by some radioactive substance (α-particles are particles endowed with charge $+2e$ and mass four times that of the H atom: today we know they are nuclei of He^4, atoms with $Z = 2$ and $A = 4$) against a very thin (thickness $\simeq 10^{-4}$ cm) golden foil ($Z = 79$) and observed how the α particles were scattered off their original trajectory. Now, on the one hand, it is obvious that the α particles, being subject to the electrostatic forces due to the presence of charges in the atoms, should somehow be scattered; on the other hand, what called both Geiger's and Rutherford's attention was the fact that, on the average, one α particle over 10^4 was scattered at angles greater than 90°: this fact was in contrast with Thomson model. Let us give a qualitative explanation of this fact. In the Thomson model the electric field attains its maximum e/a^2 at the boundary of the atom and decreases linearly down to 0 as the distance from the center decreases down to 0. In the Rutherford model, on the contrary, since the nuclear charge is considered pointlike (i.e. much smaller than the atomic size), the electric field goes like $1/r^2$ and there is no limit on the intensity of the force the α particle may be subject to, provided it passes close enough to the nucleus. In the experiment the α particles had a somewhat high speed $v \simeq 10^9$ cm/s (i.e. kinetic energy $E_k \simeq 10^6 \div 10^7$ eV $\equiv 1 \div 10$ MeV) so that, in order to produce a remarkable deviation, a remarkable force – i.e. a very strong electrostatic field – was needed. Let us note, in particular, that for a particle, shot with sufficient kinetic energy (i.e. exceeding $2 \times \frac{3}{2}(Z\,e^2/a) \simeq 3.2 \times 10^3$ eV $= 3.2$ keV) head-on the center of the atom, there is no deflection according to the Thomson model; on the contrary, according to the Rutherford model, the same particle is deflected even backwards.

Of course the analysis of Geiger–Marsden experiments is rather complex: for example, one must show that the observed large scattering angles are not due to multiple collisions – what Rutherford did by introducing, perhaps for the first time, statistical methods in the interpretation of experimental data.

To summarize, the situation was the following: α particles scattered at large angles and with a non-negligible frequency were observed; such large deviations were necessarily produced in single collisions and only the Rutherford model was able to provide a reasonable explanation for this fact. One was then entitled to assume the Rutherford model as the 'right' one.

1.2 The Problems of Atomic Size and Radiative Collapse

In the very moment Thomson model was abandoned by the scientific community in favour of Rutherford's, several difficulties stemmed from the latter model that classical physics – the physics that was consolidated by the beginning of the 20-th century – could not explain.

The first problem was that relative to the *stability of the sizes* of atoms: even by 1910 it was known that all atoms, even relative to different elements, had sizes of the order of 10^{-8} cm . From a classical point of view one knows that the size of the orbit of an electron (think of the simplest case of the H atom) does depend on the energy and cannot, therefore, be determined without knowing the value of the latter: no reason can be envisaged to maintain that all the H atoms in different conditions and obtained by the most diverse methods (e.g by dissociating molecules of H_2, or molecules of H_2O, etc.) should have about the same energy. Furthermore, even if it were so, *which* energy should they have? The same argument can be extended to molecules and lattice spacings in crystals: it should be evident that there must exist a fundamental quantity with dimensions of a length and of the order of magnitude of some ångström.

From a formal point of view, in the equations that rule the motion of the electrons around the nucleus (we mean Newton's $\vec{F} = m\,\vec{a}$ in the non-relativistic approximation) only the electron charge e and mass m_e enter as parameters: the solution of the problem may be very complicated, but if – regardless of the initial conditions – there must be some 'natural' size of the system, this may only depend on e and m_e. The point is that no length can be constructed by only e and m_e. If one takes into account relativistic corrections (but in light atoms the velocity of electrons is about $1/100$ of the velocity of light c), then in the equations of motions also c intervenes and it is possible to construct a quantity with the dimensions of a length, the so called *classical radius of the electron*:

$$r_e = \frac{e^2}{m_e\,c^2} \simeq 2.8 \times 10^{-13}\,\mathrm{cm}$$

which simply is too small to be useful to explain a size that is five orders of magnitude larger.

It should be noted that, from this point of view, Thomson atom has no such problem, since the atomic size is the same as the size of the positive charge distribution, and the latter, *by assumption*, is of the order of 10^{-8} cm, practically the same for all atoms. One could even say that Thomson model was made on purpose to 'explain' the stability of atomic sizes.

The second problem was that of *radiative collapse*: it shows up when one tries to treat the emission of electromagnetic radiation by the atom in the framework of classical electrodynamics. It is known that an accelerated charge radiates, i.e. it emits electromagnetic waves thus loosing energy. This should be so also for the electrons of an atom: they should loose energy down to a complete stop, but – inasmuch as in the Rutherford model the only stable equilibrium position is that in which all the electrons stick to the nucleus – all the electrons should fall on the nucleus and the stable size of the atom would be that of the nucleus, namely about 10^{-13} cm.

This would not be a real difficulty, provided the time needed to the atom to reach this condition is very long (e.g. several billions of years): it would mean we are observing atoms that did not yet appreciably suffer from the 'collapse' and their sizes would be 'practically' stable.

Let us then try to estimate the time of collapse in the simplest case of the H atom. We shall make two assumptions useful to simplify the calculation, by means of which we only want to obtain an order of magnitude.

1. The initial conditions are such that, in absence of radiation, the electron would move on a circular (instead of elliptic) orbit.
2. Due to energy loss by radiation, the electron will go a spiral: in order to perform the calculation we shall approximate at any point the orbit of the electron with a circle.

The second approximation is legitimate if the energy lost by the electron in a turn is much less then the energy it possesses. The power W radiated by an accelerated charge e (in this case an electron) is

$$W = \frac{2}{3}\frac{e^2}{c^3}|\vec{a}|^2 = -\frac{dE}{dt} \tag{1.7}$$

where E is the energy of the atom, and since we approximate the orbit as a circle:

$$E = -\frac{1}{2}\frac{e^2}{r}, \qquad \frac{dE}{dt} = \frac{e^2}{2r^2}\frac{dr}{dt}. \tag{1.8}$$

Since $|\vec{a}| = e^2/m_e r^2$, by (1.7) and (1.8) one has:

$$\frac{2}{3}\frac{e^2}{c^3}\frac{e^4}{m_e^2 r^4} = -\frac{e^2}{2r^2}\frac{dr}{dt}, \tag{1.9}$$

whence:

$$dt = -\frac{3}{4}\frac{m_e^2 c^3}{e^4}r^2\, dr. \tag{1.10}$$

The time of collapse τ is obtained by integrating (1.10) from R (initial radius of the orbit) to r_0 (nuclear radius); one obtains:

$$-\frac{3}{4}\frac{m_e^2 c^3}{e^4}\int_R^{r_0} r^2\, dr = \int_0^\tau dt \quad \Rightarrow \quad \tau \simeq \frac{1}{4}\frac{m_e^2 c^3}{e^4}R^3 \quad (R \gg r_0). \tag{1.11}$$

Upon inserting numerical values into (1.11) (with $R \simeq 10^{-8}$ cm) one finds $\tau \simeq 10^{-10}$ s. The time of collapse of an atom is therefore extremely short and, in addition, in this short time interval the atom should radiate an energy of the order of $e^2/r_0 \simeq 1\,\mathrm{MeV}$!

A further problem, not independent of the previous one, is that of the emission of light by atoms and of the spectral features of the emitted light: if the atoms of a gas are excited in some way (e.g. by means of collisions), each of them emits radiation on some discrete and precise frequencies (**spectral lines**) that are characteristic of that particular atom (or molecule) one is examining and are independent of the treatment the gas has undergone. The set of such frequencies constitutes the **spectrum** of that atom and it is a feature of the type of atom (hydrogen, helium, ...) one is considering: a kind of 'barcode' of the atom that enables us to reveal its presence in a substance by means of spectroscopic analysis. The case of helium (He) is emblematic: in 1868, during an eclipse, P.J.C. Jenssen, while investigating the radiation emitted by the solar crown, spotted a spectral line that could not be attributed to any of the known atoms (helium, quite abundant in the Universe, is rather rare on the Earth). He was so led to conjecture the existence of a new element on the Sun (Helios, in ancient Greek), whence its name.

In the framework of classical electrodynamics, the emission of waves of a given frequency requires that the electric dipole moment of the system that is emitting varies with time according to a harmonic law. Now an atom has a finite number of degrees of freedom, so its possible 'normal modes' (those motions of the electrons such that the atomic electric dipole moment does vary with harmonic law) also are in finite number. For light atoms such as hydrogen, helium, lithium etc. their number is of the order of unity, whereas also the spectra of these atoms (as well as those of heavier atoms) are very rich, i.e. they contain a large (possibly infinite) number of spectral lines. Furthermore, the frequencies of the spectral lines are not in simple numerical ratios among them, as it is for a fundamental frequency and its harmonics (which, in the emission of an oscillating dipole, are always there – even if with intensities much weaker than that of the fundamental line): the spectral lines of any atom are distributed in such a complicated way that no explanation has been found within classical electrodynamics.

1.3 Difficulties Related with Heat Capacities

The kinetic theory of gases makes a well defined prediction about the values of molar heat capacities at constant volume C_V for all gaseous substances.

For example, for monatomic gases (He, Ne, Ar, ...) $C_V = \frac{3}{2}R$, where $R = 1.93\,\mathrm{cal/mol\,K}$ is the ideal gas constant, for diatomic gases (H_2, N_2, O_2, $\cdots$) $C_V = \frac{5}{2}R$. In addition, for monatomic crystalline solids (as e.g. metals), $C_V = 3R$ ($\simeq 6\,\mathrm{cal/mol\,K}$: Dulong–Petit law).

So, according to classical physics, molar heat capacities do not depend on the temperature T. Before examining how the above results are derived from classical statistical mechanics, let us have a glance at the situation from the experimental point of view.

1. For monatomic gases the agreement between the theoretical prediction $\frac{3}{2}R$ and the experimental situation is very good practically for all temperatures, provided the conditions are such that the gas can be considered an ideal gas.

2. For diatomic gases, always in the conditions of ideal gases, the agreement is good for a wide range of temperatures around room temperature, let us say for T from $10\,\mathrm{K}$ to $500\,\mathrm{K}$. For low temperatures C_V decreases with decreasing T and tends to the value $\frac{3}{2}R$. For high temperatures C_V increases with increasing T. The dependence of C_V on the temperature is particularly noticeable in the case of H_2, where $C_V = \frac{5}{2}R$ only for $T \simeq 300\,\mathrm{K}$.

3. For metals the experimental situation is less favourable: only in a few cases $C_V = 3R$ at room temperature and, as a rule, the experimental value is smaller than the theoretical one. In Fig. 1.1 the experimental data relative to lead, copper and diamond (carbon) are reported. Note that all $C_V \to 0$ for $T \to 0$.

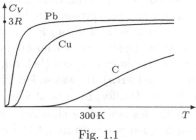

Fig. 1.1

Let us now describe how classical physics is able to make a prediction for the C_V of the several substances discussed above. By definition:

$$C_V = \left(\frac{\partial \mathcal{U}}{\partial T}\right)_V \qquad (1.12)$$

where $\mathcal{U}$ is the internal energy of a gram-molecule (or gram-atom) of the substance one is examining. It is necessary to know $\mathcal{U}$, which is provided by statistical mechanics. Let us consider a thermodynamic system, for example a gas, consisting of N identical subsystems (the molecules of the gas) that interact weakly with one another, i.e. let us suppose that the gas is almost ideal. Let f be the number of degrees of freedom of each molecule, $q_1, \ldots, q_f$ a set of Lagrangian coordinates for the single molecule, $p_1, \ldots, p_f$ the set of corresponding canonically conjugate momenta. If the system is in the condition of thermodynamic equilibrium at a given temperature T, the number $dn(q,p)$ (better: the most probable number) of molecules whose coordinates and momenta respectively have values between q_1 and $q_1 + dq_1, \ldots, q_f$ and $q_f + dq_f$; p_1 and $p_1 + dp_1, \ldots, p_f$ and $p_f + dp_f$ is given, according to classical statistical mechanics, by the Maxwell-Boltzmann distribution:

$$dn(q,p) = B\,e^{-\beta E(q,p)}\,dq\,dp\,, \qquad dq\,dp \equiv dq_1 \cdots dq_f\,dp_1 \cdots dp_f \qquad (1.13)$$

where $E(q,p)$ is the energy of a single molecule whose coordinates and momenta are $q_1, \ldots, q_f$ and $p_1, \ldots, p_f$ (i.e. $E(q,p)$ is the Hamiltonian of the molecule), $\beta = 1/k_B T$, $k_B = 1.38 \times 10^{-16}\,\mathrm{erg/K}$ is the Boltzmann constant and T is the absolute temperature. B is a constant that is determined by requiring:

$$\int dn(q,p) = N \quad \Rightarrow \quad B = \frac{N}{\int e^{-\beta E}\, dq\, dp} \,. \tag{1.14}$$

From (1.13) it immediately follows that

$$\mathcal{U} = \int E(q,p)\, dn(q,p) = N\, \frac{\int E(q,p)\, e^{-\beta E}\, dq\, dp}{\int e^{-\beta E}\, dq\, dp} \,. \tag{1.15}$$

Note that the numerator in (1.15) is, apart from its sign, the derivative of the denominator with respect to β. By defining the **partition function**:

$$Z(\beta) \overset{\text{def}}{=} \int e^{-\beta E}\, dq\, dp \tag{1.16}$$

one can write:

$$\mathcal{U} = -N\, \frac{\partial \ln Z(\beta)}{\partial \beta} \,. \tag{1.17}$$

Let us now assume that the energy E of the subsystems be a positive quadratic form of its arguments:

$$E(q,p) = \sum_{i=1}^{f} (a_i\, p_i^2 + b_i\, q_i^2) \,, \qquad a_i,\, b_i \geq 0 \tag{1.18}$$

(in the case of a gas the quadratic terms in the coordinates correspond to the potential energy of the normal modes of vibration of the molecule about the equilibrium configuration and the number of $b_i \neq 0$ is less than f). Then:

$$\mathcal{U} = N \times \frac{1}{2} k_B T \times \mathsf{v} \quad \Rightarrow \quad C_V = n\, \frac{\mathsf{v}}{2}\, R\,, \qquad n \equiv \frac{N}{N_A} \tag{1.19}$$

where v it the total number of nonvanishing a_i and b_i ($f \leq \mathsf{v} \leq 2f$).

This result, that we are now going to demonstrate, is known as

Theorem of Equipartition of Energy: *each quadratic term in the Hamiltonian of the whole system provides a contribution $\frac{1}{2} k_B T$ to the internal energy.*

If the subsystems are weakly interacting, the Hamiltonian of the whole system is approximated with the sum of the Hamiltonians of the single subsystems. Then, by (1.16) and (1.18), one has:

$$Z(\beta) = \int \exp\left(-\beta \sum_{1}^{f} (a_i\, p_i^2 + b_i\, q_i^2)\right) dq_1 \cdots dp_f \,. \tag{1.20}$$

Restricting to the variables q_i and p_i that appear with nonvanishing coefficients in (1.18), let us perform the change of variables $q_i' = \sqrt{\beta}\, q_i,\ p_i' = \sqrt{\beta}\, p_i$ that makes the dependence of the integrand on β disappear and brings a factor $\beta^{-\mathsf{v}/2}$. Thanks to the fact that the integrand is Gaussian with respect to each of these variables, the limits of integration can in any event be taken from $-\infty$ to $+\infty$ and so are not modified by the change of variables. The integration over the coordinates that do not appear in (1.18) (e.g. the coordinates of the center-of-mass of the molecule) gives a constant factor (also: finite, since the volume available for the system is finite), so in conclusion:

$$Z(\beta) \propto \beta^{-\nu/2} \quad \Rightarrow \quad \mathcal{U} = N \times \frac{\nu}{2\beta} = N\frac{\nu}{2}k_B T = n\frac{\nu}{2}RT . \tag{1.21}$$

The proof can be extended to the case where the a_i depend on the q's, as it happens when some of the q's are not Cartesian coordinates.

In particular, since the kinetic energy of a system always is a quadratic form of the p's, one may say that the average *kinetic* energy associated to each degree of freedom of the system is $\frac{1}{2}k_B T$.

Applications

Monatomic gas. The atoms are schematized as pointlike (i.e. their internal structure is ignored); for a mole:

$$f = \nu = 3, \quad E = \frac{1}{2m}(p_1^2 + p_2^2 + p_3^2) \quad \Rightarrow \quad \mathcal{U} = \frac{3}{2}RT, \quad C_V = \frac{3}{2}R .$$

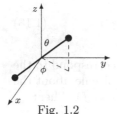

Fig. 1.2

Diatomic gas. The molecule is schematized as two (pointlike) atoms at a *fixed* distance, then $f = 5$ (three translational and two rotational degrees of freedom):

$$E = \frac{1}{2M}(P_1^2 + P_2^2 + P_3^2) + \frac{1}{2I}\left(p_\theta^2 + \frac{p_\phi^2}{\sin^2\theta}\right) \tag{1.22}$$

where M is the total mass of the molecule, θ and ϕ are the polar angles of Fig. 1.2 and I is the moment of inertia with respect to an axis passing through the center of mass and orthogonal to the axis of the molecule ($p_\theta^2 + p_\phi^2/\sin^2\theta = \vec{L}^2$, the square of the angular momentum). Note that in this case the coefficient of p_ϕ does depend on θ. Therefore:

$$\mathcal{U} = \frac{5}{2}RT, \qquad\qquad C_V = \frac{5}{2}R .$$

Solids. Most solids consist of several micro-crystals randomly oriented. In a crystal the ions, due to their mutual interactions, are arranged in a regular manner at the vertices of a lattice and, as long as the amplitude of their motions around their equilibrium positions can be considered sufficiently small, one can think that they effect harmonic oscillations. According to this picture (Einstein model), each ion is looked upon as a three-dimensional harmonic oscillator that oscillates independently of the other ions: the solid is therefore an 'ideal gas of harmonic oscillators'. For each atom one has:

$$E = \frac{1}{2m}(p_1^2 + p_2^2 + p_3^2) + \frac{1}{2}(k_1\, q_1^2 + k_2\, q_2^2 + k_3\, q_3^2) \tag{1.23}$$

where k_i are the elastic constants. If the solid is not monatomic, as e.g. sodium chloride NaCl, we shall consider both E_{Na} and E_{Cl}, which however have the same form, even if with different values of m and of the k_i's. So for a solid $f = 3n$, n being the number of atoms that make up the molecule, and

$$\mathcal{U} = \frac{6}{2} n\,RT, \qquad C_V = 3n\,R\;.$$

Note that this time also the potential energy contributes to $\mathcal{U}$, as much as kinetic energy does. For metals $n = 1$ and $C_V = 3\,R$. For NaCl, $n = 2$ and $C_V = 6\,R$, etc.

We have already examined the experimental situation. We shall know make some comments of general character.

1.   The dependence of C_V on T (with the exception of monatomic gases) takes place *as if*, with the decrease of T, also the number of degrees of freedom, contributing to the calculation of $\mathcal{U}$, decreased – or, alternatively, *as if* at low temperatures a (non-integer!) number of degrees of freedom were 'frozen' (i.e. did not contribute to $\mathcal{U}$). For example, as for a diatomic gas $C_V \to \frac{3}{2} R$ for $T \to 0$, it is as if one had a gradual freezing of the rotational degrees of freedom of the molecule. It should be clear that speaking, in this context, about the freezing of degrees of freedom only is a figurative way to visualize the experimental situation – in no way is it an attempt to explain the mechanism responsible for the dependence of C_V on T.

2.   We have seen that for certain temperatures there is a good agreement between theoretical values and experimental results. However, also in this case, we have very good reasons to be astonished about this agreement. Indeed, think about a diatomic gas: the molecule has been schematized as two atoms at a *fixed* distance, i.e. rigidly connected.

Now in Nature such rigid connections do not exist, it is more realistic to state that the two atoms interact via a potential of the type represented in Fig. 1.3 endowed with a minimum when the atoms are at a certain distance d – of the order of a few ångström – from each other. The molecule can effect oscillations about this equilibrium position. Such oscillations, as long as one can consider them of small amplitude, will

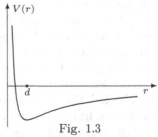

Fig. 1.3

be harmonic. Therefore it seems more appropriate to schematize the diatomic molecule as a pair of atoms held together by a spring: the rigid connection of the previous schematization is now replaced by a very 'hard' spring. But one is then confronted with a difficulty: no matter how 'hard' the spring is (provided it is not rigorously rigid), the Hamiltonian of the molecule is of the type:

$$E = \frac{1}{2m_1}\vec{p}_1^{\,2} + \frac{1}{2m_2}\vec{p}_2^{\,2} + \frac{1}{2}k(r-d)^2 \qquad (1.24)$$

where $\vec{p}_1$ and $\vec{p}_2$ are the momenta of the two atoms, r is the relative distance and k is the elastic constant of the spring between the two atoms. Note that now $f = 6$ and that there is one more quadratic term – the potential – so that, for any value of k (i.e. for any hardness of the spring), one should have $\mathcal{U} = \frac{7}{2}RT$ and $C_V = \frac{7}{2}R$.

At this point the problem is the following: the model with the spring certainly is more plausible than the previous schematization; so how is it that for a large interval of temperatures the experimental value of C_V is $\frac{5}{2}R$? Indeed we know also that, for large T, $C_V > \frac{5}{2}R$, so it seems that the vibrational degree of freedom is frozen up to $T \simeq 500\,\mathrm{K}$ and that, when T is further increased, it starts unfreezing.

Another difficulty of the type of the previous one is the following: we have schematized the atoms as pointlike – which indeed is not the case. For this reason, for example, in the discussion of the diatomic gas, we should have kept into account also the degree of freedom of rotation around the axis of the molecule, the axis that joins the two atoms. This degree of freedom should bring a further contribution $\frac{1}{2}R$ to the molar heat capacity, about which there is no trace in Nature. More to it: in metals, in addition to the ions that make up the crystal lattice, there are some completely free electrons (conduction electrons) in a number that is quite comparable to that of ions. Each electron should contribute $\frac{3}{2}k_B$ to the heat capacity, so for a gram-atom of a metal they should add a $\frac{3}{2}R$ to the C_V: to sum up, one should have $C_V \simeq 3R + \frac{3}{2}R = \frac{9}{2}R$, whereas we have seen that, most times, not even $3R$ is reached!

3.   The core of the previous discussion is the following: according to classical physics, the calculation of C_V is based upon the counting of the number of degrees of freedom. Now the concept of degree of freedom is somewhat loose – it rather is a mathematical abstraction, much as that of point mass is. So, if we ask ourselves how many degrees of freedom a lead pellet has got, we may legitimately answer three – if we are interested in a ballistic problem. But, if we are instead interested in the motion of the pellet on an inclined plane, then the number of degrees of freedom is six. In the case we would like to compute its heat capacity, then the number becomes $\simeq 10^{24}$! And one could go even further, by counting the number of electrons in the Pb atom, how many nucleons in its nucleus, how many quarks in each nucleon, and so on. The classical theory of heat capacities is therefore in a very unsatisfactory situation: it hinges upon a concept that, from the point of view of physics, is not well defined and, nonetheless, it has the pretention to predict numbers to be compared with experimental data!

1.4 Photoelectric Effect

Among the several phenomena that do not find an explanation in the framework of classical physics, let us finally consider the *photoelectric effect*. It will be the starting point, thanks to Einstein's ideas, to come in touch with the profound modifications one has to effect on classical physics in order to overcome the difficulties one was confronted with in those days.

First of all, let use examine the phenomenology of photoelectric effect: if a beam of light of short wavelength is let to strike a surface (e.g. that of a metal), the expulsion of electrons is observed. From a quantitative standpoint, the experiments show that the following three laws hold.

1. *For every substance there exists a minimum frequency ν_0 (photoelectric threshold) such that the effect is observed only with radiation whose frequency ν is greater than ν_0.*
2. *The maximum kinetic energy of the extracted electrons is independent of the intensity of the incident radiation, but depends linearly on the frequency ν:*
$$E_k^{\max} = h\left(\nu - \nu_0\right), \qquad h > 0 . \qquad (1.25)$$
3. *For a fixed frequency ν the rate of extraction of electrons (number of electrons per unit time) is proportional to the intensity of the incident radiation.*

The device by which the above laws can be verified is sketched in Fig. 1.4. It represents a photocell; light of a given frequency ν is sent on the photo-cathode, that is polarized positively with respect to the anode. Therefore the extracted electrons find a counteracting field that slows them down: only those endowed with kinetic energy $E_k > e \times V$ (V is the difference of potential between cathode and anode) will pass. The potentiometer allows one to

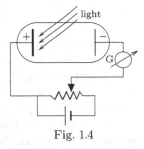

Fig. 1.4

vary V until the galvanometer G stops indicating a flow of current. In this way one is certain that all the photoelectrons have a kinetic energy $E_k < e \times V$, so the maximum kinetic energy exactly is $e \times V$.

Upon changing the frequency of the light and repeating the measurement one obtains the linear dependence expressed by (1.25) (see Fig. 1.5).

   In order to verify the third law, the intensity of light is varied and the intensity of the current flowing through G is measured: the latter is proportional to the number of electrons extracted from the pho-tocathode.

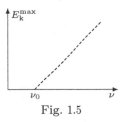

Fig. 1.5

   Let us now see why classical physics is unable to explain the laws of photoelectric effect. Let us consider the case of a metal. We already know that in a metal there are ions at almost fixed positions: in this lattice conduction electrons move almost freely. It is known from thermionic effect that a certain energy (**work function**) W of the order of some electronvolt must be provided to an electron, in order that it can be removed from the metal:

$$W_{\text{Na}} = 2.7\,\text{eV}, \quad W_{\text{Fe}} = 3.2\,\text{eV}, \quad W_{\text{Cu}} = 3.4\,\text{eV}, \quad W_{\text{Pb}} = 6\,\text{eV} .$$

In the photoelectric effect this energy is evidently provided by the incident radiation. What one expects is that the energy absorbed by the electrons is, for fixed frequency, proportional to the intensity of the incident radiation so that, for any fixed frequency and provided the intensity is sufficiently high, the emission of electrons should show up. In addition, the kinetic energy of the photoelectrons, having to equal the absorbed energy less the work function W, should be a linear function of the intensity. In this way neither the existence of a threshold nor the second law are explained.

Let us consider another aspect of the problem. Imagine to put the photocell 1 m away from a 2 watt bulb. Let us make the (very optimistic) assumption that all the light incident on the photocathode is absorbed by the conduction electrons close to the surface of the photocell; since there is roughly one conduction electron per atom and each atom occupies a volume about 10^{-24} cm³, each electron can absorb, at most, all the energy arriving on an area about 10^{-16} cm². As the bulb radiates uniformly in all directions, the energy absorbed by one electron in a second is

$$E = (2 \times 10^7) \times \frac{10^{-16}}{4\pi \times 10^4} = 1.6 \times 10^{-14}\, \text{erg s}^{-1} = 10^{-2}\, \text{eV s}^{-1} . \tag{1.26}$$

As we know that the work function W of an electron is of the order of some electronvolt, the result is that one should wait about 10^2 seconds before the photoelectric effect is detected (meanwhile the electrons would loose the acquired energy in several collisions!). From the experimental point of view, photoelectric effect practically is instantaneous ($10^{-8} \div 10^{-9}$ s).

In conclusion, classical physics, besides being unable to explain photoelectric effect, leads to conclusions that are in blatant contradiction with experiment.

Chapter 2

From Einstein to de Broglie

2.1 Photons

According to classical physics, the energy associated with a monochromatic electromagnetic wave is proportional to its intensity; the intensity can have any value above zero, and can therefore be varied with continuity. Furthermore this energy is distributed in space in a continuous way.

In order to explain the laws of photoelectric effect, in 1905 Einstein formulated an hypothesis in open contradiction with what we have reported above: an electromagnetic wave of frequency ν carries energy in 'packets' of energy E proportional to the frequency:

$$E = h\nu .\tag{2.1}$$

In other words, the energy associated to such a wave can only take the values $0,\ h\nu,\ 2h\nu,\ \cdots,\ nh\nu,\ \cdots$. That is to say that the energy of a wave of frequency ν is "quantized". These energy packets, or "quanta", or **photons**, are indivisible entities: the 'half photon' does not exist. For what regards the spatial and temporal distributions of such photons, Einstein makes no assumptions about their regularity: photons are randomly distributed.

The constant h had been introduced a few years before (1900) by M. Planck, whose name it brings, to interpret the spectrum of the radiation emitted by a black body. From Einstein relation (2.1) it appears that h has the dimensions of an action, namely an energy times a time, or equivalently of an angular momentum, or also of a linear momentum times a length: in general, of a coordinate times its canonically conjugate momentum. The numerical value of h taken from experiments is:

$$h = 6.6 \times 10^{-27} \, \mathrm{erg\,s} .$$

How much is the energy of a photon? Obviously this depends on the radiation one considers, namely on ν. Just to have an idea, let us consider radiation from the visible spectrum, i.e. with its wavelength satisfying

$$4000\,\text{Å} \le \lambda \le 7000\,\text{Å} \qquad \text{(visible spectrum)}.$$

© Springer International Publishing Switzerland 2016
L.E. Picasso, *Lectures in Quantum Mechanics*, UNITEXT for Physics,
DOI 10.1007/978-3-319-22632-3_2

Let us take for example $\lambda = 4000\,\text{Å}$. Then $(1\,\text{eV} = 1.6 \times 10^{-12}\,\text{erg})$:

$$E = h\nu = h\frac{c}{\lambda} = \frac{(6.6 \times 10^{-27}) \times (3 \times 10^{10})}{(4 \times 10^{-5}) \times (1.6 \times 10^{-12})}\,\text{eV} \simeq 3\,\text{eV}$$

that is of the same order of the energies occurring in the photoelectric effect, e.g. the work function W.

Let us now see how the laws of photoelectric effect are explained by means of the Einstein hypothesis. According to the classical mechanism of interaction among charges and the electromagnetic field, the radiation can transfer an arbitrarily small quantity of energy to the electrons. Now, instead, according to the quantum hypothesis, an electron is allowed either not to absorb energy at all or to absorb an entire quantum (the probability it absorbs two or more quanta, we shall see, is extremely small, so we will neglect this possibility). Then it is clear that, if $h\nu < W$, no electron can jump out of the metal and that the minimum frequency for which the effect is observed is $\nu_0 = W/h$. If $\nu > \nu_0$, the electrons leave the metal with a maximum kinetic energy $E_k = h(\nu - \nu_0)$ (maximum because, in leaving the metal, the electrons may loose energy by collisions). So, incidentally, h can be determined by measuring the slope of the straight line in Fig. 1.5. Furthermore, for a fixed frequency, the intensity of the radiation is proportional to the number of photons that arrive on the metal per unit time and, therefore, to the number of extracted electrons.

Let us reconsider, in the light of this interpretation, the example of the bulb discussed at the end of Sect. 1.4. If we keep on accepting the somewhat optimistic assumptions that have led to (1.26) and if we suppose that the frequency of the radiation emitted by the bulb corresponds to photons with energy of the order of some eV, then (1.26) must be reinterpreted by saying that in a second there arrives, on the average, one photon every 10^2 atoms and, as a consequence, every second only one electron over a hundred receives the energy necessary to jump out of the metal (indeed much less than this, for part of the energy is taken by the lattice).

Thanks mainly to the assumptions made to arrive at (1.26), this discussion is somewhat superficial. What one wants to emphasize is the following: according to classical physics, owing to the fact that the energy of a monochromatic wave is uniformly distributed, the electrons share 'democratically' the incoming energy and each of them takes too little of it to be able to jump out of the metal; according to the quantum hypothesis, instead, some (a few) electrons take the incoming energy and most of them take nothing at all. These few electrons give rise to the photoelectric effect.

The example of the bulb also shows how unlikely it is that an electron may absorb two ore more quanta of energy: the arrival of another photon should take place in the time interval one quantum is absorbed ($10^{-8}\,\text{s}$), but from (1.26) it follows that in $10^{-8}\,\text{s}$ there arrives one photon every 10^{10} atoms!

There are many phenomena that depend on the photoelectric effect and that, therefore, exhibit a threshold. One is in the old – i.e. pre-digital – process of taking a photograph, particularly the formation of the latent image. The photographic plate is covered with a layer of AgBr crystals: the primary effect that (according to a mechanism we shall not describe) gives rise to the formation of the latent image exactly is the photoelectric effect triggered by the incident light. In this way one understands how it is that either infrared or red light does not impress the plate, whereas this happens with a light of lesser wavelength, even if of weak intensity.

It may be helpful to have at one's disposal a mnemonic rule to obtain quickly the photon energy associated to a given radiation and viceversa. If, as customary in atomic physics, we characterize radiation by means of its wavelength λ expressed in ångström and express energy in electronvolt, then it suffices to recall that $hc = 12400\,\mathrm{eV} \times \text{Å}$, so that from $E = hc/\lambda$ one obtains:

$$E[\mathrm{eV}] = \frac{12400}{\lambda[\text{Å}]} \qquad \text{(photons)} . \qquad (2.2)$$

Photoelectric effect does not prove that an electromagnetic wave carries energy in quanta, it rather proves that its energy is absorbed in quanta. It is natural to wonder what happens with emission, namely whether electromagnetic energy is emitted either in quanta or with continuity.

Let us consider the production of X-rays ($\lambda \simeq$ 1 Å): electrons emitted by means of thermionic effect are accelerated by a difference of potential about $10^4\,\mathrm{V}$ (Fig. 2.1) and then are let to hit a metal plate. Penetrating the metal, the electrons are suddenly decelerated and radiate. The spectrum of the emitted radiation – *bremsstrahlung* (deceleration radiation) spec-

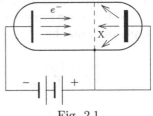

Fig. 2.1

trum – is a continuous spectrum on which some spectral lines are superposed, their position depending on the metal the plate is made of. The interesting thing is that, no matter what metal is used, the continuous spectrum has an upper limit at a certain frequency ν_0, that depends only on the accelerating potential V, indeed it is proportional to it. The interpretation of this fact is straightforward, if one admits that also emission of electromagnetic radiation takes place by quanta: each electron looses the kinetic energy $E_k \le e \times V$ by emitting one or more photons, whose frequencies are therefore such that $h\nu \le e \times V$, namely $\nu \le e \times V/h$ i.e. limited from above. The frequency $\nu_0 = e \times V/h$ is emitted when an electron looses all its energy by emitting only one photon. From the classical point of view, the spectrum should extend up to infinity, because the decelerated motion of the electron is not periodic.

2.2 Compton Effect

It is known that, in classical physics, an electromagnetic wave carries linear momentum, according to the relationship $p = E/c$. Since we now know that energy is carried by photons, it is natural to wonder whether these photons bring a linear momentum $p = E/c = h\nu/c$. This conclusion can be reached by analyzing the Compton effect: if a beam of "hard" X-rays (wavelength $\lambda_0 \lesssim 1\,\text{Å}$) is shot against whatever substance, one observes that X-rays scattered at an angle θ with respect to the direction of the incident radiation are endowed with a wavelength $\lambda(\theta)$ slightly higher than λ_0. The variation of the wavelength is independent of the particular substance off which X-rays are scattered and depends on θ according to:

$$\lambda(\theta) - \lambda_0 = 0.024\,(1 - \cos\theta)\,\text{Å}\,. \tag{2.3}$$

Compton effect can be easily explained by attributing it to the interaction between radiation and electrons and by interpreting this interaction as the collision process between a photon, of energy $h\nu$ and momentum $h\nu/c$, and a free electron (the binding energy of electrons, at least the outer ones, is of the order of some eV, whereas that of photons is $\simeq 10^5\,\text{eV}$). In the photon-electron collision total linear momentum and total energy are conserved, so it is clear that the larger the angle θ by which the photon is scattered, the higher the momentum and the energy transferred to the electron, and the larger the wavelength $\lambda(\theta)$ of the photon itself. Indeed, let $\vec{k}_0$ be the momentum of the incident photon and $\vec{k}$ and $\vec{p}$ respectively those of the photon and the electron after the collision (the initial momentum of the electron $\simeq \sqrt{2m\,E_k}$ is neglected, since typically it is two orders of magnitude smaller than that of the photon), one has:

$$\begin{cases} c\,k_0 + m_e\,c^2 = c\,k + \sqrt{m_e^2\,c^4 + c^2 p^2} \\ \vec{k}_0 - \vec{k} = \vec{p} \quad \Rightarrow \quad p^2 = k_0^2 + k^2 - 2k_0\,k\,\cos\theta \end{cases} \quad \Rightarrow$$

$$(k_0 - k + m_e\,c)^2 = m_e^2\,c^2 + p^2 = m_e^2\,c^2 + k_0^2 + k^2 - 2k_0\,k\,\cos\theta \quad \Rightarrow$$

$$m_e\,c\,(k_0 - k) = k_0\,k\,(1 - \cos\theta) \quad \Rightarrow \quad \frac{1}{k} - \frac{1}{k_0} = \frac{1}{m_e\,c}\,(1 - \cos\theta)$$

and, if $k = h\nu/c \equiv h/\lambda$, one finally obtains:

$$\lambda(\theta) - \lambda_0 = \frac{h}{m_e\,c}\,(1 - \cos\theta)\,. \tag{2.4}$$

The quantity $\lambda_C \equiv h/m_e\,c$ (that has the dimension of a length) is called the electron **Compton wavelength** and its numerical value is about $0.024\,\text{Å}$, in agreement with (2.3). We can then conclude that photons behave as corpuscles of energy $h\nu$ and momentum $h\nu/c$. According to special relativity, the relationship $p = E/c$ is typical of massless particles: photons are therefore corpuscles of vanishing mass.

2.3 General Features of Spectra. Bohr Hypotheses

Any substance can absorb and emit only electromagnetic radiations of well defined frequencies, typical of the substance itself. The frequencies a substance can absorb give rise to its **absorption spectrum**, those it can emit give rise

to its *emission spectrum*. We shall mainly be concerned with the spectra
of gaseous substances and vapours.

A device suitable to obtain absorption
and emission spectra of a substance is
schematized in Fig. 2.2: white light hits a
cell containing the gas or vapour one is interested in; a spectrograph, schematized
as the prism P_1, breaks the light transmitted by the gas up into its monochromatic components before they impress
the photographic plate L_1: a uniform il-
lumination (due to the continuous spectrum transmitted by the gas) will appear

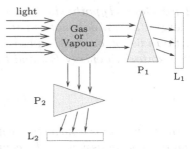

Fig. 2.2

on L_1, with the exception of some dark lines in correspondence with the frequencies absorbed by the gas. These dark lines make up the absorption spectrum. In order to obtain the emission spectrum it is necessary to provide
energy to the gas: this energy will be re-emitted as electromagnetic radiation.
This can be done in several ways, for example either by heating the gas by a
flame, or provoking an electric (arc) discharge, or even sending light on the
substance, as in Fig. 2.2.

For the sake of simplicity, we shall refer to the last method of excitation
of the gas or vapour. If one wants to analyze the emission spectrum of the
substance, one will observe the light emitted in a direction orthogonal to the
direction of the incident light: in this way the latter will not disturb the observation. The prism P_2 separates the monochromatic components that impress
the plate L_2. On L_2, on a dark background, there will appear some bright
lines in correspondence with the frequencies emitted by the gas. These lines
make up the emission spectrum. The first thing one notes is that the emission
and absorption spectra of a substance are different from each other: the emission spectrum is richer that the absorption one, with all the absorption lines
appearing in the emission spectrum. If, in addition, the incident radiation has
a spectrum consisting of frequencies ν higher than a given $\bar{\nu}$, in the emission spectrum of the gas there appear also lines corresponding to frequencies
smaller than $\bar{\nu}$.

In order to explain emission and absorption spectra, in 1913 N. Bohr put
forward some hypotheses that, as Einstein hypothesis, are in contrast with
classical physics which, by the way, is unable to explain spectra.

1-st Bohr Hypothesis: For any atom only a discrete set of energies, starting
from a minimum value on, $E_0 < E_1 < \cdots < E_n < \cdots$, are allowed.

These energies are called *energy levels* of the atom. Note that, from the
classical point of view, the allowed energies for a system *always* make up a
continuous set.

The first hypothesis concerns the *bound states* of an atom, namely those
states that correspond to orbits limited in space (think e.g. of ellipses or
circles in the hydrogen atom). For unbounded – or ionization – states, all the

classically possible energies are allowed. The state of a system corresponding to the minimum allowed energy is called the **ground-state** of the system.

2-nd Bohr Hypothesis: *When the system is in one of the above mentioned energy states, it does not radiate; either emission or absorption of electromagnetic radiation only takes place in the transition from a state with energy E_n to a state with energy E_m and the frequency of the radiation is:*

$$\nu_{nm} = \frac{|E_n - E_m|}{h}. \tag{2.5}$$

If the transition is from a state with higher energy to a state with lower energy, there occurs emission; in the contrary case one has absorption.

In the next section we will enunciate the third Bohr hypothesis that allows one to calculate the allowed energies $E_0 < E_1 < \cdots < E_n < \cdots$ in simple cases. Let us now examine the consequences of the first two Bohr hypotheses. This discussion mainly concerns the hydrogen atom.

1. It s clear that spectra are made out of lines whose frequencies are given by (2.5) with all possible E_n and E_m. Any substance may also exhibit a continuous spectrum that corresponds to transitions from a bound state to an ionization state (the latter are also called **continuum states** since for them the hypothesis of discrete energies does not apply).

2. Bohr hypotheses are compatible with Einstein hypothesis. Rather it is probably more correct to state that Bohr was led to the formulation of his hypotheses as a consequence of Einstein's. Indeed, if an atom radiates quanta, when it either emits or absorbs a photon of frequency ν, its energy varies by $h\nu$. As it was known that an atom either emits or absorbs only certain frequencies, it looks reasonable to assume that only certain energies are allowed for the atom, i.e. those for which the observed frequencies are given by (2.5).

3. One can understand why emission spectra are richer than absorption spectra. Indeed (we shall come back to this point) at room temperature the great majority of the atoms are in their ground state. As a consequence, only the lines with frequencies $\nu_{0n} = (E_n - E_0)/h$ are observed in absorption: the latter correspond to the transitions from the lowest energy level E_0 to a generic level E_n. After the atom has been excited to the level E_n, in a very short time ($10^{-7} \div 10^{-9}$ s) it re-emits radiation effecting one or more transitions to lower energy levels until the ground state is reached. In this way one observes in emission the entire spectrum given by (2.5).

4. It follows from the above discussion that all the frequencies of the emission spectrum are obtained by making all the possible differences of the frequencies of the absorption spectrum:

$$|\nu_{0n} - \nu_{0m}| = \left| \frac{E_n - E_0}{h} - \frac{E_m - E_0}{h} \right| = \frac{|E_n - E_m|}{h} = \nu_{nm}$$

This fact was already known as Ritz **combination principle**.

5. Bohr explanation of Ritz combination principle is consequence also of the fact that the energies E_n can be derived from the knowledge of the absorption spectrum. Indeed, Planck constant h being known, the absorption spectrum enables one to determine $E_1 - E_0$, $E_2 - E_0$, ..., $E_n - E_0 \to E_\infty - E_0$ i.e. all the energies up to the constant E_0. If we set to zero the energy corresponding to the ionization threshold, namely to the limiting value E_∞, one obtains $E_0 = -h\nu_\infty$, where ν_∞ is the limiting frequency of the spectrum of lines: for higher frequencies one has a continuous spectrum, corresponding to the ionization of the atom.

6. The above discussion suggests an experimental verification of Bohr hypothesis. Indeed $|E_0|$, measured by the absorption spectrum, is the ionization energy E_{I} of the atom. But the atom can be ionized by collisions with electrons or other atoms, instead of absorbtion of light. One can then measure the energy necessary to ionize an atom by collisions and compare the result with the value measured (or predicted) by spectroscopic way.

7. Spectroscopists had succeeded in grouping the lines of a spectrum in *series*, in such a way that the frequencies, or better the **wavenumbers** $1/\lambda = \nu/c$ corresponding to the lines of a spectrum could be expressed as differences between **spectroscopic terms**: $1/\lambda = \nu/c = T(n) - T(m)$ (n, m integers ≥ 0), and each series is identified by the value of n. So, for example, the first series, that is the absorption series, corresponds to the wavenumbers:

$$1/\lambda_m = T(0) - T(m), \qquad m \geq 1.$$

Now, according to Bohr, the spectroscopic terms $T(n)$ are nothing but the energy levels divided by $h\,c$: $T(n) = -E_n/(h\,c)$ and each series corresponds to the transitions that (in emission) have in common the arrival level; the different series are identified by the arrival energy level.

2.4 Hydrogen Energy Levels According to Bohr

The problem that has been left open is that of determining the energy levels of a system. In this section we will limit our discussion to the case of hydrogen and hydrogen-like atoms. Hydrogen-like atoms are atoms ionized one or more times in such a way that only one electron is left, for example He^+, Li^{++}, Be^{+++} etc. In all such cases one deals with systems with only one electron in the field of a nucleus of charge Ze, where Z is the atomic number.

In general the orbits are ellipses (bound states!), and it is known from classical physics that energy only depends on the major axis of the ellipse. So, stating that only some energies are allowed is the same as saying that only certain orbits are allowed: the problem is to calculate which ones. Let us limit ourselves, as Bohr did, to circular orbits.

The relationship between the energy and the radius of the orbit is (see (1.4) and (1.8))

$$E = \frac{1}{2}V = -\frac{1}{2}\frac{Z e^2}{r}. \tag{2.6}$$

In order to determine the allowed energies, i.e. radii, Bohr put forward the following hypothesis:

3-rd Bohr Hypothesis: *The allowed orbits are those for which the angular momentum is an integer multiple of $h/2\pi$.*

This rule can be enunciated in a more general (Bohr-Sommerfeld) form in the following way:

$$\int_{\text{orbit}} p\,dq = n\,h\,, \qquad n = 1, 2, \cdots \qquad (2.7)$$

where q and p are canonically conjugate variables and the integral is to be taken along one orbit. In any event, for circular orbits, the third Bohr hypothesis requires that:

$$\mu_e\, v\, r = n\,\hbar \qquad (2.8)$$

where $\mu_e = m_e M/(m_e + M)$ is the reduced mass of the electron-nucleus two-body system. As it is customary in quantum mechanics, we have set:

$$\hbar \equiv \frac{h}{2\pi} = 1.05 \times 10^{-27}\,\text{erg s}\,.$$

Squaring (2.8) and recalling that $E_k = \frac{1}{2}\mu_e v^2 = -\frac{1}{2}V$, one obtains

$$\frac{1}{2}\mu_e\, v^2 = \frac{n^2\hbar^2}{2\mu_e\, r^2} = \frac{Z\,e^2}{2\,r} \qquad (2.9)$$

whence:

$$r_n = \frac{n^2\hbar^2}{\mu_e\, Z\, e^2} = \frac{n^2}{Z}\,(m_e/\mu_e)\,a_B\,, \qquad a_B \equiv \frac{\hbar^2}{m_e\, e^2}\,. \qquad (2.10)$$

where a_B is the **Bohr radius**. Equation (2.10) says which are the allowed orbits, according to Bohr. By inserting (2.10) into (2.6) one finally obtains the allowed energies:

$$E_n = -\frac{1}{2}\frac{Z^2 e^2\,(m_e/\mu_e)}{n^2\, a_B}\,, \qquad n = 1, 2, \cdots\,. \qquad (2.11)$$

Putting:

$$R = \frac{e^2\,(m_e/\mu_e)}{2a_B\, h\, c} = \frac{\mu_e\, e^4}{4\pi\,\hbar^3\, c} \qquad (2.12)$$

one can write:

$$E_n = -Z^2\,\frac{R\,h\,c}{n^2}\,. \qquad (2.13)$$

The reason why we have written (2.11) in the form (2.13) is in the fact that spectroscopists had already experimentally determined the spectroscopic terms of hydrogen and had found $T(n) = R_H/n^2$ with $R_H = 109677.6\,\text{cm}^{-1}$ (**Rydberg constant** for hydrogen). The fact that R_H was known with seven significant digits is due to the high precision by which spectroscopic experiments were made even in that time. Presently Rydberg constant is known with a precision higher than one part over 10^{11}: if R_∞ stands for Rydberg constant with $M = \infty$ ($\mu_e = m_e$), one has $R_\infty = 109737.31568508(65)\,\text{cm}^{-1}$ ("2014 CODATA recommended values").

It is worth calculating the value of R_H predicted by (2.12) to realize the excellent agreement with the experimental value: we leave this exercise to the reader. We can conclude that the spectroscopic terms deduced from Bohr theory are in very good agreement with the corresponding experimental values.

Let us now examine in detail the consequences of the obtained results. Let us start with

Hydrogen: $Z = 1$

If the difference between reduced mass and electron mass is neglected ($m_e/M_H \simeq 0.5 \times 10^{-3}$), one has:

$$r_n = n^2 a_B \,, \qquad a_B = \frac{\hbar^2}{m_e\, e^2} = 0.53\,\text{Å} \,, \qquad E_n = -\frac{1}{2}\frac{e^2}{n^2\, a_B} \qquad (2.14)$$

where a_B is the radius of the first allowed orbit for the atom in its ground state. The value of a_B is satisfactory because it is in agreement with the value expected for atomic size. Note that now, having at one's disposal the Planck constant, it is possible to form a length with the right order of magnitude. Furthermore, the atom in its first orbit cannot collapse, exactly because it is in the state of minimum energy.

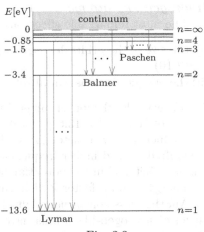

The energy levels given by (2.14) are negative and decrease, in absolute value, towards 0: the position of these level is illustrated in Fig. 2.3 (in scale). The energy E_1 of the ground state is, apart from

Fig. 2.3

its sign, the ionization energy of the atom (since the minimum value of the **quantum number** n is in the present case 1, we prefer to call E_1 instead of E_0 the energy of the ground state). One has $E_1 = -e^2/2a_B = -13.6\,\text{eV}$. The energies of the different levels, expressed in eV, are therefore:

$$E_n[\text{eV}] = -\frac{13.6}{n^2} \,. \qquad (2.15)$$

So $E_2 = -3.4$, $E_3 = -1.5$, $\cdots$.

How does the hydrogen spectrum look like? The first series (**Lyman series**) is that consisting of all transitions to level $n = 1$ (Fig. 2.3): the lines get denser and denser in the neighborhood of a

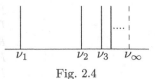

Fig. 2.4

limiting line (Fig. 2.4). The first line corresponds to a transition with $\Delta E = E_2 - E_1 = 10.2\,\mathrm{eV}$ and has therefore a wavelength $\lambda_1 = 12400/10.2 \simeq 1200\,\text{Å}$; the limiting line corresponds to $\Delta E = 13.6\,\mathrm{eV}$ and $\lambda_\infty \simeq 900\,\text{Å}$: both λ_1 and λ_∞ are in the ultraviolet, and so is all the series. Since this is the series that is observed in absorption, one understands why hydrogen is transparent: it does not absorb in the visible part of the electromagnetic spectrum.

The second series (**Balmer series**) corresponds to all transitions from levels with $n > 2$ to the first **excited level** ($n = 2$) (Fig. 2.3). The structure of this series still is of the type reported in Fig. 2.4, but in comparison with Lyman series it is shifted towards lower frequencies. Indeed:

$$\lambda_1 = \frac{12400}{1.9} \simeq 6500\,\text{Å}, \qquad \lambda_\infty = \frac{12400}{3.4} \simeq 3700\,\text{Å}$$

namely λ_1 is in the orange and λ_∞ is in the near ultraviolet so that almost all the series is in the visible. As for the third series (**Paschen series**), it all lies in the infrared.

Let us now consider the case of

Hydrogen-like atoms: $Z > 1$

The most remarkable effects are

(i) the size of the orbits decreases with increasing Z, due to the $1/Z$ factor in (2.10);
(ii) the energies increase, in absolute value, like Z^2.

So, for example, the ionization energy of He^+ (also called **second-ionization** energy of He) is $4 \times 13.6 = 54\,\mathrm{eV}$. As a consequence also all the energetic jumps increase by a factor Z^2 with respect to H and all the spectral lines are shifted toward higher frequencies. Note that the factor Z^2 has a twofold origin: a factor Z in the potential energy – and, as a consequence of (2.6), in the energy – and a factor Z^{-1} in the allowed radii.

Another, less conspicuous but important, effect is due to the fact that the nuclei of hydrogen-like atoms have different masses: this entails differences, although small, in the reduced masses of the systems upon which the Rydberg constant (2.12) and, in the very end, the energy levels (2.13) depend.

As $m_e \ll M$, one can write $\mu_e = m_e/(1 + m_e/M) \simeq m_e(1 - m_e/M)$. The term m_e/M, that for hydrogen is about $1/2000$, for deuterium is half such value (deuterium D is an isotope of hydrogen: the nucleus of deuterium consists of a proton and a neutron), for He is four times smaller.

Thus we see that the Rydberg constant slightly varies for the different atoms we have considered:

$$
\begin{aligned}
R_H &= 109677\,\mathrm{cm}^{-1} \\
R_D &= 109707\,\mathrm{cm}^{-1} \\
R_{He^+} &= 109722\,\mathrm{cm}^{-1} \\
R_{Li^{++}} &= 109728\,\mathrm{cm}^{-1} \\
R_\infty &= 109737\,\mathrm{cm}^{-1}
\end{aligned}
$$

with differences smaller than 0.5% with respect to R_∞ (the limiting case of nucleus with infinite mass), as had to be expected. No matter how small, the variation of the Rydberg constant is well detectable, given the precision of spectroscopic measurements.

However the most important fact is that the dependence of R – and therefore of the spectrum – on the nuclear mass allows one to distinguish the different isotopes of a single element. As well known, one calls *isotopes* those elements that have the same atomic number Z but different atomic weight A, due to the different number of neutrons in the nucleus, as is the case of hydrogen and deuterium.

So, for example, if one analyzes a mixture of hydrogen H ($Z = 1$, $A = 1$) and deuterium D ($Z = 1$, $A = 2$), or He^3 and He^4 ($Z = 2$, $A = 3$, 4 respectively), one finds a spectrum of doubled lines, as in Fig. 2.5. In the case of H and D the resolution $\Delta\nu/\nu = \Delta\lambda/\lambda$ necessary to distinguish

Fig. 2.5

the lines of the doublets is $\Delta\mu/\mu \simeq 2.8 \times 10^{-4}$. The relative intensity of the spectral lines enables one to find the composition of the mixture (for example, the percentage of heavy water D_2O in a sample of water).

Such effect is known as *isotopic effect* and, therefore, allows one to establish either the presence or the absence of an isotope in a given substance.

The fact that the energy levels of even order in the spectrum of He^+ coincide, up to corrections of reduced mass, with the levels of H had induced the interpretation of the spectrum of some stellar radiations as due to hydrogen: only thanks to Bohr theory it was correctly attributed to He^+.

In conclusion Bohr theory had explained several questions left unsolved by the Rutherford atomic model: stability of atomic sizes, emission of line spectra, absence of radiative collapse. All this made it possible that the theory had been very favourably welcome since, by postulating something new, provided the explanation of many facts.

We shall see in the forthcoming sections that the solution of other problems and further experimental verifications reinforce even more the validity of this theory, even if (not only) from a conceptual point of view there still remain many open problems.

2.5 Energy Levels for an Oscillator and for a Particle in a Segment

After the successful application of the Bohr quantization condition to the hydrogen atom, we feel encouraged to apply it to other systems and to explore its consequences. Both for its simplicity and, mainly, for the interest it has in almost all the fields of physics, we shall now apply the quantization condition to the harmonic oscillator. In order to emphasize the importance this system has in physics, we recall that we have already considered harmonic oscillators with regard to the heat capacities of both solids and diatomic gases.

In order to determine the energy levels of a harmonic oscillator, we shall use the Bohr-Sommerfeld condition (2.7). Recalling that $p = m\dot{q}$ and $dq = \dot{q}\,dt$,

(2.7) becomes:

$$\int_{\text{period}} m\,\dot{q}^{\,2}\,\mathrm{d}t = n\,h \tag{2.16}$$

the integral being taken over a period $\tau = 2\pi/\omega$. It is known (and the verification is straightforward) that for the harmonic oscillator the average (over one period) of the kinetic energy equals a half of the total energy E (virial theorem), so from (2.16) one obtains:

$$\int_{\text{period}} m\,\dot{q}^{\,2}\,\mathrm{d}t = \frac{2\pi}{\omega}\,E \quad \Rightarrow$$

$$E_n = n\,\hbar\,\omega\,, \quad n \geq 0 \quad \text{(harmonic oscillator)} \tag{2.17}$$

so the energy levels of a harmonic oscillator are equidistant and their distance is $\hbar\,\omega$.

One can ask why in this case n may also be 0, whereas for the hydrogen atom this is not the case: indeed, on this point Bohr theory provides no answer. In this particular case starting from either $n = 0$ or $n = 1$ makes no difference because, the energy levels being equidistant, it is the same as shifting them all by the additive constant $\hbar\,\omega$ (it is however the case to anticipate that the result provided by quantum mechanics will be $E_n = \hbar\,\omega\,(n + \frac{1}{2})$, $n \geq 0$).

If the oscillator carries an electric charge, it can either emit or absorb radiation. Due to the second Bohr hypothesis, the frequency of the either emitted or absorbed radiation equals the frequency ν_0 of the oscillator, provided the transition takes place between two adjacent energy levels; otherwise it will be an integer multiple of ν_0, namely it will correspond to a higher harmonic. Also according to classical physics an oscillator emits all the frequencies that are integer multiples of the fundamental frequency, however – when the dipole approximation holds (velocity of the oscillator $v(t) \ll c$) – higher harmonics are much less intense. One of the flaws of Bohr theory just is its incapability at predicting the intensity of the radiation emitted in the various transitions.

Let us examine one more example of application of the quantization rule (2.7).

Consider a particle constrained to move in a segment of length a (in three dimensions this would be the problem of a particle constrained within a box). From a physical point of view such a problem is encountered when considering the motion of a particle subject to a potential $V(x)$ that can be approximated

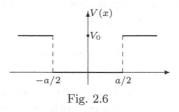

Fig. 2.6

by a constant in a region of size a and rapidly grows to another constant of higher value out of the considered region: in such cases the potential is schematized as a **potential well**, as shown in Fig. 2.6. There are several examples: the molecules of a (perfect) gas within a box; a free electron within a metal; a free electron within a crystal cell (size: a few ångström) that makes up an impurity in a given material (colour centers); ... .

In this case (2.7) gives

$$n\,h = \oint p\,dq = 2a\,p \qquad \Rightarrow \qquad p = n\,\frac{h}{2a}$$

and, since $E = p^2/2m$, one obtains the following energy levels:

$$E_n = \frac{h^2 n^2}{8m\,a^2} \qquad n \geq 0\ (?) \qquad \text{(potential well)} \qquad (2.18)$$

(in the present case, according to quantum mechanics, $n \geq 1$).

We will complete the present section with a few observations.

1. The dependence on n of the energy levels is different for each system ($\propto n^{-2}$ for the hydrogen atom, $\propto n$ for the oscillator, $\propto n^2$ for the particle in the potential well);

2. the quantization condition has been applied, and can be applied, only if the orbit is finite; consider, for example, the case of the particle in a segment: if $a \to \infty$, the integral $\oint p\,dq$ diverges and therefore, in this case, there is no quantization condition. A different way to look upon the same thing is the following: the distances among the energy levels given by (2.18) are proportional to a^{-2}; if $a \to \infty$ they tend to 0, i.e. one finds the continuum of energies from 0 to ∞ of classical physics. Therefore, for example, in the case of a potential such as that of Fig. 2.6, the result expressed by (2.18) holds only for $E_n < V_0$, whereas for energies higher than V_0 all the values are admissible (much as for the ionization states of the hydrogen atom).

3. Thanks to Stokes' theorem the integral on the left hand side of (2.7) equals the area enclosed by the orbit of energy E_n, therefore (2.7) means that the number of states (i.e. energy levels) in a (two-dimensional) volume Ω of the phase-space is given by Ω/h. In general, if $2f$ is the dimension of the phase-space, the quantization condition implies that h^f is the phase-space volume per state.

2.6 Einstein and Debye Theories for the Heat Capacities of Solids

If one accepts Bohr hypothesis about the existence of discrete energy levels, there are several things that must be re-examined from this point of view: first of all, the theory of heat capacities. Indeed, the theory of heat capacities was based on the calculation of the average energy (theorem of equipartition of energy) that is obtained from the expression $\int E(q,p)\,dn(q,p)$ by integrating over all the phase space, i.e. over all the classically admitted states. Now, according to the first Bohr hypothesis, only some states with energies E_n are allowed and, as a consequence and in the first place, the very Boltzmann distribution (1.13)

$$dn(q,p) = B\,e^{-\beta\,E(q,p)}\,dq\,dp$$

is meaningless and needs to be reformulated. Instead of asking oneself which is the (most probable) number of subsystems with coordinates and momenta between q and $q + dq$, and p and $p + dp$, one must ask which is the (most probable) number n_i of subsystems with energy E_i, where the E_i are the allowed energies. This problem is solved by the same reasoning that has led to (1.13), so a formally analogue result will obtain

$$n_i = B\,e^{-\beta\,E_i}$$

where again B can be found by requiring $\sum n_i = N$. Finally:

$$n_i = N\,\frac{e^{-\beta\,E_i}}{\sum_j e^{-\beta\,E_j}} \tag{2.19}$$

namely Boltzmann distribution still applies to the number of subsystems in the energy level E_i (the n_i are also called **populations** of the levels E_i). Indeed, (2.19) is not quite correct, it must be modified according to

$$n_i = N\,g_i\,\frac{e^{-\beta\,E_i}}{\sum_j g_j\,e^{-\beta\,E_j}} \tag{2.20}$$

where the g_i are integer numbers (normally of the order of a few units), called **degree of degeneracy** of the level E_i, whose origin is not easy to explain now: they represent the number of states corresponding to the level E_i; we shall clarify in the sequel the exact meaning of this statement. They can be ignored in the discussion of the forthcoming sections.

Before proceeding any further, we wish to briefly discuss (2.19), from which it follows that, for a given temperature, the average number of systems that are in the energy level E_i decreases exponentially with the increase of the energy. It follows that, if $E_r < E_s$, always one has $n_r > n_s$, so the higher the level the lower the population. It is interesting to evaluate the ratio of the populations relative to two energy levels E_r and E_s: for temperature T and putting $\Delta E = E_r - E_s$, one has:

$$\frac{n_r}{n_s} = \frac{B\,e^{-E_r/k_B T}}{B\,e^{-E_s/k_B T}} = e^{-\Delta E/k_B T} . \tag{2.21}$$

Let us evaluate, for example, the ratio n_2/n_1 between the populations of the first excited and the lowest energy level for an atomic system. The energy differences are of the order of some eV (an exception is provided by hydrogen, for which $E_2 - E_1 \simeq 10\,\text{eV}$): this is proven by the fact that, normally, the absorption spectrum lies in the visible region ($1.5 \div 4\,\text{eV}$). Now for such differences of energy one realizes that, at room temperature, the excited level practically is unpopulated. Indeed, for $T_0 \simeq 300\,\text{K}$, one has:

$$k_B T_0 = \frac{(1.38 \times 10^{-16}) \times 300}{1.6 \times 10^{-12}} \simeq \frac{1}{40}\,\text{eV} . \tag{2.22}$$

Both this relationship and the equivalent:

$$k_{\mathrm{B}}T = 1\,\mathrm{eV} \qquad \Rightarrow \qquad T \simeq 12000\,\mathrm{K} \qquad (2.23)$$

are very useful as they allow for a quick evaluation of $k_{\mathrm{B}}T$ for any temperature.

In conclusion, for $\Delta E \simeq 1\,\mathrm{eV}$ and $T \simeq 300\,\mathrm{K}$, the ratio between the populations, according to (2.21), is

$$\mathrm{e}^{-\Delta E/k_{\mathrm{B}}T} = \mathrm{e}^{-40} \simeq 10^{-16} \qquad \left(\mathrm{e}^{x} \simeq 10^{0.4x}\right) \,.$$

In order to have an appreciable population also for the excited levels, $k_{\mathrm{B}}T \simeq \Delta E$ will be needed. As $\exp(-\Delta E/k_{\mathrm{B}}T) \to 1$ for $T \to \infty$, one can say that only at very high temperatures the energy levels tend to become all equally (un)populated.

The fact that, at room temperature, practically all the atoms are in the ground state explains why the atoms of the same type all have the same size, whatever their past history (stability of atomic sizes), and also explains the fact that in the absorption spectrum of a substance (in normal conditions) only the first series is observed, namely only the transitions that start from the ground state.

It should be said that (2.19) applies to systems that only possess discrete energy levels (as the harmonic oscillator); for the systems that also possess a continuum of energies (as atoms, molecules ...), no quantization condition existing for them, (1.13) still holds. So we shall have:

$$\begin{cases} n_i = B\,\mathrm{e}^{-\beta\,E_i} & \text{(discrete levels)} \\ \mathrm{d}n(q,p) = B\,\mathrm{e}^{-\beta\,E(q,p)}\,\mathrm{d}q\,\mathrm{d}p/h^f & \text{(continuous levels)} \end{cases} \qquad (2.24)$$

where

$$B = N\left(\sum_j \mathrm{e}^{-\beta\,E_j} + \int \mathrm{e}^{-\beta\,E(q,p)}\,\mathrm{d}q\,\mathrm{d}p/h^f\right)^{-1}$$

the factor h^f dividing $\mathrm{d}q\,\mathrm{d}p$ is there because, as discussed at the end of Sect. 2.5, $\mathrm{d}q\,\mathrm{d}p/h^f$ is the number of quantum states in the cell $\mathrm{d}q\,\mathrm{d}p$.

Equation (2.24) may be useful, for example, in the calculation of the degree of ionization of a gas in the condition of thermal equilibrium: indeed, observe that, even if the population fast decreases as E increases, the degree of ionization can nonetheless be appreciable, owing to the great number of continuum states.

Let us now go back to the problem of the heat capacities. Having now (2.19) at one's disposal, the expression for the average energy of a subsystem (with only discrete levels) is

$$\overline{E} = \frac{\sum_i E_i\, n_i}{\sum_j n_j} = \frac{\sum_i E_i\, \mathrm{e}^{-\beta\,E_i}}{\sum_j \mathrm{e}^{-\beta\,E_j}} = -\frac{\partial}{\partial \beta} \ln Z(\beta) \qquad (2.25)$$

where

$$Z(\beta) = \sum_i \mathrm{e}^{-\beta\,E_i} \,.$$

Not always the series that defines the partition function can be explicitly summed.

In the case of the harmonic oscillator, as $E_n = n\,\hbar\omega$, the thing is simple since one is dealing with a geometric series

$$Z(\beta) = \sum_{n=0}^{\infty} e^{-\beta E_n} = \sum_{n=0}^{\infty} \left(e^{-\beta\,\hbar\omega}\right)^n = \frac{1}{1 - e^{-\beta\,\hbar\omega}}$$

so, by use of (2.25), the average energy for a quantum oscillator reads:

$$\overline{E} = \frac{\hbar\omega}{e^{\hbar\omega/k_B T} - 1}. \tag{2.26}$$

In Fig. 2.7 both $\overline{E}$ as the function of T given by (2.26) and (dashed line) the graph of the classical expression $(\overline{E}_{cl} = k_B T)$ are reported; we have put $T_c \equiv \hbar\omega/k_B$ (T_c is called **characteristic temperature**). Note that, contrary to the classical result, the average energy $\overline{E}$ depends on the frequency of the oscillator: now the quantum theory is able to distinguish 'harder' oscillators from the less 'hard'

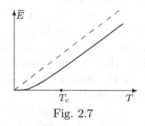

Fig. 2.7

ones; at high temperatures, (2.26) is well approximated by the classical value, up to an irrelevant additive constant. This happens when $k_B T \gg \hbar\omega$: indeed, in such a case, one can expand in series (2.26) and obtain:

$$\overline{E} = \frac{\hbar\omega}{1 + \hbar\omega/k_B T + \cdots - 1} \simeq k_B T \qquad (T \gg T_c)$$

(if the next term in the expansion of the exponential is kept, also the constant term is recovered: $\overline{E} \simeq k_B T - \frac{1}{2}\hbar\omega$).

The result could have been predicted, indeed $\hbar\omega$ is the jump between two adjacent energy levels of the oscillator: saying that $k_B T \gg \hbar\omega$ is the same as saying that the levels are very close to one another (with respect to thermal energy scale $k_B T$) and the quantization does not show up; the result is therefore the same one has classically (it could also be said that for $\hbar\omega \ll k_B T$ the series in (2.25) can be approximated by integrals and one is back to the classical case).

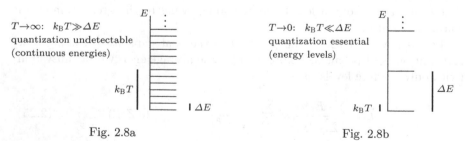

Fig. 2.8a   Fig. 2.8b

For $T \to 0$, (2.26) becomes $\overline{E} = \hbar\omega\,\exp(-\hbar\omega/k_B T)$ that tends very fast to 0, as is seen in Fig. 2.7. The two limiting cases ($T \gg T_c$ and $T \ll T_c$) are shown in Figs. 2.8a and 2.8b. Let us now see the consequences of the above discussion for what concerns the heat capacities of solids.

As we have already seen in Sect. 1.3, according to the Einstein model, a solid (for the sake of simplicity we shall deal with a monatomic solid) is made out of many identical oscillators, with frequency ν and non-interacting with one another ('ideal gas of oscillators'). As the oscillators are three-dimensional, the internal energy of a gram-atom of substance is given by $\mathcal{U} = 3N_A\overline{E}$, with $\overline{E}$ given by (2.26). The atomic heat capacity is therefore:

$$C_V = \frac{\partial \mathcal{U}}{\partial T} = 3N_A k_B \left(\frac{\hbar\omega}{k_B T}\right)^2 \frac{e^{\hbar\omega/k_B T}}{(e^{\hbar\omega/k_B T} - 1)^2}. \qquad (2.27)$$

It is easy to verify that, for $T \to 0$, $C_V \to 0$ whereas, for $T \to \infty$, $C_V \to 3R$. In Fig. 2.9 we have reported the graph of C_V as a function of the temperature: the figure shows that C_V reaches its classical value – up to less than 10% – already for $T = T_c \equiv \hbar\omega/k_B$, while at temperatures one order of magnitude smaller, $T \simeq T_c/10$, C_V is practically vanishing.

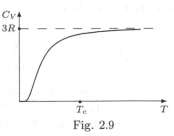

Fig. 2.9

Furthermore, we can examine how things change if we consider different solids, i.e. solids made out of oscillators with different frequencies. First of all, note that the C_V given by (2.27) is a universal function of the ratio T/T_c, where only $T_c = \hbar\omega/k_B$ depends on the considered solid (through $\omega = 2\pi\nu$), so that for different solids the curve is the same, apart from the scale of abscissae.

Let us compare, for example, lead and diamond; their physical features (e.g. the melting points) lead us to think that the oscillators the diamond consists of are harder (namely have an elastic constant with a higher value) than those of Pb: it follows that $\nu_{Pb} < \nu_C$ and consequently $T_c^{Pb} < T_c^C$. But we have seen that for $T = T_c$ the atomic heat capacity is practically equal to the classical Dulong–Petit value, so the curve relative to Pb should reach the value $3R$ at a lower temperature than that of diamond: this is exactly what happens experimentally (see Fig. 1.1).

In conclusion, the Einstein model – together with the quantization of the energy levels of the harmonic oscillator – accounts fairly well for the heat capacities of solids. Fairly well but not quite, because while (2.27) predicts that, for $T \to 0$,

$$C_V = 3R \left(\frac{h\nu}{k_B T}\right)^2 e^{-h\nu/k_B T} \qquad (2.28)$$

(recall that $\hbar\omega = h\nu$), namely that $C_V \to 0$ with all its derivatives, the curves relative to experimental data indicate that $C_V \to 0$ only as T^3. This depends on the fact that the schematization of the solid, typical of the Einstein model, is rather inaccurate. Indeed, as we have said, Einstein identifies the solid with a collection of identical and independent oscillators. One should instead think that the interactions among the oscillators are not negligible: it

is sufficient to recall that hitting a metal gives rise to a sound, which proves
the propagation of the perturbation along the whole piece of metal, i.e. the
existence of elastic waves and, in turn, the existence of interactions among
the oscillators.

It was P. Debye who proposed a model more complicated, but more plausible, assuming that the oscillators are not independent, but elastically bound
to each other: this amounts to considering the small vibrations of the lattice
ions around their equilibrium positions. Let $V(x_1, \cdots, x_{3N})$ the potential
energy of the system consisting of N ions ($3N$ degrees of freedom). By expanding V in Taylor series around the equilibrium positions x_i^0 and keeping
only the first significant terms of the expansion (namely restricting to small
vibrations) one has:

$$V(x_1, \cdots, x_{3N}) = \frac{1}{2} \sum_{i,j=1}^{3N} \frac{\partial^2 V}{\partial x_i \, \partial x_j}\bigg|_{x_i^0, x_j^0} (x_i - x_i^0)(x_j - x_j^0) + \cdots \quad (2.29)$$

that is the potential energy of a system of N *coupled* oscillators. Since the right
hand side of (2.29) is a quadratic form, there always exists a suitable change
of variables that takes such a system to a system of $3N$ one-dimensional *independent* oscillators, whose frequencies we shall denote by ν_i, $i = 1, \ldots, 3N$.

As an example, consider the system consisting of only
two coupled oscillators (Fig. 2.10a): it is straightforward to realize that the possible motions of such a
system are linear combinations of two fundamental
harmonic motions, in which the oscillators vibrate
either with the same phase (Fig. 2.10b) or with opposite phases (Fig. 2.10c), with different frequencies

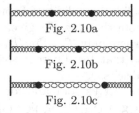

Fig. 2.10a

Fig. 2.10b

Fig. 2.10c

ν_1 and ν_2. It is in this sense that one says that two identical, elastically bound
oscillators are equivalent to a system consisting of two independent, but different (i.e. different frequencies) oscillators.

In the case at hand, one is dealing with many interacting oscillators that,
similarly, are equivalent to as many independent oscillators with different frequencies that start from a certain minimum ν_{min} up to a maximum ν_{max}.
The frequencies ν_i are no longer frequencies relative to the single ions: they
rather are collective oscillations (as in the case of Fig. 2.10), i.e. they are
frequencies (**characteristic frequencies**) of the crystal as a whole. One understands that ν_{min} must be of the order of magnitude of acoustic vibrations,
i.e. $\simeq 10^3 \, \text{s}^{-1}$.

From the point of view of classical theory, since $\overline{E}$ does not depend on
ν, there is no difference between Einstein and Debye models. Instead, applying the quantization condition to the oscillators, the Debye model leads to a
different result.

Indeed, the average energy of the i-th oscillators is given by (2.26):

$$\overline{E}_i = \frac{h\nu_i}{e^{h\nu_i/k_{\mathrm{B}}T} - 1}$$

and the total energy is $\mathcal{U} = \sum_{i=1}^{3N} \overline{E}_i$ (sum over all the oscillators).
For a fixed temperature T, put $\nu_T \equiv k_B T/h$
and assume that $\nu_{min} < \nu_T < \nu_{max}$: the oscillators whose frequency is $\nu_i < \nu_T$ – given
that for them $h\nu_i < k_B T$ – contribute to
the total energy with $\overline{E}_i \simeq k_B T$ (the classical value). For this reason such oscillators are

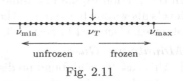

Fig. 2.11

said 'unfrozen' (Fig. 2.11). On the contrary, the oscillators for which $\nu_i > \nu_T$,
having $h\nu_i > k_B T$, give a contribution to the energy – and to the heat capacity – that tends to 0 as ν increases: they are said 'frozen'. It follows that,
by increasing T, and therefore ν_T, all the oscillators tend to unfreeze until
the classical value $3R$ for the atomic heat capacity is obtained. Instead, by
decreasing T, more and more oscillators become frozen and $C_V \to 0$. The
temperature such that all the oscillators are unfrozen, called **Debye characteristic temperature** and denoted by Θ, is defined by:

$$\Theta \overset{def}{=} \frac{h\nu_{max}}{k_B} \tag{2.30}$$

and gives an indication of the temperature at which the specific heat reaches
the classical value of the Dulong–Petit law.

So, at a certain temperature T, the solids for which $T > \Theta$ certainly will
follow the Dulong–Petit law, those for which instead $T \ll \Theta$ will still be far
away from the classical behaviour. Experimentally one has:

$$\Theta_{Fe} = 450\,K, \qquad \Theta_{Cu} = 315\,K, \qquad \Theta_{Pb} = 90\,K, \qquad \Theta_C = 2000\,K$$

(the last datum refers to diamond carbon).

The values given above for Θ and (2.30) enable one to obtain that ν_{max} is
of the order of $10^{14}\,s^{-1}$.

It is interesting to note that ν_{max} depends on the velocity v of the sound
in the solid, according to the following relationship:

$$\nu_{max} \simeq v \sqrt[3]{\frac{N}{V}} \tag{2.31}$$

where N is the number of atoms contained in the volume V. The origin of
(2.31) lies in the fact that, since a solid is not a continuous medium, the
minimum wavelength admissible for the vibrations of the solid is of the order
of magnitude of the interatomic distance, i.e. $(V/N)^{1/3}$, and $\nu_{max} \simeq v/\lambda_{min}$.
Also (2.31) leads, for ν_{max}, to values of the order of $10^{13} \div 10^{14}\,s^{-1}$.

The difference between Einstein model and Debye's is therefore in the fact
that, while for the first all the oscillators – being identical – either freeze or
unfreeze all together (this is the reason why $C_V \to 0$ too fast for $T \to 0$), in
the second model the oscillators freeze gradually when $T \to 0$, which entails
a vanishing of C_V slower ($\propto T^3$) than in the former case.

2.7 Heat Capacities of Gases

In this section we shall examine in which way the hypothesis of discrete energy levels allows one to overcome the difficulties associated with the heat capacities of gases, discussed in Sect. 1.3.

Monatomic Gases

In this case we had found no discrepancy between classical theory and experimental data. Indeed, the atoms of the gas – schematized as pointlike particles – are free particles and therefore Bohr quantization condition does not apply: the classical treatment needs no modification. One could object that the gas is closed in a vessel and the atoms are bound to move within precise limits, so that also in this case one will have energy levels.

To provide an answer to this objection, it is convenient to have a glance at the situation from a quantitative point of view. Let us take, for the sake of simplicity, the energy levels of a particle in a segment, given by (2.18). By choosing a (size of the vessel) $= 1\,\mathrm{cm}$ and $m = 10^{-24}\,\mathrm{g}$, one has:

$$E_n = n^2 E_1 \; ; \qquad E_1 = \frac{h^2}{8m\,a^2} \simeq 10^{-18}\,\mathrm{eV} \; . \tag{2.32}$$

We must now ask what is the relationship between $k_\mathrm{B}T$ and the distance among the energy levels. However this time, contrary to the case of harmonic oscillators, the distance between the energy levels is not constant but grows with n; indeed, for $n \gg 1$:

$$E_{n+1} - E_n \simeq 2n E_1 \; . \tag{2.33}$$

Of course we are interested in comparing $k_\mathrm{B}T$ with the distances among the levels that are appreciably populated, namely those levels for which $E_n \lesssim k_\mathrm{B}T$. Equation (2.32) implies that this happens for $n \lesssim \sqrt{k_\mathrm{B}T/E_1}$. Thanks to (2.33), the energy distances we are interested in are of the order of

$$\Delta E \lesssim \sqrt{E_1 k_\mathrm{B}T} \; .$$

The conditions $\Delta E \ll k_\mathrm{B}T$ or $\Delta E \gg k_\mathrm{B}T$ take respectively the form:

$$\sqrt{E_1} \ll \sqrt{k_\mathrm{B}T} \qquad \text{or} \qquad \sqrt{E_1} \gg \sqrt{k_\mathrm{B}T} \; . \tag{2.34}$$

Recalling now that, for example, for $T = 1\,\mathrm{K}$, $k_\mathrm{B}T = 1/12000\,\mathrm{eV}$, (2.32) implies that *for any temperature* $\sqrt{k_\mathrm{B}T/E_1} \gg 1$ (e.g. He becomes liquid at about $4\,\mathrm{K}$), and therefore quantization is undetectable and the atomic heat capacity C_V always is $\frac{3}{2}R$.

Diatomic Gases

In the discussion of Sect. 1.3 we had, initially, schematized the molecule as made out of two (pointlike) atoms at a fixed distance d. The Hamiltonian is given by (1.22): it can be written as

$$H = H_\mathrm{tr} + H_\mathrm{rot}$$

namely the sum of the first term that represents the translational motion of the center-of-mass, and of the second term that represents the energy associated with rotations.

The discussion made about monatomic gases teaches us that, as for the part relative to the translational degrees of freedom, quantization does not show up and, as a consequence, H_{tr} contributes $\frac{3}{2}R$ to the molar heat capacity C_V. Let us now examine the contribution of H_{rot} to C_V.

In order to determine the energy levels associated to the rotational degrees of freedom, it is convenient to recall the physical meaning of H_{rot}. For the sake of simplicity we shall restrict to the case of identical atoms: one has

$$H_{\mathrm{rot}} = \tfrac{1}{2}m\,v_1^2 + \tfrac{1}{2}m\,v_2^2 \; .$$

In the center-of-mass frame, however, $v_1 = v_2$ so that

$$H_{\mathrm{rot}} = m\,v^2 = \frac{(m\,v\,d)^2}{m\,d^2} = \frac{L^2}{2I}$$

where L is the angular momentum of the system and I is the moment of inertia with respect to an axis passing through the center of mass and orthogonal to the segment joining the two atoms.

Thanks to the Bohr quantization condition $L = n\,\hbar$ already used for the hydrogen atom (see (2.8)), one finds:

$$E_n = \frac{\hbar^2\,n^2}{2I} \qquad\qquad \text{(rigid rotator)} \qquad\qquad (2.35)$$

(actually, the energy levels we shall find later by quantum mechanics are given by $E_n = (\hbar^2/2I)\,n(n+1)$; the difference with the levels given by (2.35) decreases with the increase of the quantum number n).

To discover under which conditions the effect of quantization is detectable – recall the discussion made for monatomic gases (note that also in that case $E_n \propto n^2$) that has led to (2.34) – we must compare $k_{\mathrm{B}}T$ with $\hbar^2/2I$. In other words we can, for every gas, define a characteristic temperature $T_c^{\mathrm{rot}} \equiv \hbar^2/(2I\,k_{\mathrm{B}})$: for $T \gg T_c^{\mathrm{rot}}$ we expect to find again the classical value R as contribution to C_V by the two rotational degrees of freedom, while for $T \ll T_c^{\mathrm{rot}}$ we expect that the effect of quantization shows up with a $C_V^{(\mathrm{rot})}$ tending to 0 for $T \to 0$. Indeed, in order to calculate $\mathcal{U}$, and therefore C_V, at low temperatures for *whatever* system endowed with discrete energy levels, we may limit ourselves to consider only the first two levels (in the case the third is not too close to the second): the lower the temperature, the lesser the contributions to the average energy of the excited levels.

So, putting $\Delta E = E_2 - E_1$, one has

$$\mathcal{U} = N\,\frac{E_1\,\mathrm{e}^{-\beta\,E_1} + E_2\,\mathrm{e}^{-\beta\,E_2}}{\mathrm{e}^{-\beta\,E_1} + \mathrm{e}^{-\beta\,E_2}} = N E_1 + N\Delta E\,\frac{\mathrm{e}^{-\beta\,\Delta E}}{1 + \mathrm{e}^{-\beta\,\Delta E}}$$

whence:

$$C_V = R \left(\frac{\Delta E}{k_{\mathrm{B}} T} \right)^2 \mathrm{e}^{-\Delta E/k_{\mathrm{B}} T}, \qquad\qquad T \ll \Delta E/k_{\mathrm{B}} \qquad (2.36)$$

(as a confirmation of the generality of this result, note that (2.36) coincides with (2.27) derived directly from (2.26)).

Let us now give the values of T_C^{rot} for some gases:

$$\mathrm{H}_2: \ 85\,\mathrm{K}, \qquad \mathrm{N}_2: \ 3\,\mathrm{K}, \qquad \mathrm{O}_2: \ 2\,\mathrm{K}, \qquad \mathrm{NaCl}: \ 15\,\mathrm{K}$$

and it appears from these data that the effect of quantization, in practice, is detectable only for H_2: in this case $\Delta E = \hbar^2/2I \simeq 10^{-2}\,\mathrm{eV}$, whereas for the other gases ΔE is smaller by a factor $10 \div 100$.

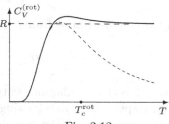

Fig. 2.12

In the present case, contrary to the case of oscillators, it is not possible to calculate $\mathcal{U}(T)$ analytically, and the same is true for $C_V(T)$; the numerical calculation is however possible. The solid curve in Fig. 2.12 has been obtained by making use of the correct quantum expression for the energy levels, including the degeneracy factors g_n that, in this case, are $g_n = 2n + 1$; only five levels have been taken into consideration (the lowest and the first four excited energy levels) and we have verified that, at least for $T \leq 2T_c^{\mathrm{rot}}$, and up to 1‰, the curve is not modified by the inclusion of 20 further levels: as it is evident, for $T \geq 0.5\,T_c^{\mathrm{rot}}$, $C_V \simeq R$.

The dashed curve is the one obtained by taking into consideration only the first two levels: in agreement with (2.36) the curve reproduces very well the behaviour of C_V at low temperatures, whereas for high temperatures tends to 0, given that for a system with a finite number of energy levels the internal energy tends to a constant that corresponds to the equipopulation of the levels.

Let us now consider the problem of the other degrees of freedom of the diatomic molecule: in the first place let us consider the degree of freedom relative to the rotation around the line that joins the nuclei. Also for this rotational degree of freedom we shall have energy levels of the type:

$$E_n = \frac{\hbar^2}{2\,I'}\, n^2$$

in which, in comparison with (2.35), only the moment of inertia is changed. A diatomic molecule has the structure reported in Fig. 2.13: the nuclei are heavy, but small ($10^{-12} \div 10^{-13}$ cm), so the main contribution to I' is given by the electrons that are light, but rotate at a distance of about 10^{-8} cm from the nuclei. So, for example, in H_2, since the ratio between the masses of nuclei and electrons is about 2000, it follows that $I' \simeq I/2000$.

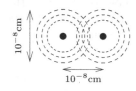

Fig. 2.13

As a consequence, in this case, the ΔE and the characteristic temperature are about 2000 times higher than in the previous case: $\Delta E \simeq 10\,\mathrm{eV} \Rightarrow T_c \simeq 10^5\,\mathrm{K}$. Therefore this degree of freedom does not contribute to C_V inasmuch as permanently frozen, even at temperatures of some thousand degrees.

Finally, to conclude, let us consider the vibrational degree of freedom of the diatomic molecule. As of now, it will be clear that, in order to understand whether this degree of freedom contributes to C_V or not at a given temperature, one must compare the $k_B T$ we are interested in with the ΔE that, dealing with a harmonic oscillator, equals $\hbar\omega$. Clearly $\hbar\omega$ depends on the 'hardness', i.e. on the elastic constant of the 'spring' that binds the two atoms.

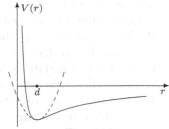

Let us therefore try to get an idea of how much is ω for a molecule. We know the potential $V(r)$ that binds the atoms is of the type reported in Fig. 1.3. In the approximation of small oscillations around the equilibrium point, we are interested in approximating $V(r)$ around the point $r = d$ as (Fig. 2.14):

Fig. 2.14

$$V^{\mathrm{harm}}(r) = \frac{1}{2} k\,(r-d)^2 - V_0\,. \qquad (2.37)$$

Let us try to estimate k by exploiting the following information:

i) an energy of the order of some electronvolt is needed to dissociate a molecule, so $V_0 \simeq 1\,\mathrm{eV}$;
ii) the distance between the atoms is of the order of the ångström; so also the distance among the points where $V^{\mathrm{harm}}(r) \simeq 0$ will be of the same order of magnitude.

These data are sufficient to determine k in (2.37) and, apart from numerical factors, it turns out that (M is the mass of the nuclei; better: the reduced mass):

$$k = \frac{V_0}{d^2} \quad \Rightarrow \quad \omega = \sqrt{\frac{k}{M}} = \sqrt{\frac{V_0}{M\,d^2}} \simeq 10^{14}\,\mathrm{s}^{-1}\,. \qquad (2.38)$$

The result is reasonable, the fact that a molecule radiates in the infrared being known. In conclusion: $\Delta E = \hbar\omega \simeq 10^{-1}\,\mathrm{eV}$.

It appears that at room temperature ($k_B T = 1/40\,\mathrm{eV}$) this degree of freedom practically is frozen, but at temperatures even not so higher ($T \simeq 10^3\,\mathrm{K}$) it may give an appreciable contribution to the C_V. We report, as an example, the characteristic temperatures $T_c^{\mathrm{vib}} \equiv \hbar\omega/k_B$ relative to the vibrational degrees freedom for some molecules:

$$\mathrm{H_2}:\ 6100\,\mathrm{K}\,, \qquad \mathrm{N_2}:\ 3340\,\mathrm{K}\,, \qquad \mathrm{O_2}:\ 2230\,\mathrm{K}\,.$$

Let us summarize in a table the results we have found, reporting for the several degrees of freedom of a diatomic molecule the typical values of ΔE and the relative characteristic temperatures T_c.

translations		$\Delta E \simeq 0,$	$T_c \simeq 0$
rotations	(diagram)	$\Delta E: \; 10^{-4} \div 10^{-2}\,\text{eV},$	$T_c: \; 1 \div 100\,\text{K}$
vibrations		$\Delta E \simeq 10^{-1}\,\text{eV},$	$T_c \simeq 10^3\,\text{K}$
(electrons		$\Delta E \simeq 1\,\text{eV},$	$T_c \simeq 10^4\,\text{K}$)
(rotations	(diagram)	$\Delta E \simeq 10\,\text{eV},$	$T_c \simeq 10^5\,\text{K}$)

Here is how the hypothesis of energy levels solves the problem of the ambiguity in the counting of the degrees of freedom: let us attribute to the system all the degrees of freedom we wish (electrons in the atoms, nucleons in the nuclei, quarks in the nucleons ...), but in the end, in order to decide whether they do contribute to the C_V, we must check how much is the distance between the energy levels associated with such degrees of freedom; for the electrons we shall find a ΔE of the order of some eV, for nucleons of some MeV, ... and in all such cases we discover that these degrees of freedom can be ignored at any reasonable temperature.

2.8 Wave-like Behaviour of Particles: the Bragg and Davisson–Germer Experiments

In the previous sections of this chapter we have introduced a certain number of hypotheses:

1. the existence of photons;
2. the Bohr hypothesis of quantization;
3. the Bohr hypothesis that relates the frequency radiated in a transition with the energy jumps.

These hypotheses have then been used with a remarkable success to explain several experimental facts. At this point it is natural to ask what survives of classical physics: are the above hypotheses a sort of 'amendments' to the laws of classical physics that, however, keeps its structure substantially untouched or, rather, the success of the above hypotheses hints at the necessity of profoundly modifying the concepts of classical physics? At the end of the present section there will be no doubt about which the answer should be.

In 1923, about ten years after Bohr had proposed his theory, L. de Broglie put forward the following problem. The electromagnetic radiation, to which always and correctly a wave-like nature had been attributed, also behaves in a totally different way, i.e. in a corpuscular way (photons). The relationship between the two aspects, wave-like $\rightarrow$ corpuscular, is provided by Planck's constant h. But h also enters, through Bohr–Sommerfeld quantization condition, problems where particles are dealt with. Furthermore, in the quantization condition integer numbers n do intervene: any physicist knows by experience that the presence of integer numbers often indicates one is dealing with undulatory phenomena (for example: interference fringes, standing waves on a string, ...). In conclusion: could it not be that, through h, one should make the contrary step, namely look for a wave-like behaviour in what has always been considered corpuscular (particles)?

Let us consider, for example, a particle going a certain orbit (an electron in a circular orbit in hydrogen, or a particle in a segment): then the Bohr–Sommerfeld condition $\oint p \, dq = n \, h$ entails:

$$p L = n h \qquad\qquad (L = \text{length of the orbit}) \qquad\qquad (2.39)$$

If we rewrite (2.39) in the form

$$L = n \frac{h}{p} \qquad\qquad (2.40)$$

and recall that for a photon $h/p = \lambda$, then (2.40) suggests the following interpretation

de Broglie hypothesis: *a wave is associated with each particle. The relationship between the wavelength λ and the momentum p is given, as for photons, by $\lambda = h/p$. The allowed orbits are those that contain an integer number of wavelengths.*

Bohr quantization condition is therefore looked upon in a totally new light.

But so far we are in the field of hypotheses. Is de Broglie's hypothesis only a new terminology or there is something more in it? What kind of wave is one talking about? We know two types of waves: elastic waves and electromagnetic waves. Certainly one is not dealing with elastic waves and, presumably, not even with electromagnetic waves, inasmuch as the de Broglie hypothesis does not necessarily refer to charged particles. It is likely that we shall have to deal with a third type of waves. In this case intuition is of little help: it is convenient to ask oneself which are the general features of waves. Waves are entities that can be summed with each other or, in loose terms, waves are things that *interfere*. So, if these de Broglie waves really exist, they must be revealed by means of interference phenomena.

Before thinking of an experiment that emphasizes the wave-like aspect of particles, it is convenient to realize the size of the wavelengths one has to deal with. Let us consider free particles: from the relations $\lambda = h/p$ and $p = \sqrt{2m\,E}$ one obtains the **de Broglie wavelength**:

$$\lambda = \frac{h}{\sqrt{2m\,E}} \, . \qquad\qquad (2.41)$$

If, as customary, we express λ in Å and E in eV, for an *electron* one has

$$\lambda[\text{Å}] = \frac{12.4}{\sqrt{E[\text{eV}]}} \qquad\qquad (\text{for electrons}) \qquad\qquad (2.42)$$

i.e. for an electron of energy $1\,\text{eV}$, $\lambda = 12.4\,\text{Å}$.

Note that, while for photons λ is proportional to $1/E$, for massive (non-relativistic) particles λ is proportional to $1/\sqrt{E}$. In addition, the higher the mass, the lower – for a given energy – the wavelength λ. In conclusion, for electrons with energy of the order of $100\,\mathrm{eV}$, λ is of the order of the ångström, i.e. of the same order of magnitude of the X-ray wavelengths.

A way to show that X-rays are electromagnetic waves consists in analyzing the reflection of X-rays by a crystal (***Bragg reflection***). Let us consider a crystal; the regularity of the arrangement of the atoms (or ions) determines many families of lattice planes: in Fig. 2.15 one of these families is represented, for example that parallel to a face of the crystal. The distance d between

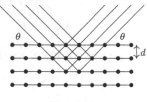

Fig. 2.15

the lattice planes (of a family) is called ***lattice spacing***: d normally is of the order of a few ångström.

If we let some monochromatic X-rays impinge on the crystal, at an angle θ with respect to the crystal face, and observe the specularly reflected (i.e. emerging at the same angle θ) radiation, one notes that the radiation is reflected only for certain particular values of θ : $\theta_1, \theta_2, \cdots$. More precisely, very sharp maxima in the intensity of the reflected radiation are observed in correspondence of the angles $\theta_1, \theta_2, \cdots$, alternating with minima in which the intensity is practically vanishing.

The explanation of this fact is the following: X-rays are reflected by the various lattice planes (Fig. 2.15): the waves reflected by two adjacent planes (Fig. 2.16) go optical paths whose difference is $2d\sin\theta$ (the darker segments in Fig. 2.16). When $2d\sin\theta$ is an integer multiple of the wave-

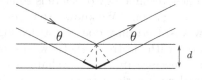

Fig. 2.16

length λ, all the reflected waves are in phase with one another and there occurs a constructive interference, i.e. a maximum in the reflected intensity. If, on the contrary, $2d\sin\theta$ is an odd integer multiple of $\lambda/2$, the waves reflected by two adjacent lattice planes are in opposition of phase and there occurs destructive interference, i.e. no reflected radiation is observed. In the intermediate cases there occurs destructive interference among the waves reflected by several lattice planes, which results in an almost vanishing intensity. In conclusion the maxima in the intensity of the reflected radiation occur for those angles θ that satisfy the ***Bragg condition***:

$$2d\sin\theta = n\lambda\,. \tag{2.43}$$

The number of maxima observed when θ is let to vary from 0 to $\pi/2$ is the maximum integer contained in $2d/\lambda$: this explains why, in order to easily observe the phenomenon with X-rays ($\lambda \simeq 1\,\text{Å}$), it is necessary that d be of the same order of magnitude of λ, i.e. why crystals are used in this type of experiments. It is also clear that the knowledge of the angles θ satisfying (2.43) allows for the determination of λ, given d; or viceversa of d, given λ.

The same experiment was made by C. Davisson and L. Germer in 1927 replacing X-rays with a sharp beam of (as much as possible) monoenergetic *electrons*.

It was observed that in this case also the electrons were specularly reflected only for some values of θ in agreement with (2.43): this proves that also electrons possess a wave-like behaviour. Furthermore, the Davisson–Germer experiment allows one to determine λ from (2.43) and to verify the relationship (2.41) between λ and E.

We add that, along with the development of experimental techniques, Davisson–Germer experiment has been repeated also with different particles: protons, neutrons, He atoms, ions; and in all these cases the Bragg condition has been confirmed for matter particles.

So the de Broglie idea of associating a wave with each particle was not simply a smart theoretical speculation, but a well precise physical reality corresponded to it.

Davisson experiment first, and Davisson–Germer experiments later had an accidental origin: as a consequence of the explosion of a bottle of liquid air in the laboratory where Davisson was making experiments that used electron beams, a vacuum tube with an electrode of poly-crystalline nickel went broken; owing to the thermal treatment the electrode underwent in order to be fixed, nickel crystallized in coarse grains and this led to results that, for what regarded the angular distribution of the electrons reflected by the electrode, were completely different from the previous ones.

Chapter 3

Introduction to the Postulates of Quantum Mechanics

3.1 Introduction

In the previous chapter we have exposed some of the fundamental steps of what is now referred to as "Old Fashioned Quantum Mechanics" and that, from 1905 until 1925, had obtained several important results, particularly in the framework of atomic physics, in addition to those we have already told. For example Sommerfeld, hinging on a refinement of the Bohr quantization condition, had been even able to calculate the relativistic corrections to the energy levels of hydrogen.

However, the set of hypotheses and rules that allowed to obtain such results scarcely could be considered a 'theory'. Too many questions had no convincing answer, one in particular: the dualism wave/corpuscle, required by both the notion of photon and by the de Broglie hypothesis, still was a very misty concept. Above all, putting all these ideas together in a well arranged and consistent theoretical scheme was still lacking. It was quite clear that many paradigms of classical physics had to be abandoned, but the new paradigms were not yet in sight.

In this chapter we will critically analyze some simple experiments (many of which really performed), the so called 'one photon experiments' (or one neutron experiments). The purpose will be to emphasize the contradictions between the concepts of classical physics – according to which a particle is a particle and a wave is a wave – and the experimental facts: in this way we shall understand how we must change our schemes of reasoning, in order to reconcile the existence of both the wave-like and the corpuscular aspect belonging to one single entity. The discussion we shall make has the twofold purpose of both introducing in an inductive way the postulates of quantum mechanics and, also, of 'psychological preparation' to its abstract formulation.

The 'key' according to which the next sections should be read is the following: we will expose the facts and try to draw some conclusions; the latter have to be regarded upon as the first steps to arrive at the interpretation of the facts, namely to build up a 'theory'. The conclusions we shall propose (maybe not always the only possible, but – without any doubt – the simplest

© Springer International Publishing Switzerland 2016
L.E. Picasso, *Lectures in Quantum Mechanics*, UNITEXT for Physics,
DOI 10.1007/978-3-319-22632-3_3

and, in some sense, the most 'natural' ones) are those that gave rise to the so called **Copenhagen interpretation** of quantum mechanics. To the present day (2014), such interpretation still is under discussion, but it has not yet been replaced by another one. The Copenhagen Institute of Physics, since 1921 directed by Niels Bohr, was – together with that in Göttingen, where Max Born worked – the center of all the discussions that regarded the then springing quantum mechanics. Many among the greatest physicists, like P.A.M Dirac, W. Heisenberg and E. Schrödinger spent there some of their time.

The next section stands, is some sense, for itself: it contains a short description of the Mach–Zehnder interferometer, to which we shall refer in the sequel. The reason to illustrate it in advance is not to interrupt the discussion of Sect. 3.3.

3.2 The Mach–Zehnder Interferometer

Mach–Zehnder interferometer, represented in Fig. 3.1, consists of four mirrors parallel to one another, arranged at the vertices of a rectangle and forming angles of 45° with the edges of the rectangle. Two such mirrors (s_2 and s_3) are normal mirrors, whereas the other two (s_1 and s_4) are semi-transparent mirrors: a beam of light that hits one of them is par-

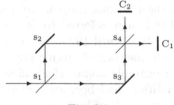

Fig. 3.1

tially reflected and partially transmitted (beam splitters): we shall assume that 50% of the incident intensity is reflected and 50% transmitted.

A beam of light, coming from the left, hits s_1: the transmitted beam is reflected by s_3 and hits s_4, where it undergoes a partial reflection towards the counter C_1 and a partial transmission towards the counter C_2. The beam reflected by s_1 is reflected by s_2, and is finally split by s_4, a part towards C_1, the other part towards C_2. So on each of the counters there arrive both part of the beam coming from the arm $s_2 - s_4$ and part of the beam coming from the arm $s_3 - s_4$: these two beams interfere with each other and the intensities I_1 and I_2 recorded at the two counters are the results of the interference between the two beams that arrive at each of them. In the case one of the two beams (no matter which) should be intercepted, e.g. by interposing an absorbing material in one of the two paths, the two counters C_1 and C_2 would both record the same intensity, the 25% of the intensity of the incident beam.

Let us take the x-axis in the direction of the incident beam. Let us assume that the incident light is a plane wave that, for the sake of simplicity, we shall assume linearly polarized (although this assumption is not strictly necessary). In this way the wave can be written as:

$$E(x,t) = E_0 \cos(k\,x - \omega\,t) \tag{3.1}$$

where $E(x,t)$ is the only nonvanishing component of the electric field.

Since the intensity of the wave is quadratic in the amplitude E_0, the amplitudes of both the waves transmitted and reflected by the beam splitter are reduced by a factor $\sqrt{2}$ with respect to the incident wave. So the wave that arrives at C_1 following the path $s_1 \to s_2 \to s_4$ is

$$\frac{1}{2} E_0 \cos(\omega t + \varphi_1) \qquad (3.2)$$

and the wave that arrives at C_1 following the path $s_1 \to s_3 \to s_4$ is

$$\frac{1}{2} E_0 \cos(\omega t + \varphi_2) . \qquad (3.3)$$

If the optical paths $s_1 \to s_2 \to s_4$ and $s_1 \to s_3 \to s_4$ were exactly equal – in length – to each other, one would have $\varphi_1 = \varphi_2$. However, on the one hand, it is impossible to gauge the interferometer in such a way that the difference of the optical paths be much less than the wavelength and, on the other hand, it may be useful to introduce a phase shift between the two waves, then in (3.2) and (3.3) we shall keep $\varphi_1 \neq \varphi_2$ and put $\varphi = \varphi_1 - \varphi_2$.

So, since the intensity of the wave incident on s_1 is the time-average over a period of $(c/4\pi)E^2(x,t)$, i.e.

$$I = \frac{c}{4\pi} E_0^2 \ \overline{\cos^2(k\,x - \omega\,t)} = \frac{c}{8\pi} E_0^2 ,$$

the intensity at C_1 is

$$I_1 = \frac{c}{4\pi} \frac{1}{4} E_0^2 \ \overline{\left(\cos(\omega t + \varphi_1) + \cos(\omega t + \varphi_2)\right)^2} = \frac{1}{2} I \left(1 + \cos\varphi\right) . \qquad (3.4)$$

As a consequence, given that $I_1 + I_2 = I$, one has:

$$I_2 = \frac{1}{2} I \left(1 - \cos\varphi\right) . \qquad (3.5)$$

The intensity at each of the counters may thus vary from 0 to I as a function of the phase shift. The phase shift may itself be varied with continuity by inserting a thin slab of glass in one of the arms of the interferometer and varying its inclination in such a way as to change the thickness of the glass gone through by the beam: if n stands for the refraction index of the glass, the phase shift one introduces in this way is $\delta\varphi = k\,(n-1)\,d = 2\pi\,(n-1)\,d/\lambda$, d being the thickness gone through.

3.3 One-Photon Interference Experiments

A beam of monochromatic light hits an opaque screen on which two holes A and B (or two parallel slits), small and very close to each other, have been made. In crossing the holes, the light undergoes diffraction and, on a photo-graphic plate put some distance away from the screen, interference fringes are observed in the zone in which the two diffraction pattern produced by A and B overlap – i.e. where light arrives from *both* A *and* B (Fig. 3.2). This is the proof of the wave character of light (T. Young experiment, 1802).

How is the above result interpreted in terms
of photons? Clearly, where the fringes are
brighter there arrive many photons, where
they are darker there arrive less.

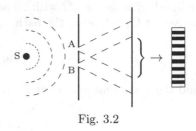

It looks as if the photons interacted with
one another in such a way as to give rise to
the interference pattern on the plate. But
what happens if one reduces the intensity
up to the point that, on the average, there is

Fig. 3.2

just one photon at a time in the path between the source and the photographic
plate? Certainly it happens that one has to increase the exposure time of the
plate in such a way as to allow many photons to arrive on it, but in the end,
after developing the image, one finds the same interference pattern as before. If
instead the plate is exposed for a short time, many very tiny spots, arranged
randomly, are observed: the spots represent the 'points' where the photons
have arrived and have produced the chemical reaction that has impressed the
plate. If one repeats the experiment with longer and longer exposure times,
one observes that the distribution of photons on the plate more and more
looses its appearance of randomness and, more and more, reproduces the
distribution of the interference pattern.

Then the interpretation in terms of interacting photons does not hold:
photons do not interact with one another and the only 'simple' conclusion is
that interference depends on the single photon (it is as if it were 'encoded' in
the single photon).

Now, however, according to the corpuscular model, a photon that leaves the source and
arrives at the plate, *either* passes through A *or*
through B. But, this being the alternative, we
do not succeed in explaining the interference
pattern: indeed, if the photon passes through
A, it is as if B were closed, and the photon
should be able to arrive at all the points that
can be reached while forming the diffraction

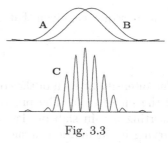

Fig. 3.3

pattern of hole A, even at those points where there are dark fringes – where,
clearly, there arrive no photons (see Fig. 3.3 where the intensity curves of
the diffraction patterns of holes A and B, as well as the intensity curve C of
the interference pattern, are reported). The same happens if we say that the
photon goes through B. So, as 50% of the photons would go through A and
50% through B, the result on the plate should be the sum of the diffraction
patterns separately produced by the holes: it is as if, after receiving the light
from A only, the plate were exposed to the light coming from B only. The
result would have nothing to do with the interference pattern C (Fig. 3.3)
that is, instead, observed.

Before proceeding any further we wish to present another experiment,
similar to the previous one, but – perhaps – more surprising: an experiment

of interference in which, instead of photons, neutrons produced in a nuclear reactor are used.

Neutron interferometer, realized for the first time in the 70's of the 20-th century, is similar to the Mach–Zehnder interferometer used for light (Figs. 3.1 and 3.4).

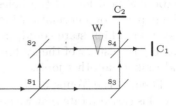

Fig. 3.4

The 'mirrors' are silicon crystals that partly transmit and partly reflect neutrons à la Bragg: for simplicity we treat s_2 and s_3 as totally reflecting mirrors. Neutrons hit the mirror s_1 and are counted by the counters C_1 and C_2. In one arm of the interferometer (the arm $s_2 - s_4$ in Fig. 3.4) a small aluminum wedge W is inserted (aluminum does not absorb neutrons and has the same task as the glass slab in the interference of light). When the wedge is pushed inside the beam, the thickness gone through by neutrons is increased – the phase shift φ is changed – and one observes that the number of neutrons registered by the two counters varies according the interference laws given by (3.4) and (3.5). In this case the problem of reducing the intensity does not show up: the intensity is so low that, when a neutron arrives at a counter, the next neutron has not yet been produced within the reactor!

So interference is not a feature of photons: also neutrons (as well as other particles like electrons, atoms, ...) give rise to the same phenomena and also in this case the partition of the neutrons between the two counters – the analogue of the partition of photons among the many fringes – cannot be ascribed to an interaction among neutrons, it rather depends on the behaviour of the single neutron.

Also in this case, according to common sense, a neutron inside the interferometer *either* goes the path $s_1 \to s_2 \to s_4$ *or* the path $s_1 \to s_3 \to s_4$: but, if thing were this way, the countings at C_1 and C_2 should be identical, regardless of the position of the wedge.

There must exists a third possibility – not conceivable from the classical point of view – and the same conclusion applies to the photons of the Young experiment.

We now introduce an important concept, that of **state** of a system. In classical physics, for example, the state of a particle at a given time is defined by position and velocity at that time. In the 'new physics', given that one cannot even say if a neutron follows either a path or another, one must be very cautious: the state of a system (photon, neutron, ...) is defined by all the pieces of information regarding the way it has been 'prepared'. For example, in the case of the Young experiment, the state of the photon is defined by saying which source has been used (i.e. which is its frequency), which polaroid sheet it has gone through (therefore which is its polarization), how far from the source is the screen with the holes located, in which way the holes have been made, how much is the distance between them, ... : in general, all what is needed to characterize completely and reproducibly the experimental situation.

Therefore, the information that defines the state of a system exclusively regards quantities under our control, i.e. that we can modify (type of source, shape and positions of the holes, ...): it is the task of the theory to specify which pieces of information indeed matter in the definition of the state of a system (which state could, in principle, even require the knowledge of the entire past history of the system – in the latter case we would better give up and look for another job!).

Then, in the Young experiment, the state **A**, defined by the fact that the hole B is closed (state that we describe by saying that the photon goes through A), obviously is different from the state **B** in which A is closed, and the state **C**, defined by the fact that both holes are open, obviously is different from both **A** and **B** (Fig. 3.3), and – less 'obviously' – it is *not*: 'sometimes **A**, sometimes **B**' (otherwise, as we have already said, the intensity on the plate should be the sum of the intensities of the diffraction patterns, namely the sum of the curves **A** and **B** of Fig. 3.3), even if – in some sense – **C** 'must have something to do' with both **A** and **B**.

In other words, photons go through neither A nor B (neutrons go neither the path $s_1 \to s_2 \to s_4$ nor the path $s_1 \to s_3 \to s_4$), but it is as if each of them went simultaneously through both A and B (... it is as if each neutron simultaneously went both the paths): clearly the state **C** has no right of citizenship in classical physics.

The words we have used to describe the state **C** ("... it is as if each neutron simultaneously went both ...") are suggested by the wave-like interpretation of the phenomenon of interference: if only hole A is open, one has a wave $A(x, y, z, t)$ between the screen and the plate; if only B is open, one has a wave $B(x, y, z, t)$ beyond the screen. If instead both holes are open, the wave is neither $A(x, y, z, t)$ nor $B(x, y, z, t)$, but it is $C(x, y, z, t) = A(x, y, z, t) + B(x, y, z, t)$.

So far, however, our conclusions are only speculative in character: we have said that the idea that the photon goes through either A or B, or that a neutron follows a well defined path, cannot be maintained. But what do experiments say? Could one set up a 'which way experiment', i.e. an experiment able to say, photon by photon (or neutron by neutron), which path has it gone?

For example, what should one expect if two photomultipliers F_1 and F_2 (counters that detect the photons by exploiting photoelectric effect) are put just after the two holes A and B (Fig. 3.5) and then a photon is let to leave the source? Should one expect that F_1 and F_2 click simultaneously?

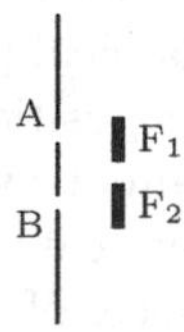

Were it so, we would succeed in splitting a photon in two parts!

Fig. 3.5

What instead happens is that each time only one photomultiplier clicks and, by repeating the experiment several times, one sees that, on the average, half of the times F_1 clicks, the other half F_2 clicks. Does this fact mean that the previous conclusion about the existence of a state **C** is wrong? And that, indeed, any photon is either in the state **A** or **B** and, in conclusion –

and contrary to what we have stated – the state **C** is 'sometimes **A**, sometimes **B**'?

Before taking this conclusion as a good one, let us observe that the insertion of the photomultipliers F_1 and F_2 does allow us to know which path has been taken by the photons, but destroys the observability of the interference fringes: indeed the photons are absorbed by the photomultipliers!

Perhaps by a smarter experiment one can both see which path is taken by the photons and observe the interference fringes.

Let us imagine the following variation on the theme of the Young experiment: just after each hole one puts two mirrors one of which (the thinner ones in Fig. 3.6) is a very light mobile mirror kept in place by a spring; when a photon crosses the screen, the observation of which mirror has recoiled enables one to know which path the photon has followed. Put in this way, it is a "Gedankenexperimente"; but, set up in a much more refined way, it can really be performed by using neutrons.

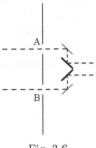

Fig. 3.6

The result is that, if we are able to detect the recoils of the mobile mirrors, the interference fringes are no longer observable: we will come back to this point at the end of this chapter. In the case of neutrons, if for examples mirrors s_2 and s_3 are mobile and their recoil can be detected, then $I_1 = I_2$, regardless of the position of the wedge.

We can conclude that the observation of interference is incompatible with the possibility of establishing which path has been followed by either the neutron or the photon.

We conclude this section by briefly summarizing the important concepts introduced in this discussion.

1. Interference fringes are observed even by sending one photon at a time: therefore it is the single photon that "interferes with itself".
2. The single photon goes through neither A nor B: there exists a third possibility, not allowed in classical physics, namely a state **C** that in some way (yet to be established) is in relationship with both **A** and **B**.
3. A measurement that enables one to know whether the photon has gone through either A or B modifies (perturbs) the state of the photon to such an extent that the interference fringes are no longer observed: either we know which path or we observe the fringes, never both the two things.

3.4 Description of the Polarization States of a Photon

In the previous section we have seen that the interference phenomena, being ascribable to the behaviour of a single particle, lead us to admit the existence of states that have no classical analogue. In order to substantiate the theory, we have to give a formal status to this concept, i.e. we have to establish which is the mathematical framework suitable to describe the states of a system and which operations can be performed on them (e.g. in classical physics the states of a system are represented by the points of the phase space).

The magic word to open this door is the word 'interference': interference means superposition principle, i.e. linearity, namely linear combinations – and this takes us to assume that the states of a system are represented by entities for which such operations are possible: the elements of a vector space.
Which kind of vector space? Real or complex? Finite or infinite-dimensional? Endowed with a scalar product or not? To be able to give an answer to these questions, let a simple example guide us: the description of the polarization states of a photon.

To describe a system means to give the pieces of information necessary and sufficient to characterize it: for example, we know very well that in an electromagnetic wave there are both the electric and the magnetic field, but for its description the electric field is sufficient; so, if one has a plane monochromatic wave that propagates in the direction of the z-axis, the wave is described by

$$\begin{cases} E_x(z,t) = E_{0x} \cos(k z - \omega t + \varphi_1) \\ E_y(z,t) = E_{0y} \cos(k z - \omega t + \varphi_2) \\ E_z(z,t) = 0 \, . \end{cases} \tag{3.6}$$

Putting $E_{0x} \equiv E_0 \cos\vartheta$, $E_{0y} \equiv E_0 \sin\vartheta$, (3.6) can be written in the form:

$$\begin{cases} E_x(z,t) = \frac{1}{2} E_0 \cos\vartheta \, (e^{i\varphi_1} e^{i(k z - \omega t)} + \text{c.c.}) \\ E_y(z,t) = \frac{1}{2} E_0 \sin\vartheta \, (e^{i\varphi_2} e^{i(k z - \omega t)} + \text{c.c.}) \\ E_z(z,t) = 0 \end{cases} \tag{3.7}$$

(c.c. stands for "complex conjugate"); if $\vec{e}_1$ and $\vec{e}_2$ denote the unit vectors of the x and y axes, one has:

$$\vec{E}(z,t) = \frac{1}{2} \left[(\cos\vartheta \, e^{i\varphi_1} \, \vec{e}_1 + \sin\vartheta \, e^{i\varphi_2} \, \vec{e}_2) \, e^{i(k z - \omega t)} + \text{c.c.} \right] . \tag{3.8}$$

As for the pieces of information on the type of wave (frequency, polarization, intensity, direction of propagation), the c.c. is superfluous: it is there only to recall us that the electromagnetic field is real. The piece of information concerning the intensity of the wave is contained in the factor E_0: so, if one is interested in the single photon, all the information is contained in the expression:

$$(\cos\vartheta \, e^{i\varphi_1} \, \vec{e}_1 + \sin\vartheta \, e^{i\varphi_2} \, \vec{e}_2) \, e^{i(k z - \omega t)} \, . \tag{3.9}$$

Indeed the description (3.9) is redundant: we can write (3.9) as

$$e^{i\varphi_1} (\cos\vartheta \, \vec{e}_1 + \sin\vartheta \, e^{i\varphi} \, \vec{e}_2) \, e^{i(k z - \omega t)} \tag{3.10}$$

where $\varphi = \varphi_1 - \varphi_2$. The phase factor $e^{i\varphi_1}$ has no role in establishing the state of the photon: already in (3.6) it possesses no physical meaning (inasmuch as it can be changed by redefining the origin of time or of the coordinates) and, as a consequence, in all the expressions that follow. Therefore, whatever the value of φ_1 may be in (3.10), the polarization state of the photon is always the same. Note that, instead, $\varphi = \varphi_1 - \varphi_2$ cannot be changed by a redefinition of the origin of time or of the coordinates: in other word φ_1 and φ_2 are defined up to the *same* additive constant.

Equation (3.10) is therefore a possible way to characterize the state of a monochromatic photon:

- from $k\,z$ one deduces that the photon propagates along the z-axis;
- from ω the frequency can be derived, then the energy of the photons (this piece of information is already contained in k: $\omega = c\,k$);
- from the vector $\cos\vartheta\,\vec{e}_1 + \sin\vartheta\,e^{i\varphi}\,\vec{e}_2$ the state of polarization of the photon is known.

So, in order to describe our photon, it would suffice to give the triple $(\vec{k}, \vartheta, \varphi)$:

$$\text{state of the photon}: \qquad (\vec{k}, \vartheta, \varphi) \equiv |\,\vec{k}, \vartheta, \varphi\,\rangle \qquad (3.11)$$

inasmuch as from this (3.10) is immediately recovered (the reason for the notation $|\cdots\rangle$ will become clear in the next chapter).

Let us now assume we are dealing with photons all having the same $\vec{k}$, i.e. the same state of motion; in this case what distinguishes the various photons is their polarization state. This is determined by the *complex* vector (from now on vectors will be written omitting the arrow):

$$e_{\vartheta\varphi} \equiv \cos\vartheta\,e^{i\varphi_1}\,e_1 + \sin\vartheta\,e^{i\varphi_2}\,e_2 = e^{i\varphi_1}\left(\cos\vartheta\,e_1 + \sin\vartheta\,e^{i\varphi}\,e_2\right). \qquad (3.12)$$

The states of linear polarization are those in which $\varphi = 0$ (namely $\varphi_1 = \varphi_2$: the components of the electric field vibrate in phase); in particular, if $e_{\vartheta\varphi} = e_1$ (e_2), the photon is polarized along the x (y) direction, whereas the states of (either right or left) circular polarization are given (up to a phase factor) by

$$e_{\sigma\pm} = \frac{1}{\sqrt{2}}\,(e_1 \pm i\,e_2)\,. \qquad (3.13)$$

One sees that the polarization states of a photon are described by the elements of a two-dimensional *complex* linear space $\mathcal{H}_2$ and that vectors proportional to each other by a phase factor represent the same polarization state. They can be expressed as linear combinations, i.e. **superposition**, of either the basis vectors e_1 and e_2, as in (3.12), or of any other basis (e.g. the vectors $e_{\sigma\pm}$).

Not only is the phase factor $\exp(i\,\varphi_1)$ in (3.10) irrelevant, but also the multiples (by an arbitrary complex factor) of $e_{\vartheta\varphi}$ do not possess any particular significance: all the possible polarization states of a *single photon* are completely described by the vectors (3.12) by varying ϑ and $\varphi = \varphi_1 - \varphi_2$. We can then adopt the following

Convention: *vectors proportional to each other (by a complex proportionality factor) represent the same state.*

The set of vectors proportional to one another constitute a **ray**: we then have a bijective correspondence between the polarization states of a photon and the rays of $\mathcal{H}_2$.

We are able to provide an answer to some of the questions we have posed at the beginning of this section. The vector space $\mathcal{H}$ by which we represent the states of a system is a complex vector space. The dimension of $\mathcal{H}$ depends on the system one is considering; normally it is infinite: in the example of the polarization states of a photon the dimension of $\mathcal{H}$ is finite because, dealing with photons all in the same state of motion, we have decided to neglect the latter. If also the states of motions are taken into consideration, the states are no longer representable in a finite-dimensional space.

The discussion of the next section will convince us that $\mathcal{H}$ must be endowed with a Hermitian scalar product.

3.5 Discussion of Experiments with Polaroid Sheets

The light emitted by a monochromatic source, e.g. by a sodium lamp (yellow light), is a *statistical mixture* of photons (because any atom emits independently of the others) with the same frequency, but with different polarizations (there exist different atomic states that belong to the same energy level and the polarization of the emitted photon depends on both the initial and the final state involved in the transition between two energy levels). Therefore a beam of natural light never is polarized.

There exist in Nature – but they are also commercially available – some plastic substances (polaroid sheets) that allow one to obtain linearly polarized light: such substances possess the feature of being totally transparent for the light polarized in a given direction (called *optical axis*), to absorb completely the light polarized in a direction orthogonal to the optical axis, and to have a *linear* behaviour: the outgoing field depends linearly on the ingoing field (what is being described here is the *ideal* polaroid: transparency and absorption never are complete, and the same is true as far as linearity is concerned).

Let us consider a plane wave propagating in the direction of the z-axis, and let us position a polaroid sheet orthogonal to the direction of propagation and take, for example, the x axis parallel to the direction of the optical axis. The plane wave is described by (3.6) and can be considered as the superposition of two polarized waves, the first $\vec{E}_1(z, t) = (E_{0\,x}, 0, 0) \cos(k\,z - \omega\,t + \varphi_1)$ parallel to the x-axis, the second $\vec{E}_2(z, t) = (0, E_{0\,y}, 0) \cos(k\,z - \omega\,t + \varphi_2)$ parallel to the y-axis.

Thanks to the linearity of the polaroid sheet, the first wave is completely transmitted, the second is completely absorbed; so the outgoing wave is polarized in the direction of the x-axis, namely parallel to the optical axis of the polaroid sheet.

If we denote by I the intensity of the ingoing wave, by I_{tr} that of the transmitted wave and by $\vec{E}_0$ the vector whose components are $(E_{0\,x}, E_{0\,y}, 0)$, one has:

$$I = \frac{c}{8\pi}\,\vec{E}_0^{\,2}\,, \qquad I_{\mathrm{tr}} = \frac{c}{8\pi}\,\vec{E}_{0\,x}^{\,2} \quad \Rightarrow \quad I_{\mathrm{tr}} = I\cos^2\vartheta \qquad (3.14)$$

where ϑ is the angle formed by $\vec{E}_0$ and the optical axis.

Equation (3.14) is known as **Malus Law**.

If unpolarized light hits the polaroid, the transmitted light is polarized parallel to the optical axis: it is in this way that, at the expense of the intensity, a beam of polarized light can be obtained. This can be shown experimentally having at one's disposal two polaroid sheets with parallel optical axes and verifying that all the light transmitted by the first sheet is transmitted also by the second (or, alternatively, if the optical axes are orthogonal to each other, that the system consisting of the two sheets is completely opaque).

Let us now imagine we have at our disposal a source of linearly polarized light and a polaroid sheet put in place in such a way that its optical axis makes an angle ϑ with the direction of the polarization of the light. Let us reduce the intensity to such an extent as to have, on the average, only one photon at a time in the path between the source and the polaroid sheet. Downstream of the sheet we put something (e.g. a photomultiplier) that enables us to count the photons that cross the sheet. After many photons have been sent on the polaroid sheet we note the following facts:

1. if $\vartheta = 0°$, all the photons cross the sheet;
2. if $\vartheta = 90°$, no photon crosses the sheet;
3. for generic ϑ, $N\cos^2\vartheta$ photons are transmitted and $N\sin^2\vartheta$ are absorbed, N being the total number of the sent photons.

The first two points confirm that the concept of polarization applies also to the single photon, not only to a beam consisting of many photons: it therefore makes sense to talk about polarized photons.

The third point, on the one side, confirms that Malus law holds for photons as well; on the other side it poses a problem: when a photon polarized in the direction $\vartheta \neq 0°, 90°$ hits the polaroid sheet, what will it do? Will a fraction $\cos^2\theta$ of it go through? If it were so and if Einstein relation (2.1) (relating the energy of the photon with its frequency) holds, the transmitted photon should have the frequency $\nu' = \nu\cos^2\vartheta$. That things are not in this way is witnessed by the obvious observation that if the ingoing light is yellow (or red or green), the light that emerges from the polaroid sheet is yellow (or red or green), namely it has the same frequency as the ingoing light. Here is another example of the conflict between the wave nature of light – expressed by the superposition principle – and the corpuscular nature introduced by Einstein relation (2.1).

The best thing to do is to control, photon by photon, when the photomultiplier clicks: it happens that sometimes the photon passes through, sometimes no. In the beginning, until only few photons have been sent, it looks as if the fact that a photon either crosses or not the sheet is totally random; in the long run, instead, one sees that about $N\cos^2\vartheta$ have gone through and, the higher N, the more this result is precise.

For any photon that arrives at the polaroid sheet we cannot know whether it will pass or not. All that we can say, by the definition of **probability**, is that one has a probability $\cos^2\vartheta$ the photon will pass and a probability $\sin^2\vartheta$ it will be absorbed: indeed, the probability of an event is defined, repeating many time the experiment (i.e. sending many times a photon), by the ratio [number of favourable events]/[number of possible cases] (statistical definition of probability). In the experiment described here the number of possible cases is N and the number of favourable events is $N \cos^2 \vartheta$.

Only in the two cases when either $\vartheta = 0°$ or $\vartheta = 90°$ we know a priori and with certainty how the photon sent on the polaroid sheet will behave; in all the other cases we can only say how much is the probability it will either pass or not.

We have then made a ***measurement*** on the photons: indeed we are using instruments, the polaroid sheet and the photomultiplier, able to provide an answer to a precise question: "does the photon go through the polaroid or not?" and the conclusion of the above discussion is that, even knowing everything about the photon (polarization state, frequency, etc. – i.e. its state) and about the measuring apparatus, the result of such a measurement *cannot be predicted a priori* (apart from the cases $\vartheta = 0°$, $\vartheta = 90°$), but it is only ***statistically*** determined.

The same thing happens in the experiment of Fig. 3.5: we know that, in the long run, 50% of the times F_1 clicks, 50% of the times F_2 clicks; but every time we send a photon, we do not know whether F_1 or F_2 will click. (Usually a measurement provides a numerical result: no problem, it is sufficient to have at one's disposal a display showing (whatever) number if F_1 clicks and a different number if F_2 clicks.)

The fact that the result of a measurement, even knowing the state of the system, cannot be predicted with certainty, but only statistically, is a new fact that has no analogue in classical physics. It is true that, if a die is thrown, one cannot predict the number that will come out, but can only say that the probability for a given number is $1/6$. However this is a consequence of the insufficient information at one's disposal: according to the classical schemes, if one knew with great precision the density distribution of the die, the forces exerted in the throwing, the effect of friction etc., then, for given initial conditions, one would be able to determine a priori the number that will come out. The difference between the case of the photon and that of the die (i.e. between the quantum and the classical points of view) lies exactly in this: in the case of the die the statistical character of predictability is eliminable, in the case of the photon it is *ineliminable*, i.e. it is an intrinsic fact.

In the Young experiment the photographic plate that records the arrival position of each photon is a measurement instrument (in the experiment with neutrons the instrument consists of the two counters): also in this case we cannot know a priori in which point of the plate will the photon be absorbed, but, since in the long run we will find the whole interference pattern on the plate, we know that the probability a photon will arrive at a certain point is proportional to the intensity of the interference pattern in that point. Evidently, the probabilistic information regarding the points of the plate where the photons can go – and where they must not go – is written in the state of the photons (all the photons are in the same state, since they have been "prepared" in the same way).

In the 'variation on the theme' of Young experiment with mobile mirrors the measurement instrument consists of the mirrors themselves plus, of course, the device that records the momentum transferred to the mirrors, and the result of the measurement is that 50% of the times the photon hits A, 50% of the times B.

The important thing that emerges in this case (we have already hinted at in Sect. 3.3) is that *the measurement perturbs the state* of the photons: indeed, without the mirrors (or with all mirrors fixed) the interference pattern does form on the plate, with mobile mirrors it does not. The same fact we find again in the experiment with polaroid sheets: we know that, after crossing a polaroid sheet, all the photons are polarized in the direction parallel to its optical axis, so the polaroid sheet changes the polarization state of the photons initially polarized in the direction $\vartheta \neq 0°$ and $\vartheta \neq 90°$.

The polaroid sheet is not the ideal object for this discussion, for it has the flaw it makes the system (i.e. the photons) disappear in all the cases they do not cross it.

Let us instead consider a birefringent crystal: it is a crystal that has different refraction indices for light linearly polarized in a given direction and for light polarized in the orthogonal direction; so, as depicted in Fig. 3.7, a beam of light, in crossing the crystal, is split in an *extraordinary* ray E,

Fig. 3.7

polarized parallel to the direction of the optical axis, and an *ordinary* ray O, polarized in the orthogonal direction.

All we have seen for the polaroid sheets applies to the birefringent crystal; in particular Malus law holds: if the incident photons are all polarized in a direction that forms the angle ϑ with the optical axis of the crystal, $N \cos^2 \vartheta$ of them will be revealed to emerge in the extraordinary ray and are polarized parallel to the optical axis, $N \sin^2 \vartheta$ of them will emerge in the ordinary ray and are polarized orthogonally to the optical axis. The only difference with the polaroid sheet is that in the latter the ordinary ray is absorbed and only the extraordinary one emerges.

The reason why we prefer to deal with the birefringent crystal instead of the polaroid sheet is the following: in discussing what happens while making a measurement on a system (the photon, in the present case), we are interested in that the instrument perturbs the system as little as possible. The flaw of the polaroid sheet is that it even makes the photons disappear – all those do not go through it. This kind of perturbation can be eliminated just by replacing the polaroid sheet with the crystal. The type of perturbation we shall *never* succeed in eliminating is the one already discussed: if a photon polarized in the direction ϑ with respect to the optical axis is sent, it emerges either in the direction of O or in the direction of E (Fig. 3.7). In both cases its polarization state is different from the initial one: also in this case the measurement – that

---

consists in detecting, e.g. by means of mobile mirrors, whether the photon emerges in either O or E – perturbs the system. If, instead, mobile mirrors are not used, it is possible to recombine the two rays that emerge from the crystal and, by suitably adjusting the optical paths, operate in such a way that each photon will be in the very end in a given polarization state (e.g. the initial one): in this case the state of the photon has not been perturbed, indeed we have not made a measurement: the system has been allowed to *causally* evolve in another state. We shall come back on this point (that is a very important one) in the next chapter, when discussing the difference between a statistical mixture and a pure state.

Note that, in the case a measurement is made, after the measurement the photon can be in either one or the other of two *precise* states of polarization, that parallel and that orthogonal to the optical axis. So the measurement perturbs the system forcing it to go in either one or the other (according to some probabilities) of particular states connected with the measurement one is making (if one used a birefringent crystal with its optical axis put in a different direction, the states of the photons after the measurement would, accordingly, be different).

This happened already in the different situations discussed in Sect. 3.3: for example, in that described in Fig. 3.6, the photon in the state **C** was forced, as consequence of the measurement with mobile mirrors, to go either in the state **A** or in the state **B**.

Denoting by e_1 and e_2 the vectors that represent the polarization states respectively parallel and orthogonal to the optical axis of the crystal, i.e. the only states that are possible after the measurement, the (linear) polarization state of the incident photons is

$$e_\vartheta = \cos\vartheta\, e_1 + \sin\vartheta\, e_2 \,. \tag{3.15}$$

The theory must enable us to calculate the probabilities that, after the measurement, the system be in either the state e_1 or the state e_2: if in $\mathcal{H}_2$ we introduce the Hermitian scalar product defined by

$$(e_1\,,e_1) = 1 = (e_2\,,e_2)\,, \quad (e_1\,,e_2) = 0\,; \qquad (u\,,v) = (v\,,u)^* \tag{3.16}$$

such probabilities are given by

$$\begin{aligned} p_1 &\equiv P(e_\vartheta \to e_1) = \cos^2\vartheta = |(e_\vartheta\,,e_1)|^2 \\ p_2 &\equiv P(e_\vartheta \to e_2) = \sin^2\vartheta = |(e_\vartheta\,,e_2)|^2 \end{aligned} \tag{3.17}$$

and this shows the necessity that in $\mathcal{H}$ a scalar product be defined, inasmuch as it is by means of it that it will be possible to express the probabilities of the several result a measurement may yield.

The discussion of this section has emphasized that in quantum mechanics the concept of measurement plays a fundamental role that has no counterpart in classical physics: the reason lies in the fact, although even in classical physics it is true that any measurement perturbs the system, it is understood that such a perturbation can be made as small as one wishes by improving the measurement apparatus. This is not possible in quantum mechanics: the perturbation induced by the measurement is an intrinsic fact and cannot be eliminated.

Einstein, to whom we owe the birth of quantum mechanics, never accepted the idea that the laws of physics might be intrinsically probabilistic and considered this fact a signal of the incompleteness of the theory: quantum mechanics in the Copenhagen interpretation was for Einstein a provisional version of the definitive theory, which necessarily had to be a deterministic theory, much as classical physics is: his statement "God doesn't play dice" is famous. To the present day a way of thinking inspired by Einstein ideas receives much attention and activity: roughly speaking, the idea is that the (polarization) state is not completely determined by the vector $e_{\vartheta\varphi}$, but also by an additional variable ε (called *hidden variable*) that can assume certain values (e.g. between 0 and 1): any photon 'is born' with a *well determined* value of ε and the latter determines whether the photon will end in either the ordinary or the extraordinary ray. It seems that the experiments exclude this possibility, at least in the form told above. We shall come back to this point in the last chapter.

Let us summarize the fundamental concepts introduced in the present section.

1. The result of a measurement, apart from particular cases ($\vartheta = 0°$ and $\vartheta = 90°$ in the case illustrated above), is not determined a priori, we can only know the probability of the various results.
2. Apart from particular cases (again $\vartheta = 0°$ and $\vartheta = 90°$) a measurement perturbs the state of the system. The perturbation caused by the measurement cannot be eliminated by using 'more refined' instruments.
3. After a measurement, the system can be in one and only one from a precise set of states.

3.6 Compatible and Incompatible Measurements: Heisenberg Uncertainty Relation

Suppose one wants to determine both the polarization state and the frequency – i.e. the energy – of a photon. After determining the polarization state in some way (e.g. by means of a birefringent crystal), in order to determine its frequency it will suffice to let it go through a prism. Since the prism does not change the polarization state, we can simultaneously know polarization and frequency of the photon. One then says that the two measurements, of polarization and frequency, are *compatible* with each other.

It is then clear that in the quantum scheme there may exist measurements incompatible with one another: this happens when the second measurement perturbs the state of the system in such a way that the information, acquired with the first measurement, gets lost. In this case the two quantities one wants to measure cannot be simultaneously known.

A typical example of quantities that cannot be simultaneously known is that of position and velocity of a particle.

Suppose one has an electron whose velocity is known. We want now to determine the position of the electron. To this end it is necessary to observe it: we therefore send some monochromatic light on it and collect the light scattered by the electron by means of a lens (microscope); this light will be focalized in a point P' of a photographic plate (image point) and from the knowledge of the position

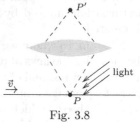

Fig. 3.8

of this point one can find the position of the electron (Fig. 3.8). This is the basic idea to measure the position of the electron (**Heisenberg microscope**).

Let us now examine how the microscope works. First of all, let us recall that any optical instrument has a finite resolution power: this means that, given two points P_1 and P_2 a distance δ apart from each other (Fig. 3.9), there exists a $\delta_{\min}$ such that, for $\delta < \delta_{\min}$ the instrument (lens) is not able to distinguish them. Why does this happen? A lens 'cuts' a portion of the wave front of the incident light, the size of this portion being equal to that of the lens – much as a hole in a screen – and so gives rise to the phenomenon of diffraction: we could say a lens is a "hole of glass in the vacuum".

So the images of points are not points, but diffraction patterns produced by the lens, i.e. small spots. Since, due to diffraction, any ray incident on the lens gives rise to a cone of rays, whose aperture is $\sin \phi \simeq \lambda/d$ (Fig. 3.9), d being the diameter of the lens, it follows that the image of a point is a spot whose linear size is $l_2 \sin \phi = l_2 \lambda/d$.

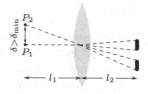

Fig. 3.9

The images of points P_1 and P_2 will be distinguished if the relative spots do not overlap: this happens if $l_2 \lambda/d < \delta l_2/l_1$, i.e. if $\delta/l_1 > \lambda/d$.

If the source – in the present case the electron – emits (or scatters) only one photon, its arrival will be recorded in a precise point of the photographic plate; according to the discussion of Sect. 3.3, the photon could have arrived in whatever point of the small 'image spot': this means that, from the knowledge of the point the photons arrives at (image point), one can trace back the position of the source up to an uncertainty $\delta_{\min} = l_1 \lambda/d$.

Note that, in order to improve the resolution power of an instrument, it is necessary either to decrease λ or to increase d, or both. Going back to our experiment, one realizes that the position of the electron is known with an uncertainty $\Delta x \simeq l_1 \lambda/d = \lambda/(2 \sin \phi)$, in which 2ϕ is the angle under which the electron 'sees' the lens (Fig. 3.10).

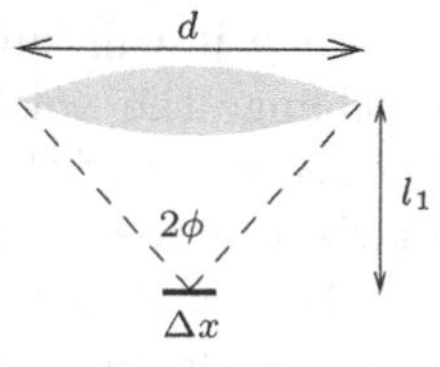

It is our choice to make this Δx big or small, according to our wish, by varying the parameters l_1, λ, d.

Fig. 3.10

Once the position has been measured with the desired precision, let us see what happened to the velocity of the electron.

If the electron has been seen by the microscope, it means that at least one photon has been scattered by the electron and has gone through the lens: we record its arrival at P' (Fig. 3.8), but we do not know in which point it has crossed the lens, namely in which direction it has been scattered. This entails that we cannot precisely know the momentum of the scattered electron and, therefore, the momentum exchanged between electron and photon. Quantitatively: the x component of the momentum h/λ of the scattered photon is determined only up to $\Delta p_x = 2(h/\lambda) \sin \phi$ which, by momentum conservation, is also the uncertainty by which we know the momentum of the electron after the position measurement. Also Δp_x can be made either big or small as much as one wishes, by varying l_1, λ, d, but at the expense of Δx. Indeed:

$$\Delta x \times \Delta p_x \simeq h \qquad (3.18)$$

that is known as **Heisenberg uncertainty relation.**

The consequences of (3.18) are evident: not only it is impossible to know simultaneously position and momentum of a particle ($\Delta x = 0$ and $\Delta p_x = 0$), but the better we know one of the two, the worse we know the other.

It should be clear from the above discussion that $\Delta p_x \neq 0$ has been caused by the perturbation that the measurement of position has produced on the electron. Let us then see why this perturbation can be eliminated in the classical scheme, whereas it cannot in the quantum description. Indeed, while we can make Δx as small as we wish – for example by decreasing λ –, we can simultaneously (according to the classical scheme) decrease the intensity of the radiation used to illuminate the electron, and this makes the momentum transferred to the electron decrease as much as one wants. The reason why this description is not realistic is that it does not take into account photons, i.e. the existence of the quantization, that prevents one from decreasing 'as much as one wishes' the intensity of the radiation, once λ has been fixed.

From the historical point of view, (3.18) has played a fundamental role in the passionate discussions among, on the one side, the supporters of the Copenhagen interpretation and, on the other side, those (as Einstein himself) who did not adhere to it. It is exactly because of this role that in some contexts it is dubbed as *indeterminacy principle*: indeed, within the quantum theory, it is not a principle, but a theorem (it follows from other 'principles') and will therefore be referred to as **uncertainty relation**.

When, in the next chapter, we shall demonstrate the uncertainty relation, we shall see that its correct form is

$$\Delta x \times \Delta p_x \geq \frac{1}{2} \hbar. \qquad (3.19)$$

The inequality sign and the fact that the right hand side is $\frac{1}{2} \hbar$, instead of h, do not alter the conceptual meaning of (3.18) that we have derived without paying too much attention to numerical factors.

We have now the means to understand, at least qualitatively, why in the Young experiment with the mobile mirrors (Fig. 3.6) the formation of the interference pattern is incompatible with the observation of the recoil of the mirrors. Indeed, two conditions should be satisfied: first, the formation of the interference pattern requires that the uncertainty in the length of optical path of the photons must not exceed λ, and this puts an upper limit to the uncertainty Δs of the position of the mobile mirrors; second, the possibility of observing the recoil of the mirrors (i.e. the momentum transferred by the photons whose momentum is h/λ), puts an upper limit to the uncertainty Δp_s of the momentum of the mirrors. These two conditions turn out to be incompatible with the uncertainty relation (3.18) (or (3.19)) applied to the mirrors.

Actually (3.19) has a more precise (and general) meaning than that emerging from the way we have derived (3.18) (which is Heisenberg's), and that can give rise to misunderstandings (as witnessed by the discussions we have hinted at): indeed, the arguments we have used to derive (3.18) may induce one to think that both the position and the momentum uncertainties arise from the instrumental impossibility of knowing where does the photon start from and which direction does it take, but that the latter – in spite of the fact that we cannot know them – are well determined.

It is then understandable that somebody would like to maintain the following point of view: 'the fact that no experiment will enable one to exactly determine position and velocity of a particle does not prevent from thinking that the particle has at any given time a well determined position and a well determined velocity'. One is of course free to think whatever he wishes, however this attitude can be criticized for two reasons. The first is that in this way one will end up in considering reality as something different from what can be measured – this is however a problem we shall leave to philosophers. The second, for us more important, is the following: it is as if, concerning the two-slit experiment, we insisted in maintaining that the photon goes through either A or B, even if no measurement will ever enable us to control whether this is true or not. But we know that in this way we are not able to explain how is it that the interference fringes, not the superposition of the diffraction patterns, are observed. Similarly: if the electron at any time 'has' a well determined position and velocity, then in the Davisson–Germer experiment the electron necessarily will be reflected by one (and only one) lattice plane: we do not know which one, the electron does. But again, as in the just discussed Young experiment, we are not able to explain why specular reflection occurs only for those angles that satisfy the Bragg condition (2.43). Also in this case we must admit that the electron is simultaneously reflected by many lattice planes – which is clearly incompatible with the idea of the electron as a corpuscle, i.e. endowed with well determined position and velocity. More: if the neutron in the interferometer of Fig. 3.4 ...

Chapter 4

The Postulates of Quantum Mechanics

4.1 The Superposition Principle

Most of the fundamental concepts that are at the basis of quantum mechanics have already emerged during the discussion of the previous chapter. Now the problem is to formalize these concepts within a coherent structure, namely to go 'from words to facts'. We will try to proceed in this process of formalization both by making good use of the already discussed examples and by arriving at the formal aspects of the theory as much as possible from the physical point of view (for example, this is the way we shall succeed in establishing that observable quantities are represented by operators on the space of states).

In Sect. 3.4 we have already recognized the necessity to describe the states of a system by means of the vectors of a complex vector space: this is the first fundamental principle at the basis of quantum mechanics, known as

Superposition Principle: the states of a system are represented by the elements (vectors) of a vector space $\mathcal{H}$ over the complex field. Vectors proportional to one another (by a complex factor) represent the same state. Therefore the states are in correspondence with the rays of $\mathcal{H}$.

Moreover, in Sect. 3.5, we have seen that – in order to give a predictive character to the theory – it is necessary that the space $\mathcal{H}$ be endowed with a Hermitian scalar product.

The dimension of $\mathcal{H}$ depends on the system under consideration: normally it is infinite.

The space $\mathcal{H}$, that we shall assume complete, is therefore a **Hilbert space**. We also assume that $\mathcal{H}$ be separable (namely it admits a countable orthonormal basis).

As far as the (infinite) dimension and the separability are concerned, these – in the case of one or more particles – will be deduced by the quantization postulate to be introduced in the sequel.

One can now ask oneself if the correspondence between states of the system and rays of $\mathcal{H}$ is a bijective one, namely if to any vector (or ray) of $\mathcal{H}$ there always corresponds a state of the system. It is likely that this assumption is

© Springer International Publishing Switzerland 2016   61
L.E. Picasso, *Lectures in Quantum Mechanics*, UNITEXT for Physics,
DOI 10.1007/978-3-319-22632-3_4

not true and not even necessary; on this point, that is rather marginal, we shall come back later and we shall see that, probably, it is sufficient to assume that the physical states are in correspondence with a dense, algebraically closed (i.e. closed under finite linear combinations) subset of $\mathcal{H}$. However there is no doubt that the hypothesis that $\mathcal{H}$ be a complete space is extremely useful from the mathematical point of view – indeed it is an hypothesis one cannot simply give up: that is why we shall assume it is fulfilled. This (probably) means that the vectors representative of physical states are immersed in a larger (closed) ambient. We have used a dubitative form because, usually, one takes for granted that the correspondence states/rays is bijective.

The vectors of $\mathcal{H}$ will be denoted by the symbol $| \cdots \rangle$, called "ket" (Dirac notation). The scalar product between two vectors $|A\rangle$ and $|B\rangle$ is denoted by the 'bra'-'ket':

$$\langle B \mid A \rangle = \langle A \mid B \rangle^* \qquad\qquad (4.1)$$

and is linear in $|A\rangle$ and antilinear in $|B\rangle$; the scalar product between $\alpha|A\rangle$ and $\beta|B\rangle$ (α, β complex numbers) is $\beta^*\alpha\langle B \mid A \rangle$: it is the same as saying that one takes the scalar product between the "ket" $\alpha|A\rangle$ and the "bra" $\beta^*\langle B|$ (by the way we note that $|A\rangle$ and $\alpha|A\rangle$ represent the same state).

The important physical aspect of the superposition principle is that it expresses the fact that the states, being represented by vectors, may 'interfere' with each other: if $|A\rangle$ and $|B\rangle$ represent two states, then

$$\alpha|A\rangle + \beta|B\rangle \qquad\qquad \text{for any } \alpha,\ \beta \in \mathbb{C} \qquad\qquad (4.2)$$

still represents a state different from both $|A\rangle$ and $|B\rangle$ – just as for the polarization states of light (represented by the vectors (3.12)) and, in general, for waves.

Note that the states (4.2) are (varying α and β) ∞^2, not ∞^4 (α and β are complex numbers!) because α and β can be multiplied by the same complex factor: $\alpha|A\rangle + \beta|B\rangle$ and $\alpha'|A\rangle + \beta'|B\rangle$ represent the same state if and only if $\alpha : \alpha' = \beta : \beta'$.

In the example discussed in Sect. 3.3 $|A\rangle$ and $|B\rangle$ represent two states of the photon: $|A\rangle$ ($|B\rangle$) represents the state one has when only the hole A (B) is open: when both holes are open the state of the photon is represented by

$$|C\rangle = |A\rangle + |B\rangle . \qquad\qquad (4.3)$$

The vague expression used in Sect. 3.3 for the state **C** that 'has to do with' the states **A** and **B** is now translated into a precise mathematical form: the vector $|C\rangle$ that represents the state **C** of the photon is a linear combination of the vectors $|A\rangle$ and $|B\rangle$ that represent the states of the photon when either only hole A or only hole B is open.

(Sometimes, in order keep the terminology quick enough, we shall simply say 'the state $|A\rangle$' instead of 'the state represented by the vector $|A\rangle$'. The abbreviation is not appropriate for two reasons: (i) there is no correspondence between states and vectors, only between states and rays; (ii) one thing is the state of the system, another thing is the way we represent it.)

Always referring to the Young experiment (or to its analogue with neutrons) we now ask how the ∞^2 states $\alpha \,|\, A \rangle + \beta \,|\, B \rangle$ can by physically realized: we may change at our wish the ratio $|\alpha/\beta|$ by varying, by means of diaphragms, the sizes of the two holes (or – in the experiment with neutrons of Fig. 3.4 – by making use, for the semi-transparent mirror s_1, of a crystal with suitable reflection and transmission coefficients); the relative phase between α and β can be varied by putting (in the Young experiment) a small glass slab of suitable thickness in front of one of the holes (it will have the same effect as either the glass slab in the Mach–Zehnder interferometer or the aluminum wedge in the neutron interferometer of Fig. 3.4). If we now send many photons and then develop the photographic plate, we see that, according to the value of $|\alpha/\beta|$, the contrast among the bright and the dark fringes of the interference pattern changes; and if either $\alpha \to 0$ or $\beta \to 0$, the latter becomes the diffraction pattern of the open hole, whereas in the second case (i.e. in presence of the phase shifter) we see that all the interference pattern has undergone a translation. There exist, indeed, ∞^2 interference patterns.

In the case of the states of polarization of light, we know that the ∞^2 states corresponding to the vectors (3.12) are the ∞^2 states of elliptic polarization of the photons.

In Sect. 3.3 we have said that the state of a system is defined by the way in which the system is prepared: the postulates we will introduce in the next sections will teach us how this information is codified in the vector $|\, A \rangle$ that represents the state.

4.2 Observables

We call **observables** the quantities that can be measured on a system: in most cases one has to do with the same quantities that can be measured according to classical physics (an exception is provided by the spin, that has no classical analogue). For example, in the case the system is a particle, energy, angular momentum, the components q_i, $1 = 1, 2, 3$ of the position of the particle, the components of its linear momentum etc. are observables (with some proviso on the last two): in general the functions $f(\vec{q}, \vec{p})$.

We already know the fundamental role played in quantum mechanics by the process of measurement; therefore we think to associate one or more instruments of measurement with each observable: for example we associate a Heisenberg microscope with the observable 'position q_1', a magnetic field spectrometer with the observable 'momentum $|\vec{p}\,|$', etc.

From now on, by the term 'observable' we shall mean both the quantity that can be measured and an instrument suitable for measuring it.

Let ξ be an observable and $\xi_1, \xi_2, \cdots \xi_i \cdots$ the possible results of the measurements of ξ on the system. The real numbers ξ_i are called the **eigenvalues** of ξ (we shall see later that the denumerability of the eigenvalues follows from the separability of $\mathcal{H}$).

For example, the possible results of the measurements we have cited in the previous chapter, when either dealing with the Young experiment modified by the presence of mobile mirrors (Sect. 3.3) or with the experiment involving the birefringent crystal (Sect. 3.5), were two (either one of the mirrors was hit), and we can couple a device to our apparatus exhibiting a number on its display, e.g. the numbers $+1$ and -1, depending on the result. In the latter case the eigenvalues are two: $\xi_1 = +1$, $\xi_2 = -1$.

More: the energies allowed for a system, i.e. its energy levels, are the eigenvalues of the observable 'energy'.

It will be a specific task of the theory to specify which are the eigenvalues relative to each observable. Indeed one of the main problems of quantum mechanics precisely is the determination of the eigenvalues for the different observables and, among these, energy will have a privileged role.

In general, if the observable ξ is measured on a system in the state $|A\rangle$, the result of the measurement is not a priori determined (see the experiments with polaroid sheets), but all the numbers ξ_i can be found as result, with probabilities p_i that depend on the state $|A\rangle$. In other words, if the measurement of ξ is made many times on the system, that *any time* is in the state $|A\rangle$ (this means that many copies of the system, all prepared in the same way, are at one's disposal), the different results ξ_i will be obtained with frequencies proportional to p_i.

We shall call **eigenstates** of ξ those particular states on which the result of measurements of ξ is determined a priori, therefore it is always the same (example: the two rectilinear polarization states of a photon, respectively parallel and orthogonal to the optical axis of a birefringent crystal, are eigenstates of the observable associated with the crystal) and we shall call **eigenvectors** of ξ the vectors of $\mathcal{H}$ that represent the eigenstates of ξ (the abuse made in using the words 'eigenvectors' and 'eigenvalues', that have a precise meaning in the framework of linear algebra, will be justified later).

An eigenstate of ξ **corresponding to the eigenvalue** ξ_i is a state for which the result of the measurement always is ξ_i (so to any eigenstate there corresponds one of the eigenvalues; we shall shortly see that also the viceversa is true): therefore for it $p_i = 1$ whereas, if $j \neq i$, $p_j = 0$; a representative vector of it is denoted by $|\xi_i\rangle$ and is called an **eigenvector of ξ corresponding to** ξ_i (we shall often improperly say 'the eigenstate $|\xi_i\rangle$' instead of 'the eigenstate of ξ represented by the vector $|\xi_i\rangle$').

We postulate that:

If ξ is measured on a system and the result is ξ_i, immediately after the measurement the system is in an eigenstate of ξ corresponding to ξ_i.

So, if the system is in a state $|A\rangle$ and a measurement of ξ is made, we cannot know a priori which result we will obtain and in which state the system will be after the measurement, but when the measurement has been made and has given the result ξ_i, we know that the system is in an eigenstate of ξ corresponding to the eigenvalue ξ_i: therefore, if immediately after the first measurement of ξ a second measurement of the same observable is made, certainly the same result will be obtained (and therefore to any eigenvalue there corresponds *at least* one eigenvector).

So, in general, a measurement perturbs the state of the system: $|A\rangle \rightarrow |\xi_i\rangle$ (an exception is provided by the case in which $|A\rangle$ itself is an eigenstate of ξ). Note that this postulate expresses in general what we have seen about the birefringent crystal in the previous chapter.

It is easy to realize that in many measurement processes this postulate is contradicted. It must therefore be understood in the following sense: it defines which are the 'ideal' instruments of measurement that correspond to observables quantities, fixing their behaviour (so, in this sense, it is a definition rather than a postulate). In addition, it postulates that for any observable there exists at least one 'ideal' instrument suitable to measure it.

There may exist *one* or *more* eigenstates corresponding to a given eigenvalue ξ_i of the observable ξ: in the first case we will say that the eigenvalue ξ_i is **nondegenerate**, while in the second we will say that the eigenvalue ξ_i is **degenerate**. The physical importance of this definition is the following: if a measurement of ξ yields the eigenvalue ξ_i as result, and if this is nondegenerate, then we know in which state the system is after the measurement. In the contrary case, if ξ_i is a degenerate eigenvalue, we only know that after the measurement the system is in an eigenstate of ξ corresponding to the eigenvalue ξ_i, but we do not know which one.

A **nondegenerate observable** is an observable such that to *any* of its eigenvalues there corresponds only one eigenstate, i.e. all of its eigenvalues are nondegenerate; in most cases the observables are degenerate.

According to the above discussion, a measurement of a nondegenerate observable always completely determines the state of the system after the measurement; if instead, in correspondence with the eigenvalue found as a consequence of the measurement there exists more than one eigenstate (degenerate eigenvalue), the information on the state of the system after the measurement is only partial.

We shall see in Sect. 4.4 that a postulate, known as **von Neumann postulate**, will enable us to determine the state immediately after the measurement even when the result is a degenerate eigenvalue. For the time being, consistently with this postulate (that we shall be able to enunciate only after proving that the set of the eigenvectors of on observable, corresponding to a given eigenvalue, is a linear manifold), we may assume that:

If a measurement of the observable ξ is made on the system in the state $|A\rangle$, the state after the measurement is univocally determined by the initial state $|A\rangle$ and by the found eigenvalue ξ_i.

We shall see that the von Neumann postulate will allow us to state that the 'arrival state', within the set of all the possible eigenstates corresponding to the found *degenerate* eigenvalue, is that for which the initial state has undergone the least possible perturbation.

Example: the birefringent crystal is a nondegenerate observable if the system is the photon regardless of its state of motion (i.e. the state space is $\mathcal{H}_2$);

if instead the system is 'all' the photon, then it is a degenerate observable: indeed it allows us to determine the polarization state of the photons, not their state of motion. In any event, von Neumann postulate allows us to state that if we take a photon in a well determined state (for example it propagates in a given direction with energy $E = h\nu$) and is linearly polarized at an angle ϑ with respect to the optical axis of a birefringent crystal, after the measurement has established whether the photon emerges either in ordinary or in the extraordinary ray, the photon is in a well determined state that depends only on the initial state and on the result of the measurement: the polarization state is either parallel or orthogonal to the axis of the crystal, according to whether the photon has emerged respectively either in the extraordinary or in the ordinary ray and, if our instrument is 'ideal' (in the above said meaning), the state of motion (i.e. energy and direction of motion) has been left unchanged.

A device that allows us to determine both the polarization and the motion state of the photons (crystal + prism + etc.) is instead a nondegenerate observable.

Provisionally we take the following statement as a further postulate, even if it will be seen to be a consequence of the other postulates we will enunciate in the sequel:

Any state is eigenstate of some nondegenerate observable.

From the point view of physics this means that any state can be 'prepared' by the measurement of a suitable observable: i.e. one uses the observable as a filter (much as the polaroid or the birefringent crystal to prepare linearly polarized photons) by making measurements on the system and accepting only the state that is obtained when the measurement yields the desired result.

It therefore emerges that a type of information sufficient to characterize the state of a system consists in knowing which nondegenerate observable has been measured on the system and the result of such a measurement: it is not necessary to know all the past history of the system.

4.3 Transition Probabilities

Let ξ be an observable. We know that a measurement of ξ on a system in the state $|A\rangle$ will give one of the eigenvalues ξ_i as a result and that, owing to the measurement process, the system makes a transition to an eigenstate $|\xi_i\rangle$. Therefore the probability p_i of finding a given eigenvalue ξ_i as a result is also called transition probability from $|A\rangle$ to $|\xi_i\rangle$. Let in general $|A\rangle$ and $|B\rangle$ be the representative vectors of two states; the probability that, owing to a measurement on the system in the state $|A\rangle$, the system goes in the state $|B\rangle$ is called **transition probability** from $|A\rangle$ to $|B\rangle$.

One postulates that the above transition probability $P(|A\rangle \to |B\rangle)$ does not depend on the ('ideal') instrument used to perform the measurement and is given by:

$$P(|A\rangle \to |B\rangle) = \frac{|\langle B \mid A \rangle|^2}{\langle A \mid A \rangle \langle B \mid B \rangle}. \qquad (4.4)$$

Definition (4.4) is a 'good definition': indeed

- P does not depend on the arbitrary factor present in the correspondence states $\leftrightarrow$ vectors: it does depend, as it must be, on the states, not on the vectors chosen to represent them. Indeed, if α and β are complex numbers, $\alpha\,|\,A\,\rangle$ and $\beta\,|\,B\,\rangle$ represent the same states as $|\,A\,\rangle$ and $|\,B\,\rangle$ and one has

$$P(\alpha\,|\,A\,\rangle \to \beta\,|\,B\,\rangle) = \frac{|\alpha|^2\,|\beta|^2\,\big|\langle\,B\mid A\,\rangle\big|^2}{|\alpha|^2\langle\,A\mid A\,\rangle\,|\beta|^2\langle\,B\mid B\,\rangle} = P(\,|\,A\,\rangle \to |\,B\,\rangle)\,.$$

- $0 \le P \le 1$: it follows from the Schwartz inequality.

Note that $P(\,|\,A\,\rangle \to |\,B\,\rangle) = 0$ if and only if $|\,A\,\rangle$ and $|\,B\,\rangle$ are orthogonal to each other: there occur no transitions among states (represented by vectors) orthogonal to one another; whereas $P = 1$ if and only if $|\,A\,\rangle$ and $|\,B\,\rangle$ represent the same state.

Usually it is convenient to represent the states by vectors normalized to 1:

$$\langle\,A\mid A\,\rangle = 1 = \langle\,B\mid B\,\rangle$$

in which case (4.4) reads:

$$P(\,|\,A\,\rangle \to |\,B\,\rangle) = \big|\langle\,B\mid A\,\rangle\big|^2 \qquad \text{(valid for normalized vectors).} \quad (4.5)$$

The transition probability between two states $|\,A\,\rangle$ and $|\,B\,\rangle$ is, from the operational point of view, always well defined inasmuch as we have postulated that for any state $|\,B\,\rangle$ there always exists (at least) one observable that has such a state as an eigenstate corresponding to a nondegenerate eigenvalue.

In any event, even if ξ_i is a degenerate eigenvalue of ξ, the probability p_i that a measurement of ξ on a state $|\,A\,\rangle$ gave ξ_i as a result is given by $\big|\langle\,A\mid \xi_i'\,\rangle\big|^2$ (normalized vectors), where $|\,\xi_i'\,\rangle$ is that eigenstate of ξ that, thanks to von Neumann postulate, is univocally determined by the initial state $|\,A\,\rangle$ and by the eigenvalue ξ_i.

Example: let us consider a photon in the linear polarization state

$$|\,e_\vartheta\,\rangle = \cos\vartheta\,|\,e_1\,\rangle + \sin\vartheta\,|\,e_2\,\rangle\,.$$

The probability that, owing to a measurement (e.g. by means of a birefringent crystal) it makes a transition to the state $|\,e_1\,\rangle$, is given by

$$P(\,|\,e_\vartheta\,\rangle \to |\,e_1\,\rangle) = \big|\langle\,e_1\mid e_\vartheta\,\rangle\big|^2 = \cos^2\vartheta\,.$$

Indeed, the two vectors $|\,e_1\,\rangle$ and $|\,e_2\,\rangle$ are orthogonal to each other (as elements of $\mathcal{H}_2$) because a measurement on photons in the state $|\,e_1\,\rangle$ will never give photons in the state $|\,e_2\,\rangle$. One is back to Malus law.

4.4 Consequences and von Neumann Postulate

Let us now analyze some consequences of the postulates so far introduced. We will firstly discuss the case of

Nondegenerate observables

Let ξ be a nondegenerate observable, ξ_i its eigenvalues and $|\xi_i\rangle$ its eigenstates that we will assume normalized to 1.

The quantity $|\langle\xi_i\mid\xi_j\rangle|^2$ is the probability that, owing to a measurement, e.g. of the observable ξ, a transition from $|\xi_i\rangle$ to $|\xi_j\rangle$ or viceversa has taken place. But, if $i\neq j$, this probability is 0 because a measurement of ξ on the state $|\xi_i\rangle$ will always give ξ_i as a result, never ξ_j. Therefore:

$$\langle\xi_i\mid\xi_j\rangle = \delta_{ij} = \begin{cases} 0 & \text{if } i\neq j \\ 1 & \text{if } i=j. \end{cases} \tag{4.6}$$

Namely: the normalized eigenvectors of an observable ξ are an **orthonormal system** of vectors. Let us demonstrate that such a system is also **complete**.

We will reason by contradiction: if a vector $|A\rangle$ orthogonal to all the $|\xi_i\rangle$ existed, one would have $p_i = P(|A\rangle \to |\xi_i\rangle) = |\langle\xi_i\mid A\rangle|^2 = 0$ for any i, which is absurd since, by definition of probability, $\sum_i p_i = 1$. Therefore: the eigenvectors of the (nondegenerate) observable ξ form an orthonormal basis.

Any vector $|A\rangle$ of the space $\mathcal{H}$ can therefore be expanded in series of the vectors of the basis (Fourier series):

$$|A\rangle = \sum_{i=1}^{\infty} a_i |\xi_i\rangle. \tag{4.7}$$

The coefficients a_i of the Fourier series are calculated by taking the scalar product of both sides of (4.7) with the generic (normalized) vector belonging to the basis:

$$\langle\xi_i\mid A\rangle = \sum_{j=1}^{\infty} a_j \langle\xi_i\mid\xi_j\rangle = a_i. \tag{4.8}$$

Equation (4.7) can then be rewritten by substituting the expression $\langle\xi_i\mid A\rangle$ to the a_i:

$$|A\rangle = \sum_{i=1}^{\infty} |\xi_i\rangle\langle\xi_i\mid A\rangle. \tag{4.9}$$

Note that, thanks to (4.6), $\langle A\mid A\rangle$ (the squared norm of the vector $|A\rangle$) is given by

$$\langle A\mid A\rangle = \sum_{i,j} a_i^* a_j \langle\xi_i\mid\xi_j\rangle = \sum_{i=1}^{\infty} |a_i|^2 < \infty \tag{4.10}$$

so, only if $\sum_i |a_i|^2 < \infty$, does (4.7) define a vector belonging to $\mathcal{H}$.

If the vector $|A\rangle$ is normalized, the coefficients a_i have a rather direct meaning: $|a_i|^2$ is the transition probability p_i from $|A\rangle$ to $|\xi_i\rangle$ (and from (4.10) one has $\sum_i p_i = 1$):

$$p_i = \left|\langle A \mid \xi_i \rangle\right|^2 = |a_i|^2 . \tag{4.11}$$

If the $|\xi_i\rangle$ are normalized, but $|A\rangle$ is not, then

$$p_i = \frac{|a_i|^2}{\langle A \mid A \rangle} .$$

The fact that only the $|a_i|^2$ and not directly the a_i have been given a physical meaning should *not* lead one to think that only the absolute value of the a_i is endowed with a physical meaning: once the basis is fixed (i.e. once the vectors $|\xi_i\rangle$ are fixed), changing the a_i by a phase factor, $a_i \rightarrow \exp(\mathrm{i}\,\varphi_i)\,a_i$, amounts to changing the state, unless the φ_i are all equal to one another, in which case $|A\rangle \rightarrow \exp(\mathrm{i}\,\varphi)|A\rangle$, and the state is not changed.

Example: the states of polarization of a photon , represented by the vectors

$$|e_\vartheta\rangle = \cos\vartheta\,|e_1\rangle + \sin\vartheta\,|e_2\rangle , \qquad |e_{\vartheta\varphi}\rangle = \cos\vartheta\,|e_1\rangle + \sin\vartheta\,\mathrm{e}^{\mathrm{i}\varphi}\,|e_2\rangle$$

are different from each other: linear polarization in the first case, elliptic in the second, even if

$$P(|e_\vartheta\rangle \rightarrow |e_1\rangle) = P(|e_{\vartheta\varphi}\rangle \rightarrow |e_1\rangle)$$
$$P(|e_\vartheta\rangle \rightarrow |e_2\rangle) = P(|e_{\vartheta\varphi}\rangle \rightarrow |e_2\rangle) .$$

If instead we want the probability of transmission through a polaroid sheet with its optical axis at 45° in the x-y plane, i.e. the transition probability to the state

$$\frac{1}{\sqrt{2}}\left(|e_1\rangle + |e_2\rangle\right),$$

the latter is different for the two states $|e_\vartheta\rangle$ and $|e_{\vartheta\varphi}\rangle$. Therefore, in general, two states represented by the vectors:

$$|A\rangle = \sum_{j=1}^{\infty} a_j\,|\xi_j\rangle , \qquad |B\rangle = \sum_{j=1}^{\infty} a_j\,\mathrm{e}^{\mathrm{i}\,\varphi_j}\,|\xi_j\rangle \tag{4.12}$$

are different, even if they behave in the same way as far as the measurements of the observable ξ are concerned (the probabilities $|A\rangle \rightarrow |\xi_j\rangle$ and $|B\rangle \rightarrow |\xi_j\rangle$ are equal to each other), but – just because they are different states – there will certainly exist some other observable η, measuring which will give rise to a different behaviour of $|A\rangle$ and $|B\rangle$.

Let us now discuss the case of

Degenerate Observables

Let us now examine how the previous results are modified if ξ is a degenerate observable. Also in this case one has that

the eigenvectors of an observable corresponding to different eigenvalues are orthogonal to one another
     and that
the set of all the eigenvectors of an observable is complete.

The proofs, that we omit, are similar to those of the nondegenerate case (it is however necessary to make use of another consequence of the von Neumann postulate: if, owing to a measurement of the observable ξ, $|A\rangle \to |\bar{\xi}_i\rangle$, then, from among all the eigenstates of ξ corresponding to the eigenvalue ξ_i, the vector $|\bar{\xi}_i\rangle$ is that for which the probability transition is a maximum).

We can no longer state that the eigenvectors of the (degenerate) observable ξ form an orthonormal basis: we are not guaranteed (and it is not true) that different eigenvectors corresponding to the same eigenvalue are mutually orthogonal. Indeed the following important theorem holds:

Theorem: *any linear combination of the eigenvectors of an observable corresponding to the same eigenvalue still is an eigenvector of the observable corresponding to the same eigenvalue.*

Let ξ_1 be a degenerate eigenvalue: the theorem states that if $|\xi_1'\rangle$ and $|\xi_1''\rangle$ represent two eigenvectors of ξ corresponding to the same eigenvalue ξ_1, any vector $\alpha|\xi_1'\rangle + \beta|\xi_1''\rangle$ represents an eigenstate of ξ corresponding to the eigenvalue ξ_1.

Proof: let us put $|A\rangle = \alpha|\xi_1'\rangle + \beta|\xi_1''\rangle$. In order to show that $|A\rangle$ is an eigenstate of ξ corresponding to the eigenvalue ξ_1, one must show that a measurement of ξ on $|A\rangle$ always gives ξ_1 as result, which is equivalent to say that a measurement of ξ will never give ξ_i, $i \geq 2$,, namely $p_i = 0$ for $i \geq 2$. Let us reason by contradiction: suppose we find ξ_i ($i \geq 2$) as a result. Then the system after the measurement is in an eigenstate of ξ corresponding to ξ_i: $|\xi_i\rangle$. The transition probability $|A\rangle \to |\xi_i\rangle$ is $p_i = |\langle A \mid \xi_i\rangle|^2$, but

$$\langle A \mid \xi_i\rangle = \alpha^*\langle \xi_1' \mid \xi_i\rangle + \beta^*\langle \xi_1'' \mid \xi_i\rangle = 0$$

for $\langle \xi_1' \mid \xi_i\rangle = \langle \xi_1'' \mid \xi_i\rangle = 0$ (orthogonality of eigenvectors corresponding to different eigenvalues), therefore $p_i = 0$, against the hypothesis.

Thanks to the continuity of the scalar product, the conclusion extends to linear combinations of whatever (either finite or infinite) number of eigenvectors and we can in conclusion state that

The set of all the eigenvectors of an observable corresponding to the same eigenvalue is a (closed) linear subspace of the Hilbert space $\mathcal{H}$.

This linear manifold is called the **eigenspace** of the observable corresponding to the eigenvalue ξ_i.

The dimension of this manifold (that can be either finite or infinite), i.e. the number of *independent* vectors it contains, is called **degree of degeneracy** of the eigenvalue. For a nondegenerate eigenvalue the degree of degeneracy is therefore 1. The g_i's of (2.20) are precisely the degrees of degeneracy of the eigenvalues of the observable energy, i.e. the number of independent states corresponding to the energy level E_i.

We have seen that, for a degenerate observable, the set of its eigenvectors is complete, but – contrary to the nondegenerate case – is not an orthonormal system: indeed, in every degenerate eigenspace of an observable obviously there exist vectors not orthogonal to one another. However, it is known that from a complete set it is always possible, by means of an orthonormalization process, to extract a complete orthonormal set. Therefore we are able to form an orthonormal basis consisting of eigenvectors of a (degenerate) observable ξ: such a basis contains all the eigenvectors corresponding to nondegenerate eigenvalues and a system of mutually orthogonal vectors corresponding to any degenerate eigenvalue, whose number equals its degree of degeneracy, i.e. a system that is complete in the considered eigenspace. The choice of the orthonormal basis is not unique, because clearly in any degenerate eigenspace of the observable infinite choices of mutually orthogonal vectors are possible. In any event, once an orthonormal basis is fixed: $|\xi_1^{(1)}\rangle, |\xi_1^{(2)}\rangle, \cdots, |\xi_2^{(1)}\rangle, \cdots$, one can still expand any vector $|A\rangle$ in terms of it:

$$|A\rangle = a_1^{(1)}|\xi_1^{(1)}\rangle + a_1^{(2)}|\xi_1^{(2)}\rangle + \cdots + a_2^{(1)}|\xi_2^{(1)}\rangle + \cdots . \tag{4.13}$$

The coefficients $a_1^{(1)}, a_1^{(2)}, \cdots, a_2^{(1)}, \cdots$ are obtained as in (4.8) and (4.10) still holds.

Two problems remain open:

1. how much is the probability p_i that a measurement of ξ on $|A\rangle$ gives the *degenerate* eigenvalue ξ_i as a result? and
2. which is the state of the system after such a measurement?

To both questions the von Neumann postulate (we already have partially enunciated and utilized) gives the answer:

von Neumann Postulate: *if a measurement of ξ on $|A\rangle$ gives the (degenerate) eigenvalue ξ_i as a result, the state after the measurement is represented by the (ray to which belongs the) vector $|\bar{\xi}_i\rangle$ that is obtained by orthogonally projecting $|A\rangle$ onto the eigenspace of ξ corresponding to the eigenvalue ξ_i.*

If $|A\rangle$ is given by (4.13), then:

$$|\bar{\xi}_i\rangle = a_i^{(1)}|\xi_i^{(1)}\rangle + a_i^{(2)}|\xi_i^{(2)}\rangle + \cdots + a_i^{(n)}|\xi_i^{(n)}\rangle + \cdots \tag{4.14}$$

(the sum extends only to the vectors $|\xi_i^{(k)}\rangle$ that correspond to the eigenvalue ξ_i and, if $|A\rangle$ is normalized to 1, in general the vector $|\bar{\xi}_i\rangle$ is not).

Now also the answer to the first problem is straightforward:

$$p_i = P(|A\rangle \to |\bar{\xi}_i\rangle) = \frac{|\langle A|\bar{\xi}_i\rangle|^2}{\langle \bar{\xi}_i|\bar{\xi}_i\rangle} = \frac{\left(|a_i^{(1)}|^2 + |a_i^{(2)}|^2 + \cdots + |a_i^{(n)}|^2 + \cdots\right)^2}{|a_i^{(1)}|^2 + |a_i^{(2)}|^2 + \cdots + |a_i^{(n)}|^2 + \cdots}$$
$$= |a_i^{(1)}|^2 + |a_i^{(2)}|^2 + \cdots + |a_i^{(n)}|^2 + \cdots \tag{4.15}$$

Therefore p_i is the sum of the transition probabilities $|A\rangle \to |\xi_i'\rangle$, $|A\rangle \to |\xi_i''\rangle, \cdots$ from $|A\rangle$ to the vectors of whatever orthonormal set of eigenvectors of ξ corresponding to the eigenvalue ξ_i, in number equal to the degree of degeneracy, i.e. complete in the considered eigenspace.

Note that, since the vector representing the state after the measurement is obtained by projection of the vector $|A\rangle$, an equivalent way of expressing the von Neumann postulate is to state that: the system effects the transition to the state, from among the eigenstates of ξ corresponding to ξ_i, for which the probability transition is a maximum; or that: the measurement has perturbed the system the least possible; indeed, the orthogonal projection is, in the manifold of the eigenvectors corresponding to ξ_i, the vector 'closest' to the initial vector $|A\rangle$.

An obvious consequence of von Neumann postulate is that, if the initial state $|A\rangle$ already is an eigenstate of ξ corresponding to the (degenerate) eigenvalue ξ_i, the state is not perturbed by the measurement: indeed the projection of $|A\rangle$ coincides with $|A\rangle$; therefore, in particular, even if $\langle \xi_i' \mid \xi_j'' \rangle \neq 0$, a measurement of ξ will never be able to induce the transition $|\xi_i'\rangle \to |\xi_i''\rangle$: the measurement of another observable η, for which $|\xi_i'\rangle$ is not one of its eigenstates but $|\xi_i''\rangle$ is, will be necessary to induce such a transition.

4.5 Operators Associated with Observables

In the process of formalization of physical concepts we are pursuing, we have represented the states of a system by means of vectors in a Hilbert space; there now remains to understand in which way we shall formalize (i.e. by means of which mathematical entities we shall represent) the observables that, from the physical point of view, are characterized by the existence of those particular states that we have called *eigenstates* and by the eigenvalues.

The terms we have used (eigenvectors, eigenvalues) have not been introduced casually, but they constituted an anticipation of what will be the result of this process of formalization. Indeed, it is possible to associate, in a quite natural way, an **operator** ξ^{op} on $\mathcal{H}$ (namely a linear application from $\mathcal{H}$ onto $\mathcal{H}$) to each observable ξ in the following way: no matter how a basis consisting of the eigenvectors $|\xi_i\rangle$ of the (possibly degenerate) observable ξ is chosen, one defines ξ^{op} on the vectors of this basis as

$$\xi^{\mathrm{op}}|\xi_i\rangle \overset{\text{def}}{=} \xi_i|\xi_i\rangle \tag{4.16}$$

and, just to start, one extends the definition by linearity to all the finite linear combinations formed with the vectors of the basis: if $|A\rangle = \sum_1^n a_i|\xi_i\rangle$,

$$\xi^{\mathrm{op}}|A\rangle \overset{\text{def}}{=} \sum_{i=1}^n a_i\,\xi^{\mathrm{op}}|\xi_i\rangle = \sum_{i=1}^n a_i\,\xi_i|\xi_i\rangle\,. \tag{4.17}$$

Then, when it is possible, the definition is extended to the infinite linear combinations: if $|A\rangle = \sum_1^\infty a_i|\xi_i\rangle$ (being $\sum_1^\infty |a_i|^2 < \infty$),

$$\xi^{\mathrm{op}}|A\rangle \overset{\text{def}}{=} \lim_{n\to\infty} \sum_{i=1}^n a_i\,\xi_i\,|\xi_i\rangle \equiv \sum_{i=1}^\infty a_i\,\xi_i\,|\xi_i\rangle\,. \tag{4.18}$$

Let us firstly check that the definition is a 'good definition', namely independent of the choice of the basis (although still consisting of eigenvectors of ξ): indeed, within any eigenspace of the observable ξ, ξ^{op} is a multiple of the identity application:

$$| \bar{\xi}_i \rangle = \sum_k \alpha_k | \xi_i^{(k)} \rangle \quad \Rightarrow \quad \xi^{\mathrm{op}} | \bar{\xi}_i \rangle = \sum_k \alpha_k \, \xi_i \, | \xi_i^{(k)} \rangle = \xi_i | \bar{\xi}_i \rangle \quad (4.19)$$

and, as a consequence, does not depend on the basis chosen in the eigenspace.

Why did we say 'when it is possible' before writing (4.18)? Since

$$\left\| \sum_{i=1}^n a_i \, \xi_i \, | \xi_i \rangle \right\|^2 = \sum_{i=1}^n \xi_i^2 \, |a_i|^2,$$

one realizes that, if the series $\sum_1^\infty \xi_i^2 \, |a_i|^2$ does not converge, by (4.10) the series $\sum_1^\infty a_i \, \xi_i \, | \xi_i \rangle$ does not define a vector in $\mathcal{H}$. Then, unless $|\xi_i| < M$ for any i (bounded operator: in this case $\sum_1^\infty \xi_i^2 \, |a_i|^2 < M^2 \sum_1^\infty |a_i|^2 < \infty$), it is not possible to define the operator ξ^{op} on all $\mathcal{H}$: the domain D_ξ of the operator ξ^{op}, namely the set of vectors on which it is defined, consists of those vectors for which $\sum_1^\infty \xi_i^2 \, |a_i|^2 < \infty$ and the latter (indeed already the set of finite linear combinations) form a set that is **dense** in $\mathcal{H}$; for the vectors that belong to D_ξ one has therefore:

$$\xi^{\mathrm{op}} | A \rangle = \sum_i a_i \, \xi_i \, | \xi_i \rangle \,, \qquad | A \rangle \in D_\xi \,, \qquad \overline{D}_\xi = \mathcal{H} \,. \qquad (4.20)$$

We will not exceedingly worry about these technical (domain) problems, that certainly are important from the mathematical point of view, *but* absolutely marginal from the physical standpoint. More to it: in principle, one should state that from the physical point of view all the observables are represented by bounded operators ($|\xi_i| < M$ for any i), since no instrument can yield results 'as large as one wishes': the scale of an instrument is always bounded both from above and from below. The only reason for which, in practice, we cannot totally forget domain problems lies in the fact that almost all the operators associated with the observables $f(q,p)$ (owing to the quantization postulate we have not yet enunciated) will exhibit an unbounded spectrum: one can say that such operators do not faithfully represent the physical observables, namely the measurement instruments, but they rather provide a mathematical schematization for them. Stated in different words, the root of the domain problems is in the mathematical schematization, not in physics.

4.6 Properties of the Operators Associated with Observables

Let us now examine the properties of ξ^{op}.

1. ξ^{op} is a self-adjoint operator: $\xi^{\mathrm{op}} = (\xi^{\mathrm{op}})^\dagger$.

Consistently with the just made statement that we do not want to be overwhelmed by domain problems, in the case of unbounded operators we give neither the definition of adjoint operator nor that of self-adjoint operator.

For bounded operators, defined on the whole $\mathcal{H}$, such definitions coincide with those of the finite-dimensional linear spaces: a bounded operator η (defined on all $\mathcal{H}$) by definition is known when, for any vector $|A\rangle$, the vector $|C\rangle \equiv \eta\,|A\rangle$ is known; but any vector $|C\rangle$ is itself known when, for any $|B\rangle$, the scalar products $\langle B \mid C \rangle$ are known (the knowledge of the scalar products between $|C\rangle$ and the elements of an orthonormal basis is sufficient: in the latter case $|C\rangle$ is given by an expression of the type (4.7)), so η is determined by the knowledge of the scalar products $\langle B \mid \eta \mid A \rangle$ for any $|A\rangle$ and for any $|B\rangle$ ($\langle B \mid \eta \mid A \rangle$ stands for the scalar product of $|B\rangle$ and $|C\rangle = \eta\,|A\rangle$).

The adjoint $\eta^{\dagger}$ of η can then be defined by the following equation:

$$\langle B \mid \eta^{\dagger} \mid A \rangle \overset{\text{def}}{=} \langle A \mid \eta \mid B \rangle^{*} \tag{4.21}$$

(normally mathematicians write η^{*} instead of $\eta^{\dagger}$, whereas we will reserve the asterisk for the complex conjugation of numbers); η is self-adjoint if $\eta = \eta^{\dagger}$.

Therefore $\xi^{\mathrm{op}} = (\xi^{\mathrm{op}})^{\dagger}$ (bounded) is equivalent to:

$$\langle B \mid \xi^{\mathrm{op}} \mid A \rangle = \langle A \mid \xi^{\mathrm{op}} \mid B \rangle^{*} \qquad \text{for all} \quad |A\rangle, |B\rangle \in \mathcal{H}. \tag{4.22}$$

If ξ^{op} is not bounded, (4.22) – holding for all the vectors $|A\rangle$, $|B\rangle$ in the domain of ξ^{op} – only expresses the fact that ξ^{op} is **Hermitian** (as D_{ξ} is dense, the correct term would be *symmetric*, however we will always use the term *Hermitian*, which is the one normally used by physicists); in order that ξ^{op} be self-adjoint, something more is needed (the domains, on which ξ^{op} and $(\xi^{\mathrm{op}})^{\dagger}$ are defined, must coincide).

We will limit ourselves to show that ξ^{op} is Hermitian (i.e. symmetric).

Proof: (use will be made of: (4.9), the linearity of ξ^{op}, the linearity of the scalar product: $\langle B \mid (\sum_{i} |A_{i}\rangle) \rangle = \sum_{i}\langle B \mid A_{i}\rangle$, its Hermiticity, expressed by (4.1), and the reality of the eigenvalues ξ_{i})

$$\langle B \mid \xi^{\mathrm{op}} \mid A \rangle = \langle B \mid \xi^{\mathrm{op}} \sum_{i} |\xi_{i}\rangle\langle \xi_{i} \mid A \rangle = \sum_{i}\langle B \mid \xi^{\mathrm{op}} |\xi_{i}\rangle\langle \xi_{i} \mid A \rangle$$

$$= \sum_{i} \xi_{i}\langle B \mid \xi_{i} \rangle \times \langle \xi_{i} \mid A \rangle = \sum_{i} \xi_{i}\langle A \mid \xi_{i} \rangle^{*} \times \langle \xi_{i} \mid B \rangle^{*}$$

$$= \left(\sum_{i} \xi_{i}\langle A \mid \xi_{i} \rangle \times \langle \xi_{i} \mid B \rangle \right)^{*} = \langle A \mid \xi^{\mathrm{op}} \mid B \rangle^{*}$$

2.   We have always used the terminology $|\xi_{i}\rangle$: eigenvectors; ξ_{i}: eigenvalues. This terminology has a precise mathematical meaning, however ours has not been an abuse of terms: indeed all the eigenvectors and eigenvalues of ξ^{op} (in the mathematical sense) are the eigenvectors and eigenvalues of the observable ξ (in the physical sense) and viceversa. The second part of the proof lies in the definition of ξ^{op} and (4.19). There remains to show that, if $\xi^{\mathrm{op}}|\mu\rangle = \mu\,|\mu\rangle$, then μ is one of the eigenvalues ξ_{i} of the observable ξ^{op} and $|\mu\rangle$ is one of the $|\xi_{i}\rangle$: however, this is an exercise we leave to the reader.

Let us summarize: every observable ξ corresponds to a self-adjoint operator ξ^{op}; the eigenvectors and eigenvalues of ξ^{op} are all and only the eigenvectors and eigenvalues of ξ. We are therefore authorized, from now on, to identify, and as a consequence to represent by the same symbol ξ, the observable and the operator associated with it: therefore ξ simultaneously represents both a physical quantity, the instrument (or the instruments) suitable to measure it and the linear operator that corresponds to it: so we shall no longer use the notation ξ^{op}.

It is natural, at this point, to ask whether any self-adjoint operator is associated with an observable.

Before answering this question we recall that the eigenvalues of a self-adjoint operator are real and that eigenvectors corresponding to different eigenvalues are orthogonal to one another: we recall the proof of these two facts, just to practice with the formalism.

By taking the scalar product of $|\eta'\rangle$ with $\eta\,|\eta'\rangle = \eta'\,|\eta'\rangle$ one obtains:

$$\langle \eta' \mid \eta \mid \eta' \rangle = \eta' \langle \eta' \mid \eta' \rangle$$

and as $\eta = \eta^\dagger$, from (4.22) it follows that $\langle \eta' \mid \eta \mid \eta' \rangle = \langle \eta' \mid \eta \mid \eta' \rangle^*$ and, as a consequence, η' is real.

Let now

$$\eta\,|\eta'\rangle = \eta'\,|\eta'\rangle\,, \qquad \eta\,|\eta''\rangle = \eta''\,|\eta''\rangle\,, \qquad \eta' \neq \eta''\,.$$

By taking the scalar product of the first with $|\eta''\rangle$ and of the second with $|\eta'\rangle$ one has

$$\langle \eta'' \mid \eta \mid \eta' \rangle = \eta' \langle \eta'' \mid \eta' \rangle\,, \qquad \langle \eta' \mid \eta \mid \eta'' \rangle = \eta'' \langle \eta' \mid \eta'' \rangle$$

and by subtracting the first from the complex conjugate of the second ($\eta = \eta^\dagger$!) one finds:

$$(\eta' - \eta'')\langle \eta'' \mid \eta' \rangle = 0 \quad \Rightarrow \quad \langle \eta'' \mid \eta' \rangle = 0\,.$$

It therefore seems that self-adjoint operators have the right properties to be considered as operators associated with observables. It is however necessary to keep in mind that, from this point of view, in a (separable) Hilbert space there may occur very unpleasant things. It may happen that a self-adjoint operator does not have enough eigenvectors as to form a basis, while we know that the set of the eigenvectors of an observable forms a basis (example: in the space $L_2(a,b)$ consisting of the square-integrable functions $f(x)$ in the interval $a \leq x \leq b$, the operator $f(x) \to x\,f(x)$ is self-adjoint and has no eigenvector: indeed the equation $x\,f(x) = \lambda\,f(x)$ has the only solution $f(x) \sim 0$).

The self-adjoint operators whose eigenvectors form a basis are the operators with purely discrete spectrum, while the others are the operators with continuous spectrum plus – possibly – a discrete component (see Fig. 2.3).

We then postulate that

Any self-adjoint operator with purely discrete spectrum is associated with an observable.

Any operator associated with an observable will itself be called an 'observable': we will identify in this way, not only in the notation but even in the name, measurement instruments and operators.

We shall however see in the sequel that it is not only advisable but even right to attribute the name 'observable' also to self-adjoint operators that do not have a purely discrete spectrum (as e.g. the energy of the hydrogen atom: there are energy levels that form the discrete component of the spectrum, and there is the continuum of the ionization states). We shall see that they can be considered as 'limits' (in a sense to be specified) of operators endowed with discrete spectrum and therefore correspond to limiting cases of bona fide observables (also in this case one deals with those 'limit' concepts so frequent in physics: point mass, instantaneous velocity, ...).

Also in this case (much as in the case of the bijective correspondence between states and rays) perhaps it is neither true nor necessary to assume that *all* the self-adjoint operators represent some observable.

What we had provisionally assumed, namely that for any state $|A\rangle$ there exists at least one observable that possesses $|A\rangle$ as an eigenstate corresponding to a nondegenerate eigenvalue, follows from the last postulate we have enunciated: indeed, it is sufficient to take the observable corresponding to the projector onto the one-dimensional linear manifold generated by the vector $|A\rangle$ (the definition of projection operator will be recalled in the next section). Anyway, in Sect. 6.3 we will come back to this problem.

There still remains a problem: given an observable quantity (e.g. energy, angular momentum, ...) how do we know the operator that must be associated with it? This problem will be given its answer by the quantization postulate we will enunciate in Sect. 4.12.

4.7 Digression on Dirac Notation

The aim of this section only is to compare the notation used by mathematicians with Dirac's, the latter being the one used in almost all the texts on quantum mechanics. Therefore, in the following formulae, all the technical issues that concern domain problems etc., will be omitted.

The differences between the two types of notation originate from

1. the fact that mathematicians represent vectors simply by the letters of the alphabet: u, v, $\cdots$, whereas we use the letters of the alphabet (or any other symbol) boxed within the "ket" $|\ \rangle$: $|u\rangle$, $|v\rangle$; $|+\rangle$, $|-\rangle$; $|\uparrow\rangle$, $|\downarrow\rangle$; $|\odot\rangle$, $|\odot\rangle$; $\cdots$ (even these fancy symbols we will make use of);
2. the different notation for the scalar product: (u, v) for mathematicians, $\langle u | v \rangle$ for us.

Indeed, Dirac "bra" $\langle u |$ is the element of another vector space: the space dual to $\mathcal{H}$, i.e. the space of the linear and continuous functionals on $\mathcal{H}$. This is totally legitimate thanks to the Riesz theorem, according to which the space dual to $\mathcal{H}$ is isomorphic to $\mathcal{H}$ itself.

Let us now examine the main notational differences that follow from the above points: let ξ be a (not necessarily self-adjoint) linear operator; the vector that results from the application of ξ to a vector is ξu for mathematicians and $\xi | u \rangle$ (*not* $| \xi u \rangle$) for us. So we will write $\langle v | \xi | u \rangle$ that has the same meaning as $(v, \xi u)$ for mathematicians. Moreover the adjoint $\xi^\dagger$ of an operator ξ is defined by means of the equation $(\xi^\dagger v, u) = (v, \xi u)$ that we can write with our notation only after taking the complex conjugate of both sides:

$$(u, \xi^\dagger v) = (v, \xi u)^* \quad \longleftrightarrow \quad \langle u | \xi^\dagger | v \rangle = \langle v | \xi | u \rangle^* .$$

The last equation amounts to saying that the "bra" corresponding to the "ket" $\xi | u \rangle$ is $\langle u | \xi^\dagger$, and if $\xi | u \rangle = | v \rangle$, then $\langle u | \xi^\dagger = \langle v |$; in particular, if $| \xi_i \rangle$ is an "eigenket" of $\xi = \xi^\dagger$, $\langle \xi_i |$ is an "eigenbra" of ξ corresponding to the eigenvalue ξ_i: $\langle \xi_i | \xi = \xi_i \langle \xi_i |$.

Let us list some properties of the Hermitian conjugation ($\xi \to \xi^\dagger$) that immediately follow from the definition of adjoint operator:

$$\begin{aligned}
(\xi^\dagger)^\dagger &= \xi \\
(\alpha \xi)^\dagger &= \alpha^* \xi^\dagger \\
(\xi + \eta)^\dagger &= \xi^\dagger + \eta^\dagger \\
(\xi \eta)^\dagger &= \eta^\dagger \xi^\dagger .
\end{aligned} \tag{4.23}$$

Dirac notations are not favourably looked at by the mathematicians, whereas for the physicists, that have to make a large use of them, they are rather comfortable because follow rules of easy applicability: for example, to take the scalar product of $| u \rangle$ with $| v \rangle$ means to 'glue' the bra $\langle v |$ and the ket $| u \rangle$; moreover, much as the conjugate of α is α^* and the conjugate of an operator ξ is $\xi^\dagger$, the conjugate of a ket (bra) is the corresponding bra (ket); this rule, that ensues from (4.21) and (4.23), allows one to write in an almost automatic way an expression like $\langle v | \xi \eta \zeta \cdots | u \rangle^*$: it suffices to conjugate any element and reverse their order:

$$\langle v | \xi \eta \zeta \cdots | u \rangle^* = \langle u | \cdots \zeta^\dagger \eta^\dagger \xi^\dagger | v \rangle .$$

We conclude this section by recalling the definition of projection operators and showing how also these can be easily expressed with the Dirac notation.

Let $| v \rangle \in \mathcal{H}$, $\langle v | v \rangle = 1$. The projection operator $\mathcal{P}_v$ onto the (one-dimensional) manifold generated by $| v \rangle$ is defined by

$$\text{for all } u \in \mathcal{H}: \quad \mathcal{P}_v | u \rangle = | v \rangle \langle v | u \rangle$$

so, since applying $\mathcal{P}_v$ to $| u \rangle$ is equivalent to 'glue' $| v \rangle \langle v |$ to $| u \rangle$, we can write:

$$\mathcal{P}_v = | v \rangle \langle v | .$$

Let now $\mathcal{V}$ be a (closed) linear manifold of arbitrary dimension included in $\mathcal{H}$.

If $| v_i \rangle \in \mathcal{V}$, $i = 1, 2, \cdots$ is whatever set of orthonormal vectors that generate $\mathcal{V}$, the projection operator $\mathcal{P}_\mathcal{V}$ onto $\mathcal{V}$ is defined by:

$$\text{for all } |u\rangle \in \mathcal{H}: \quad \mathcal{P}_{\mathcal{V}}|u\rangle = \sum_i |v_i\rangle\langle v_i|u\rangle$$

and we can therefore write:

$$\mathcal{P}_{\mathcal{V}} = \sum_i |v_i\rangle\langle v_i|. \tag{4.24}$$

In particular, if $|\xi_i\rangle$ is an orthonormal basis in $\mathcal{H}$, the identity operator $\mathbb{1}$ –
i.e. the projector onto all $\mathcal{H}$: $\mathbb{1}|u\rangle = |u\rangle$ – can be written as

$$\mathbb{1} = \sum_i |\xi_i\rangle\langle \xi_i|. \tag{4.25}$$

Equation (4.25) is known as the **completeness relation** because it expresses
the fact the vectors $|\xi_i\rangle$ form a complete orthonormal set.

So, for example, by applying both sides of (4.25) to the vector $|u\rangle$, one
(re)obtains (4.9). Moreover:

$$\langle u|u\rangle = \langle u|\mathbb{1}|u\rangle = \sum_i \langle u|\xi_i\rangle\langle \xi_i|u\rangle = \sum_i |\langle u|\xi_i\rangle|^2$$

i.e. (4.10).

If $\mathcal{P}$ is a projection operator, it is straightforward to verify that

$$\mathcal{P} = \mathcal{P}^\dagger, \qquad \mathcal{P}^2 = \mathcal{P}. \tag{4.26}$$

If ξ is (the operator associated with) an observable and we denote by $\mathcal{P}_i$ the
projector onto the manifold consisting of the eigenvectors of ξ corresponding
to the eigenvalue ξ_i (the eigenspace of ξ corresponding to ξ_i), by (4.25) and
(4.24) one has:

$$\sum_i \mathcal{P}_i = \mathbb{1} \Rightarrow \xi \equiv \xi \times \mathbb{1} = \xi \sum_i \mathcal{P}_i = \sum_i \xi_i \mathcal{P}_i = \sum_i |\xi_i\rangle \xi_i \langle \xi_i| \tag{4.27}$$

where the sum appearing in the last expression extends to all the vectors of
an orthonormal basis of eigenvectors of ξ; so if a given eigenvalue is n times
degenerate ($n \leq \infty$), in the sum there are n eigenvectors corresponding to
that eigenvalue (in the last term of (4.27) the eigenvalue ξ_i is placed between
the bra and the ket only for aesthetical reasons).

4.8 Mean Values

If we make N measurement of the observable ξ on the system in the state $|A\rangle$
(we recall that this means to have at one's disposal N copies of the system, all
in the state $|A\rangle$, and that any measurement is made on one of such copies),
we find the eigenvalues ξ_i – each N_i times ($\sum N_i = N$) – as results.

We can define the **mean value** $\bar{\xi}$ of the observable ξ in the state $|A\rangle$ as
the mean value of the obtained results; if N is very large, one has:

$$\bar{\xi} \stackrel{\text{def}}{=} \frac{1}{N}\sum_i \xi_i N_i = \sum_i \xi_i p_i, \qquad p_i = \frac{N_i}{N}. \tag{4.28}$$

Let us take $\langle A \mid A \rangle = 1$. Due to (4.15) (von Neumann postulate), if $\mathcal{P}_i$ is (as in the previous section) the projector onto the eigenspace of ξ corresponding to the eigenvalue ξ_i, one has

$$p_i \equiv P\big(\mid A \rangle \to \mid \bar{\xi}_i \rangle\big) = P\big(\mid A \rangle \to \mathcal{P}_i\mid A \rangle\big) = \frac{|\langle A \mid \mathcal{P}_i \mid A \rangle|^2}{\langle A \mid \mathcal{P}_i \times \mathcal{P}_i \mid A \rangle}$$

(the denominator is $\langle \bar{\xi}_i \mid \bar{\xi}_i \rangle$), and as $\big($see (4.26)$\big)$ $\mathcal{P}_i^2 = \mathcal{P}_i$ and, in addition, $\langle A \mid \mathcal{P}_i \mid A \rangle = \langle A \mid \mathcal{P}_i^2 \mid A \rangle \geq 0$, one has

$$p_i = \langle A \mid \mathcal{P}_i \mid A \rangle \tag{4.29}$$

(we have rewritten (4.15) using projectors).

Therefore, owing to (4.27),

$$\bar{\xi} = \sum_i \xi_i \langle A \mid \mathcal{P}_i \mid A \rangle = \langle A \mid \sum_i \xi_i \mathcal{P}_i \mid A \rangle = \langle A \mid \xi \mid A \rangle . \tag{4.30}$$

Unless the contrary is specified, $\langle A \mid A \rangle = 1$ will be always assumed.

The quantity $\langle A \mid \xi \mid A \rangle$ usually is more correctly called "*expectation value* of the observable ξ in the state $\mid A \rangle$" because, p_i being the probabilities provided by the theory – i.e. theoretical probabilities – $\bar{\xi}$ is what, according to the theory, should be expected as mean value of the results of the measurements of ξ on $\mid A \rangle$.

Let us now consider the operator ξ^2 and let us examine its properties: all the eigenvectors $\mid \xi_i \rangle$ of ξ also are eigenvectors of ξ^2 corresponding to the eigenvalues ξ_i^2:

$$\xi^2 \mid \xi_i \rangle = \xi \times \xi \mid \xi_i \rangle = \xi \, \xi_i \mid \xi_i \rangle = \xi_i \, \xi \mid \xi_i \rangle = \xi_i^2 \mid \xi_i \rangle$$

and, as the $\mid \xi_i \rangle$ form a complete set, also ξ^2 is an observable: if the application $\xi_i \to \xi_i^2$ is injective (i.e. if ξ has no opposite eigenvalues), then ξ and ξ^2 have all and only the same eigenvectors and the observable corresponding to ξ^2 is obtained by simply changing the scale of the instrument that measures ξ; in the contrary case the instruments that measure ξ and ξ^2 are different, since there are eigenvectors of ξ^2 that are not eigenvectors of ξ.

Incidentally, in the same way we can define $f(\xi)$ as the operator that has the same eigenvectors of ξ and eigenvalues $f(\xi_i)$: if $\xi_i \to f(\xi_i)$ is injective, then $f(\xi)$ does not differ, except for the scale, from the observable ξ.

The expectation value of ξ^2:

$$\overline{\xi^2} = \langle A \mid \xi^2 \mid A \rangle = \sum_i \xi_i^2 \, p_i$$

is the quadratic mean value of the results of the measurements of ξ on $\mid A \rangle$.

The mean-square deviation $\Delta \xi = \sqrt{\overline{\xi^2} - \bar{\xi}^2}$ is therefore given by

$$(\Delta \xi)^2 = \langle A \mid \xi^2 \mid A \rangle - \langle A \mid \xi \mid A \rangle^2 \tag{4.31}$$

or equivalently by:

$$(\Delta\xi)^2 = \langle A \mid (\xi - \bar{\xi})^2 \mid A \rangle . \qquad (4.32)$$

Indeed: $\overline{(\xi - \bar{\xi})^2} = \overline{(\xi^2 - 2\xi\bar{\xi} + \bar{\xi}^2)} = \overline{\xi^2} - \bar{\xi}^2$.

The meaning of $\Delta\xi$ is well known: it represents the size of the dispersion of the results around the mean value; moreover, in the present framework one has the following

Theorem: $\Delta\xi = 0$ if and only if $\mid A \rangle$ is an eigenvector of ξ.

Indeed, if $\xi \mid A \rangle = \xi' \mid A \rangle$ one has:

$$\langle A \mid \xi \mid A \rangle = \xi'\langle A \mid A \rangle = \xi' \quad \text{and} \quad \langle A \mid \xi^2 \mid A \rangle = \xi'^2\langle A \mid A \rangle = \xi'^2$$

whence $\Delta\xi = 0$. If, viceversa, $\Delta\xi = 0$, then:

$$0 = (\Delta\xi)^2 = \langle A \mid (\xi - \bar{\xi}) \times (\xi - \bar{\xi}) \mid A \rangle$$

and, since the latter is the squared norm of the vector $(\xi - \bar{\xi}) \mid A \rangle$ (recall that $(\xi - \bar{\xi}) = (\xi - \bar{\xi})^\dagger$), it must be that

$$(\xi - \bar{\xi}) \mid A \rangle = 0 \quad \Rightarrow \quad \xi \mid A \rangle = \bar{\xi} \mid A \rangle .$$

For this reason $\Delta\xi$ is also called **uncertainty** of ξ in $\mid A \rangle$: if $\Delta\xi = 0$ the value of ξ in $\mid A \rangle$ is completely determined. As a consequence, by means of the only knowledge of the expectation values, it is possible to establish whether a given state is an eigenstate of an observable and, in the affirmative case, to know the corresponding eigenvalue. More to it: thanks to (4.29), also the transition probabilities can be expressed as expectation values.

So all the physical information that characterizes the state of a system – transition probabilities, mean values of the observables, observables of which the state is eigenstate and the corresponding eigenvalues – can be traced back to expectation values of observables, and in this sense the knowledge of a state is equivalent to the knowledge of all the expectation values in the state itself. It is even possible to reformulate quantum mechanics by defining the state of a system as the collection of the expectation values of all the observables (linear positive functionals on the algebra of observables), instead of a vector of the Hilbert space. This formulation has been proposed by J. von Neumann, I.E. Segal, R. Haag and D. Kastker, and is equivalent, for a system of particles, to the 'Hilbert' formulation due to Dirac.

4.9 Pure States and Statistical Mixtures

We have already insisted on the fact that, since the interpretation of quantum mechanics is a statistical interpretation, saying 'the state represented by the vector $\mid A \rangle$' presupposes the possibility that one can prepare many copies of the system in the same state $\mid A \rangle$: in this case one says that the system is in a **pure state**.

Let us now assume we have $N_1 \gg 1$ copies of the system in the state $\mid A \rangle$ and $N_2 \gg 1$ copies of the same system in the state $\mid B \rangle$ and that we measure, on each of these $N = N_1 + N_2$ systems, an observable ξ: the theory predicts that, if $p_i^{(A)}$ and $p_i^{(B)}$ are the probabilities of finding the eigenvalue ξ_i respectively when the system is in either the state $|A\rangle$ or the state $|B\rangle$, we will find the eigenvalue ξ_i a number $N_1 p_i^{(A)} + N_2 p_i^{(B)}$ of times, therefore with a probability

$$p_i = \frac{1}{N}\left(N_1 p_i^{(A)} + N_2 p_i^{(B)}\right) = \frac{N_1}{N} p_i^{(A)} + \frac{N_2}{N} p_i^{(B)} . \qquad (4.33)$$

Therefore the mean value of the results of the measurements is:

$$\langle\!\langle \xi \rangle\!\rangle = \sum_i \xi_i\, p_i = \frac{N_1}{N}\langle A\,|\,\xi\,|\,A\rangle + \frac{N_2}{N}\langle B\,|\,\xi\,|\,B\rangle . \qquad (4.34)$$

For example, the $N = N_1 + N_2$ states could have been obtained by sending N photons in the same polarization state on a birefringent crystal with the optical axis at an angle ϑ with respect to the polarization direction: in the latter case $p_1 \equiv N_1/N = \cos^2\vartheta$ and $p_2 \equiv N_2/N = \sin^2\vartheta$ (N_1 and N_2 are respectively the numbers of photons emerging in the extraordinary and ordinary ray) are not 'certain numbers', but are themselves probabilities that indeed have (in the present case) their origin in the probabilistic nature of quantum mechanics, but are such because we have not recorded (e.g. by means of mobile mirrors) how many photons emerged in the extraordinary ray and how many in the ordinary ray – just as if we threw the die and did not look at the result. For this reason we have preferred to use, in (4.34), a notation different from that used in (4.28) for the mean value of ξ: $\langle\!\langle \xi \rangle\!\rangle$ is a 'classical mean of quantum means'.

In general, if a collection of quantum systems consists of systems in the (not necessarily orthogonal) states $|u_1\rangle$, $|u_2\rangle$, $\cdots$, $|u_n\rangle$, $\cdots$ respectively in percentages $p_1, p_2, \cdots, p_n, \cdots$ ($\sum_n p_n = 1$), one says that the collection of our systems is a **statistical mixture:** the numbers p_n can be either 'certain percentages' or percentages due to our ignorance, namely due to the fact that we have only partial information about how the systems have been prepared. Statistical mixtures are, for example, either the set of states of a system after that on them some observable has been measured (see the above example with photons), or any thermodynamic system (e.g. a gas): it is a statistical mixture of its subsystems (the molecules).

A statistical mixture is therefore described by the set of pairs

$$\{\,|u_1\rangle,\, p_1;\; |u_2\rangle,\, p_2;\; \cdots;\; |u_n\rangle,\, p_n;\; \cdots\,\} \qquad (4.35)$$

and (4.33) and (4.34) are generalized in the form:

$$P(\text{mixture} \rightarrow |u\rangle) = \sum_n p_n |\langle u_n\,|\,u\rangle|^2; \quad \langle\!\langle \xi \rangle\!\rangle = \sum_n p_n \langle u_n\,|\,\xi\,|\,u_n\rangle \qquad (4.36)$$

namely the mean value of any observable in a statistical mixture is the 'mean of the means'.

Assume now that $|u\rangle$ and $|v\rangle$ are two orthogonal states; let us consider a system in the pure state

$$|a\rangle = c_1 |u\rangle + c_2 |v\rangle , \qquad\qquad |c_1|^2 + |c_2|^2 = 1 \qquad\qquad (4.37)$$

(the S_{p} ensemble) and the statistical mixture of states $|u\rangle$ and $|v\rangle$ with percentages $p_1 = |c_1|^2$ and $p_2 = |c_2|^2$ that we represent as

$$\{ |u_1\rangle, \, p_1 = |c_1|^2 ; \; |u_2\rangle, \, p_2 = |c_2|^2 \} \qquad\qquad (4.38)$$

(the S_{m} ensemble): we ask ourselves in which way the two ensembles can be distinguished, since in both cases the probability to find the system in the state $|u\rangle$ being p_1 and the probability to find it in the state $|v\rangle$ being p_2.

If ξ is an observable, the mean value of ξ in the ensemble S_{p} is:

$$\bar{\xi} = \left(c_1^* \langle u| + c_2^* \langle v| \right) \xi \left(c_1 |u\rangle + c_2 |v\rangle \right) =$$
$$= |c_1|^2 \langle u\,|\xi|\, u\rangle + |c_2|^2 \langle v\,|\xi|\, v\rangle + c_1^* c_2 \langle u\,|\xi|\, v\rangle + c_2^* c_1 \langle v\,|\xi|\, u\rangle \quad (4.39)$$

whereas the mean value of ξ in the ensemble S_{m} is given by (4.36):

$$\langle\!\langle \xi \rangle\!\rangle = |c_1|^2 \langle u\,|\,\xi\,|\,u\rangle + |c_2|^2 \langle v\,|\,\xi\,|\,v\rangle \qquad\qquad (4.40)$$

so that the two mean values differ by the quantity $2\,\mathcal{R}e\left(c_1^*\, c_2 \langle u\,|\,\xi\,|\,v\rangle \right)$ that appears in (4.39) because the states $|u\rangle$ and $|v\rangle$ may interfere: this is expressed by the fact that the state $|a\rangle$ is a *coherent* superposition of $|u\rangle$ and $|v\rangle$, while in the ensemble S_{m} we have an *incoherent* mixture of the states $|u\rangle$ and $|v\rangle$: in (4.37) also the phases of the complex number c_1 and c_2 are relevant (better: their relative phase) whereas in (4.38) only the absolute values intervene.

If only observables having $|u\rangle$ and/or $|v\rangle$ as eigenstates are measured, $\langle u\,|\,\xi\,|\,v\rangle = 0$ (by assumption $|u\rangle$ and $|v\rangle$ are orthogonal) and we cannot distinguish the two ensembles, but – in general – $\langle u\,|\,\xi\,|\,v\rangle \neq 0$ and the two ensembles provide different results.

For example, assume that the ensemble S_{p} consists of systems in the state $|a\rangle = \frac{1}{\sqrt{2}} \left(|u\rangle + |v\rangle \right)$ while the ensemble S_{m} is $\{ |u\rangle, \, p_1 = \frac{1}{2}; \; |v\rangle, \, p_2 = \frac{1}{2} \}$ and we ask how much is, in the two cases, the probability to find a system in the state $|b\rangle = \frac{1}{\sqrt{2}} \left(|u\rangle - |v\rangle \right)$. In the first case the probability is 0, since the two states $|a\rangle$ and $|b\rangle$ are orthogonal to each other; in the second case, instead:

$$p = p_1 \times P\big(|u\rangle \to |b\rangle\big) + p_2 \times P\big(|v\rangle \to |b\rangle\big) = \frac{1}{2} \times \frac{1}{2} + \frac{1}{2} \times \frac{1}{2} = \frac{1}{2} \, .$$

The difference between pure state and statistical mixture is well exemplified by the experiment of neutron interferometry cited in Sect. 3.3 (Fig. 3.4), in the two cases in which the mirrors s_2 and s_3 are either fixed or mobile mirrors able to detect the transit of neutrons.

Referring to Fig. 4.1, let us call $|x\rangle$ the state of a neutron that travels 'horizontally' (in the figure) and $|y\rangle$ that of a neutron that travels upwards. So the neutrons that arrive at the semi-transparent mirror s_1 are in the state $|x\rangle$ and downstream of s_1 are in the state

$$|a\rangle = \frac{1}{\sqrt{2}}\left(|x\rangle + e^{i\,\alpha}|y\rangle\right).$$

We have assumed equal transmission and reflection coefficients of s_1; the factor $e^{i\,\alpha}$ is compatible with this assumption and a priori we cannot exclude that it be introduced by the reflection (indeed, we shall see that, owing to reasons of probability conservation, it must be equal to $\pm i$). In the reflections at mirrors s_2 and s_3 $|x\rangle \rightarrow |y\rangle$

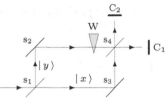

Fig. 4.1

and $|y\rangle \rightarrow |x\rangle$ (the phase factors introduced by the reflections at s_2 and s_3 are equal to each other, therefore irrelevant). So, if the mirrors s_2 and s_3 are fixed:

$$|x\rangle \xrightarrow{(s_1)} |a\rangle \equiv \frac{1}{\sqrt{2}}\left(|x\rangle + e^{i\,\alpha}|y\rangle\right) \xrightarrow{(s_2, s_3)} |b\rangle \equiv \frac{1}{\sqrt{2}}\left(|y\rangle + e^{i\,\alpha}|x\rangle\right).$$

The wedge W on the path $s_2 \rightarrow s_4$ introduces the phase shift φ:

$$|b\rangle \xrightarrow{(W)} |b_\varphi\rangle \equiv \frac{1}{\sqrt{2}}\left(|y\rangle + e^{i\,(\alpha+\varphi)}|x\rangle\right)$$

and finally the state $|b_\varphi\rangle$ hits the semi-transparent mirror s_4 where (by symmetry)

$$|x\rangle \rightarrow \frac{1}{\sqrt{2}}\left(|x\rangle + e^{i\,\alpha}|y\rangle\right); \qquad |y\rangle \rightarrow \frac{1}{\sqrt{2}}\left(|y\rangle + e^{i\,\alpha}|x\rangle\right)$$

therefore:

$$|b_\varphi\rangle \xrightarrow{(s_4)} |c\rangle \equiv \frac{1}{\sqrt{2}}\left(\frac{1}{\sqrt{2}}\left(|y\rangle + e^{i\,\alpha}|x\rangle\right) + e^{i\,(\alpha+\varphi)}\frac{1}{\sqrt{2}}\left(|x\rangle + e^{i\,\alpha}|y\rangle\right)\right)$$
$$= \frac{1}{2}\left(e^{i\,\alpha}(1 + e^{i\,\varphi})|x\rangle + (1 + e^{i\,(2\alpha+\varphi)})|y\rangle\right).$$

The probabilities p_1 and p_2 that either the counter C_1 or the counter C_2 clicks are respectively given by:

$$p_1 = |\langle c \mid x\rangle|^2 = \frac{1}{4}|1 + e^{i\,\varphi}|^2 = \frac{1}{2}(1 + \cos\varphi)$$

$$p_2 = |\langle c \mid y\rangle|^2 = \frac{1}{4}|1 + e^{i\,(2\alpha+\varphi)}|^2 = \frac{1}{2}(1 + \cos(2\alpha + \varphi))$$

from which it follows that, owing to $p_1 + p_2 = 1$, $\alpha = \pm\pi/2$, i.e. we have found again the result (3.4).

Let us now assume that s_2 and s_3 are mobile mirrors able to detect the collision of a neutron. In the latter case

$$|x\rangle \xrightarrow{(s_1)} |a\rangle \equiv \frac{1}{\sqrt{2}}\left(|x\rangle + e^{i\,\alpha}|y\rangle\right) \xrightarrow{(s_2,s_3)} \{\,|x\rangle,\, p_1 = \tfrac{1}{2};\, |y\rangle,\, p_2 = \tfrac{1}{2}\,\}$$

therefore, downstream of s_2 and s_3 the system is no longer in a pure state, but in a statistical mixture. The wedge W has no effect on the 50% of neutrons that take the path $s_2 \to s_4$ ($e^{i\,\varphi}|x\rangle \sim |x\rangle$), therefore:

$$\{\,|x\rangle,\, p_1 = \tfrac{1}{2};\, |y\rangle,\, p_2 = \tfrac{1}{2}\,\} \xrightarrow{(s_4)} \begin{cases} \{\,\frac{1}{\sqrt{2}}(|x\rangle + e^{i\,\alpha}|y\rangle),\, p_1 = \tfrac{1}{2}; \\[2mm] \frac{1}{\sqrt{2}}(|y\rangle + e^{i\,\alpha}|x\rangle),\, p_2 = \tfrac{1}{2}\,\}. \end{cases}$$

According to the first of (4.36), each of the two components of the mixture has a probability

$$\frac{1}{2} \times \left(\frac{1}{\sqrt{2}}\right)^2 = \frac{1}{4}$$

to be detected by either C_1 or C_2, i.e. – as we have already said in Sect. 3.3 – the two counters always record the same number of neutrons (probability $2 \times \tfrac{1}{4}$), independently of the position of the wedge.

In conclusion, the difference between the pure state and the statistical mixture

$$|b\rangle = \frac{1}{\sqrt{2}}(|y\rangle + e^{i\,\alpha}|x\rangle),\qquad \{\,|x\rangle,\, p_1 = \tfrac{1}{2};\, |y\rangle,\, p_2 = \tfrac{1}{2}\,\}$$

that we have downstream of the mirrors s_2, s_3, lies in that in the first case it is possible – by inserting the semi-transparent mirror s_4 in the apparatus – to make the two components $|x\rangle$ and $|y\rangle$ interfere with each other, while in the second case there is no such possibility.

4.10 Compatible Observables

In classical physics, given the state of a system, any observable has a well determined value in that state: for example, given position and velocity of a particle (i.e. its state) energy, angular momentum etc. are known. It is not so in quantum mechanics: first of all, if ξ is an observable, only in the case $|A\rangle$ represents an eigenstate of ξ it makes sense to say that ξ has a value in $|A\rangle$ (the corresponding eigenvalue); in the contrary case we can say, for example, which is the mean value of ξ in $|A\rangle$, or which is the probability that a measurement of ξ yields a given result. In summary, in this case only statistical information is available.

Furthermore, if we have two observables ξ and η and if $|\xi'\rangle$ is an eigenvector of ξ, then ξ has a value in $|\xi'\rangle$, but in general η does not possess a value in this state. If it happens that a certain state simultaneously is an eigenstate of both ξ (eigenvalue ξ') and η (eigenvalue η'), than both ξ and

η have a determined value in that state that we will accordingly represent by $| \xi', \eta' \rangle$. One has therefore, at least as far as the state $| \xi', \eta' \rangle$ is concerned, a kind of compatibility of the two observables ξ and η. If the states on which ξ and η are compatible (i.e. states that are simultaneous eigenstates of ξ and η) are enough as to form a *basis*, then we will say that ξ and η are **compatible observables**. In other words:

1-st definition: *two observables are said compatible if they admit a complete set of simultaneous eigenvectors.*

So, for, example, if the system is a photon, two birefringent crystals whose optical axes are neither parallel nor orthogonal to each other, certainly are *not* compatible observables, inasmuch as there exists no (polarization) state that is simultaneous eigenstate of both; they are instead compatible observables both in the case the optical axes are parallel and in the case they are orthogonal to each other.

We have already introduced in Sect. 3.6 the concept of compatible observables, but at first sight it does not seem that the two definitions have anything to do with each other. The connection is provided by the fact that there is *complete equivalence* between the definition given above and the following:

2-nd definition: *two observables ξ and η are said compatible if it happens that, given any eigenstate $| \xi' \rangle$ of ξ and, having made a measurement of η in it, the state after such a measurement still is an eigenstate of ξ $\big($ therefore corresponding to the eigenvalue ξ': $P\big(| \xi' \rangle \to | \xi'' \rangle\big) = 0$ if $\xi'' \neq \xi'\big)$.*

It is known that the state, immediately after the measurement of η, is an eigenstate of η: it is therefore a simultaneous eigenstate of ξ and η: $| \xi', \eta' \rangle$.

In general, if ξ' is a degenerate eigenvalue, the state $| \xi', \eta' \rangle$ after the measurement of η is different from the state $| \xi' \rangle$ in which η is measured.

More than giving demonstration of the equivalence of the two definitions (that we will however give), it is important to emphasize the physical significance of the compatibility of two observables, as it emerges from the second definition: it indeed says that two observables are compatible if they can both be measured in a state and the second measurement, even if it perturbs the state in which the system is after the first measurement, is such that the information acquired with the first measurement is not lost; more to it: if the first observable one measures is nondegenerate, the second measurement does not perturb the state. Schematically:

$$| A \rangle \xrightarrow{\text{measurement of } \xi} | \xi' \rangle \xrightarrow{\text{measurement of } \eta} | \xi', \eta' \rangle$$

and if ξ is nondegenerate, then $| \xi', \eta' \rangle = | \xi' \rangle$.

The proof of the equivalence of the two definitions, under the assumption that ξ and η are nondegenerate, is very simple: if ξ and η are compatible according to the first definition, since ξ and η have the same eigenvectors (they are nondegenerate!), the measurement of η is a repetition of the first measurement (only the scale of the instrument changes) and therefore does not perturb the state: schematically (the sign $\equiv$ means 'by assumption equal to')

$$|A\rangle \xrightarrow{\xi} |\xi'\rangle \equiv |\xi',\eta'\rangle \xrightarrow{\eta} |\xi',\eta'\rangle .$$

If instead ξ and η are compatible according to the second definition, the proof is outlined in the following diagram, where the sign $\equiv$ follows from the hypothesis of nondegeneracy of ξ, and therefore there exist only one eigenstate corresponding to the eigenvalue ξ' :

$$|A\rangle \xrightarrow{\xi} |\xi'\rangle \xrightarrow{\eta} |\xi',\eta'\rangle \equiv |\xi'\rangle .$$

We propose the demonstration of the equivalence without the nondegeneracy assumption of the observables just as an exercise.

Let us assume, according to the first definition, that ξ and η have a complete set of simultaneous eigenvectors One has

$$|A\rangle \xrightarrow{\xi} |\xi'\rangle \xrightarrow{\eta} |\eta'\rangle , \qquad |\eta'\rangle = \alpha |\xi',\eta'\rangle + \sum_i c_i |\xi_i,\eta'\rangle , \quad \xi_i \neq \xi'$$

but, since by assumption the transition probability $|\xi'\rangle \rightarrow |\eta'\rangle$ is nonvanishing and equals $|\alpha|^2 \times |\langle \xi' \mid \xi',\eta'\rangle|^2$, one must have $\alpha \neq 0$; furthermore, owing to von Neumann postulate, such probability must be a maximum, whence $|\alpha| = 1$, $c_i = 0$. In conclusion, $|\eta'\rangle = |\xi',\eta'\rangle$.

Viceversa, let us assume (second definition) that for *any* $|\xi'\rangle$

$$|\xi'\rangle \xrightarrow{\eta} |\xi',\eta'\rangle .$$

Let us consider the (closed) linear manifold $\mathcal{V}$ generated by *all* the simultaneous eigenvectors and let us assume (by contradiction) that $\mathcal{V} \neq \mathcal{H}$. Let then $|\xi'\rangle \in \mathcal{V}_\perp$ (it exists!). But then, as $|\xi',\eta'\rangle \in \mathcal{V}$, we arrive at the contradiction that $P(|\xi'\rangle \xrightarrow{\eta} |\xi',\eta'\rangle) = 0$. Then $\mathcal{V} = \mathcal{H}$.

There is now an important algebraic characterization concerning compatible observables, expressed by the following

Theorem: *Two observables are compatible if and only if*

$$\xi\eta = \eta\xi \tag{4.41}$$

holds for the operators associated with them.

In the latter case one says that ξ and η **commute** with each other. The expression:

$$[\xi,\eta] \equiv \xi\eta - \eta\xi$$

is called the **commutator** of ξ and η, then (4.41) reads:

$$[\xi,\eta] = 0 .$$

The demonstration we will give has not the status of a rigorous demonstration because, as usual, we shall ignore the problems relative to the domains of the two operators: it rather aims at emphasizing the intuitive aspects of the problem.

If ξ and η are compatible, by definition the set of their simultaneous eigenvectors $|\xi',\eta'\rangle$ is complete. One has:

$$\xi\eta\,|\,\xi',\eta'\rangle = \xi\,|\,\xi',\eta'\rangle\,\eta' = |\,\xi',\eta'\rangle\,\xi'\,\eta'$$
$$\eta\xi\,|\,\xi',\eta'\rangle = \eta\,|\,\xi',\eta'\rangle\,\xi' = |\,\xi',\eta'\rangle\,\eta'\,\xi'\,.$$

Therefore the operators $\xi\eta$ and $\eta\xi$ give the same results on the vectors $|\,\xi',\eta'\rangle$; but, since the vectors of the type $|\,\xi',\eta'\rangle$ generate the whole Hilbert space, (4.41) follows.

The viceversa needs to be shown, namely that if $[\xi,\eta] = 0$, then ξ and η have a complete set of simultaneous eigenvectors.

We start by showing a lemma that, owing to its importance and the frequent use we will make of it in the sequel, deserves to be taken out of the demonstration of the theorem.

Lemma: *if* $[\xi,\eta] = 0$ *and if* $\xi\,|\,\xi'\rangle = \xi'\,|\,\xi'\rangle$, *then* $\eta\,|\,\xi'\rangle$ *still is an eigenvector of* ξ *belonging to the eigenvalue* ξ':

$$\xi\,(\eta\,|\,\xi'\rangle) = \xi'\,(\eta\,|\,\xi'\rangle)\,. \tag{4.42}$$

Indeed:

$$\xi\eta\,|\,\xi'\rangle = \eta\xi\,|\,\xi'\rangle = \eta\xi'\,|\,\xi'\rangle = \xi'\,\eta\,|\,\xi'\rangle$$

(notice that it was not even necessary to assume that η be a self-adjoint operator).

Let us now conclude the demonstration of the theorem. Let us first consider the case in which one of the two observables, e.g. ξ, is nondegenerate. In this case, since by assumption $[\xi,\eta] = 0$, the lemma immediately takes us to the result: indeed, since $\eta\,|\,\xi'\rangle$ is an eigenvector of ξ belonging to the *nondegenerate* eigenvalue ξ', it must be a multiple of $|\,\xi'\rangle$, namely:

$$\eta\,|\,\xi'\rangle = \eta'\,|\,\xi'\rangle$$

i.e. any eigenvector $|\,\xi'\rangle$ of ξ must also be an eigenvector of η.

Let us now consider the degenerate case. Let $\mathcal{H}_i$ be the eigenspace of ξ belonging to the eigenvalue ξ_i; owing to the lemma, applying η to any vector in $\mathcal{H}_i$ always gives, as a result, a vector in $\mathcal{H}_i$, i.e. any eigenspace of ξ is invariant under η: η acts in independent ways on each $\mathcal{H}_i$. Even now it is intuitive that the restriction of η to $\mathcal{H}_i$ has a set of eigenvectors complete in $\mathcal{H}_i$; but all the vectors of $\mathcal{H}_i$ are eigenvectors of ξ, therefore in each $\mathcal{H}_i$ we have a complete set of simultaneous eigenvectors: $|\,\xi_i,\eta_j\rangle$. In this way a set of simultaneous eigenvector of ξ and η, complete for the whole $\mathcal{H}$, is obtained.

Let us now prove what we left to the intuition, namely that η has a set of eigenvectors complete in $\mathcal{H}_i$.

Let $|\,\eta_k\rangle$, $k = 1, 2, \cdots$ be a set of eigenvectors of η, complete in $\mathcal{H}$. It is known that it is possible to effect, in a unique way, the decomposition

$$| \eta_k \rangle = | \eta_k^{(i)} \rangle + | \eta_k^{(\perp)} \rangle , \quad | \eta_k^{(i)} \rangle \in \mathcal{H}_i , \ | \eta_k^{(\perp)} \rangle \in \mathcal{H}_i^{(\perp)} ; \quad \mathcal{H}_i^{(\perp)} \equiv \bigoplus_{j \neq i} \mathcal{H}_j .$$

But, from $\eta \, | \, \eta_k \rangle = \eta_k \, | \, \eta_k \rangle$, one has

$$\eta \, | \, \eta_k^{(i)} \rangle + \eta \, | \, \eta_k^{(\perp)} \rangle = \eta_k \, | \, \eta_k^{(i)} \rangle + \eta_k \, | \, \eta_k^{(\perp)} \rangle$$

and, since $\eta \, | \, \eta_k^{(i)} \rangle \in \mathcal{H}_i$ and $| \, \eta_k^{(\perp)} \rangle \in \mathcal{H}_i^{(\perp)}$, owing to the uniqueness of the decomposition of the vector $\eta \, | \, \eta_k \rangle$, one must have:

$$\eta \, | \, \eta_k^{(i)} \rangle = \eta_k \, | \, \eta_k^{(i)} \rangle , \qquad \eta \, | \, \eta_k^{(\perp)} \rangle = \eta_k \, | \, \eta_k^{(\perp)} \rangle .$$

Let now $| A \rangle \in \mathcal{H}_i$ be orthogonal to all the $| \eta_k^{(i)} \rangle$. As $| A \rangle \in \mathcal{H}_i$, it is orthogonal to all the $| \eta_k^{(\perp)} \rangle$, therefore to all the $| \eta_k \rangle$. So $| A \rangle$ is the null vector and, as a consequence, the set of all the $| \eta_k^{(i)} \rangle$ is complete in $\mathcal{H}_i$.

For the reader that has understood the theorem, the following remarks should be superfluous.

1. The theorem does *not* say that, if $[\xi , \eta] = 0$, then any eigenvector of one observable also is an eigenvector of the other; it says that there exists (i.e. one can find) a complete set of simultaneous eigenvectors. For example the identity operator $\mathbb{1}$ commutes with any operator and any vector is an eigenvector of $\mathbb{1}$, but certainly it is not true that any vector also is an eigenvector of whatever observable.

2. It is nonetheless true that, if one of the observables – e.g. ξ – is nondegenerate, then any eigenvector of ξ also is an eigenvector of η.

3. If $[\xi , \eta] \neq 0$, it may happen that ξ and η have *some* simultaneous eigenvector, certainly not as many as to form a complete set.

The results we have obtained can be generalized to the case of more than two observables: the observables $\xi, \eta, \zeta, \cdots$ are compatible if the operators associated with them all commute with one another. It can be then shown that there exists a complete set of simultaneous eigenvectors $| \xi', \eta', \zeta' \cdots \rangle$ of such observables; and viceversa: if a set of observables possesses a complete set of simultaneous eigenvectors, then they commute with one another.

Let us now introduce the concept of **complete set of compatible** (or commuting) **observables**: if $\xi, \eta, \zeta, \cdots$ are n compatible observables, any simultaneous eigenstate is identified by a n-tuple of eigenvalues $\xi_i, \eta_j, \zeta_k, \cdots$: $| \xi_i , \eta_j , \zeta_k , \cdots \rangle$. It may happen that for a given n-tuple there is more than just one simultaneous eigenstate. If instead there never are two or more simultaneous eigenstates of $\xi, \eta, \zeta, \cdots$ belonging to the same n-tuple, than we say that $\xi, \eta, \zeta, \cdots$ form a complete set of compatible observables.

This definition generalizes the concept of nondegenerate observable to the case of two or more observables. In less precise but more intuitive terms one could say that a set of compatible observables is complete when, any single observable being (possibly) degenerate, the set is globally nondegenerate.

The physical importance of this concept is the same as that of a nondegenerate observable: given the n-tuple of eigenvalues $\xi_i, \eta_j, \zeta_k, \cdots$, the state of the system is known. We leave to the reader the demonstration of the fact that, much as for a nondegenerate observable, the set of the simultaneous eigenvectors of a *complete* set of compatible observables makes up an orthogonal basis, and viceversa.

4.11 Uncertainty Relations

Let ξ and η be two (either compatible or not) observables, $\Delta\xi$ and $\Delta\eta$ the root mean squares of the results of measurements of ξ and η on a system in the generic state $|s\rangle$. The following important theorem due to H. P. Robertson, relating the product of the uncertainties $\Delta\xi$ and $\Delta\eta$ with the mean value of the commutator of ξ and η in the state $|s\rangle$, follows:

Uncertainty Relation:

$$\Delta\xi\,\Delta\eta \geq \frac{1}{2}\left|\langle s|\,[\xi,\eta]\,|s\rangle\right|. \tag{4.43}$$

Let us demonstrate (4.43). We introduce the non-Hermitian operators:

$$\alpha \equiv \xi + i\,x\,\eta, \qquad \alpha^\dagger \equiv \xi - i\,x\,\eta$$

where x is a real parameter. Let us consider the product $\alpha^\dagger\,\alpha$ (pay attention to the order of the factors, for in general ξ and η do not commute):

$$\alpha^\dagger\alpha \equiv (\xi - i\,x\,\eta)\,(\xi + i\,x\,\eta) = \xi^2 + x^2\,\eta^2 + i\,x\,[\xi,\eta]. \tag{4.44}$$

Note that, if $\langle s\,|\,s\rangle = 1$, for any operator α the inequality

$$\langle s\,|\,\alpha^\dagger\alpha\,|\,s\rangle \geq \langle s\,|\,\alpha^\dagger\,|\,s\rangle \times \langle s\,|\,\alpha\,|\,s\rangle \tag{4.45}$$

holds as a consequence of either the Schwartz inequality applied to the vectors $\alpha\,|\,s\rangle$ and $|\,s\rangle\langle s\,|\,\alpha\,|\,s\rangle$, or the completeness relation (4.25): indeed, if the vectors $|\,s_i\rangle$ make up a basis of which $|\,s\rangle$ is an element, e.g. $|\,s\rangle = |\,s_1\rangle$, then:

$$\langle s\,|\,\alpha^\dagger\alpha\,|\,s\rangle = \sum_i \langle s\,|\,\alpha^\dagger\,|\,s_i\rangle\langle s_i\,|\,\alpha\,|\,s\rangle$$

$$= \langle s\,|\,\alpha^\dagger\,|\,s\rangle\langle s\,|\,\alpha\,|\,s\rangle + \sum_{i>1}\langle s\,|\,\alpha^\dagger\,|\,s_i\rangle\langle s_i\,|\,\alpha\,|\,s\rangle.$$

In the above relation all the terms of the last sum ($i > 1$) are positive ($= |\langle s_i\,|\,\alpha\,|\,s\rangle|^2$), whence the thesis.

Then from (4.44) and (4.45) one has:

$$\langle s\,|\left(\xi^2 + x^2\,\eta^2 + i\,x\,[\xi,\eta]\right)|\,s\rangle \geq \langle s\,|\,(\xi - i\,x\,\eta)\,|\,s\rangle \times \langle s\,|\,(\xi + i\,x\,\eta)\,|\,s\rangle$$

i.e.

$$\overline{\xi^2} + x^2\,\overline{\eta^2} + x\,\overline{i\,[\xi,\eta]} \geq \overline{\xi}^2 + x^2\,\overline{\eta}^2$$

that is:

$$x^2 (\Delta \eta)^2 + x\, \overline{i\,[\xi, \eta]} + (\Delta \xi)^2 \geq 0 \,.$$

Since the sign $\geq$ must hold for any real x, the discriminant of this quadratic form in the variable x must be ≤ 0:

$$\left(\overline{i\,x\,[\xi, \eta]} \right)^2 - 4\,(\Delta \xi)^2\,(\Delta \eta)^2 \leq 0$$

from which (4.43) immediately follows.

The case in which the commutator $[\xi, \eta]$ is a multiple of the identity operator $\mathbb{1}$ is particularly important. Note that, as $\xi^\dagger = \xi$ and $\eta^\dagger = \eta$,

$$[\xi, \eta]^\dagger = (\xi \eta - \eta \xi)^\dagger = \eta \xi - \xi \eta = -[\xi, \eta]$$

(i.e. $[\xi, \eta]$ is an *anti* Hermitian operator) then, if $[\xi, \eta]$ is a multiple of $\mathbb{1}$, the multiplicative factor must be a pure imaginary number:

$$[\xi, \eta] = i\,c\,\mathbb{1}, \qquad c \in \mathbb{R}$$

(from now on the identity operator will be omitted, as $\mathbb{1}\,|\,s\,\rangle = |\,s\,\rangle$: numerical quantities and multiples of the identity operator – there is no substantial difference between the two – are often called **c-numbers**).

In the latter case, i.e. when the commutator $[\xi, \eta]$ is a multiple of the identity operator, (4.43) becomes:

$$\Delta \xi\, \Delta \eta \geq \frac{1}{2}\,|c| \qquad\qquad (\text{valid if} \quad [\xi, \eta] = i\,c\,) \qquad\qquad (4.46)$$

i.e. the product of the uncertainties is greater of or equal to a fixed quantity that does not depend on the state one is considering.

4.12 Quantization Postulate

The main difference between classical physics and quantum mechanics lies in the fact that, while in the classical scheme the observables give rise to a commutative algebra (for example: $q\,p$ and $p\,q$ are the same thing), in the quantum scheme the observables – being represented by operators – in general do not obey the commutative property.

Now, given a physical system and having established which are its observables, the (quantum) theory is complete if it enables us to find the eigenvalues and eigenvectors of the several observables, the degeneracies of the eigenvalues and the probability transitions among any two states of the system: it can be shown that all this is possible if for any pair of observables the commutator is known. We will not give a proof of the above statement, but we shall have several occasions to realize that it is true.

The quantum scheme we have discussed so far is a general scheme and makes no reference to any particular physical system. From now on the physical systems we shall be concerned with will consist of one or more particles: indeed, we have in mind to apply quantum theory to atoms and to show how it is possible to arrive at a no less than quantitative understanding of the properties of even complex atoms.

Which are the observables of a system consisting of n particles? Classically we have the positions q_i and momenta p_i ($i = 1, 2, \cdots, 3n$) and their functions (by the q_i we shall always understand the Cartesian coordinates: as in Nature there are no constraints, there is no reason not to make such a choice). We make the hypothesis that the above quantities are the observables for the system, even when it is considered from the quantum point of view. Whether this hypothesis is right or wrong will emerge from the comparison between the predictions of the theory and the experimental results: for example, we shall see that, in the case of electrons, other observables (the spin) will be needed.

According to the discussion of the previous sections, to any observable there corresponds a self-adjoint operator. So, for example, $q_i \to q_i^{\mathrm{op}}$, $p_i \to p_i^{\mathrm{op}}$ (we provisionally go back to the notation ξ^{op}).

If we now have an observable $f(q, p)$, we postulate that the operator associated with it is obtained by replacing q and p in f respectively with q^{op} and p^{op}: in other words we postulate that $f^{\mathrm{op}} \equiv f(q^{\mathrm{op}}, p^{\mathrm{op}})$. So, for example, in the case of the harmonic oscillator to the energy

$$H = \frac{1}{2m}(p^2 + m^2 \omega^2 q^2)$$

there corresponds the self-adjoint operator:

$$H^{\mathrm{op}} = \frac{1}{2m}[(p^{\mathrm{op}})^2 + m^2 \omega^2 (q^{\mathrm{op}})^2] \, .$$

Sometimes an ambiguity arises, due to the fact that, while in $f(q, p)$ the order of factors is unessential, in $f(q^{\mathrm{op}}, p^{\mathrm{op}})$ it is important, so that the resulting operator may be not Hermitian. In practice this ambiguity is not very relevant: for example one can write:

$$q\,p = \frac{1}{2}(q\,p + p\,q) \to \frac{1}{2}(q^{\mathrm{op}}\,p^{\mathrm{op}} + p^{\mathrm{op}}\,q^{\mathrm{op}})$$

that is Hermitian. From now on we will no longer write the symbol ^{op} to distinguish the quantum operators: by q, p, $f(q, p)$ we will positively denote the operators associated with the corresponding observables.

According to the above discussion , it is necessary to know the commutator of any pair of observables $f(q, p)$ and $g(q, p)$. It can be seen that such commutators are known if the following commutators:

$$[\, q_i \, , \, q_j \,], \qquad [\, q_i \, , \, p_j \,], \qquad [\, p_i \, , \, p_j \,] \qquad\qquad (4.47)$$

are known. This can be achieved by means of the following essential

Formal Properties of Commutators:

Let ξ, η, ζ be operators. One has:

1. $[\xi, \eta] = -[\eta, \xi]$,
2. $[\xi, \eta]^\dagger = -[\eta, \xi]$ if $\xi = \xi^\dagger$ $\eta = \eta^\dagger$,
3. $[\xi, \eta + \zeta] = [\xi, \eta] + [\xi, \zeta]$,
4. $[\xi, \eta\zeta] = \eta[\xi, \zeta] + [\xi, \eta]\zeta$,
5. $[\xi\eta, \zeta] = \xi[\eta, \zeta] + [\xi, \zeta]\eta$,
6. $[\xi, [\eta, \zeta]] + [\eta, [\zeta, \xi]] + [\zeta, [\xi, \eta]] = 0$.

(4.48)

Note – this observation may provide a good mnemonical rule – that 4 and 5 recall the Leibniz rule for the derivative of a product; 5 follows from 1 and 4; 6 is known as Jacobi identity.

It should be clear that, according to the above rules, the calculation of a commutator of any $f(q,p)$ and $g(q,p)$ that are either polynomials or power series in q and p is led back to the 'elementary' commutators (4.47). So, in order to complete the scheme, it is necessary and sufficient to know the commutators (4.47). Let us firstly note that not all of them can be vanishing: if it were so, all the observables would commute with one another and one would be taken back to the classical case.

In order to determine the commutators (4.47) we shall resort to an analogy between the quantum commutators and the Poisson brackets of classical mechanics – an analogy that, as it will be seen a posteriori, will enable us to recover classical mechanics as a limiting case of quantum mechanics.

As well known, in classical mechanics the Poisson Bracket $[f, g]_{\mathrm{PB}}$ between $f(q,p)$ and $g(q,p)$ is defined in the following way:

$$[f, g]_{\mathrm{PB}} \equiv \sum_{i=1}^{3n} \left(\frac{\partial f}{\partial q_i} \frac{\partial g}{\partial p_i} - \frac{\partial f}{\partial p_i} \frac{\partial g}{\partial q_i} \right) . \qquad (4.49)$$

The Poisson brackets enjoy the same formal properties of commutators listed in (4.48) (for the Poisson brackets, however, the order of factors in 4 and 5 is unessential) and play an important role e.g. in the theory of canonical transformations, or even in the equations of motions that can be written in the form:

$$\frac{\mathrm{d}}{\mathrm{d}t} f(q,p) = [f, H]_{\mathrm{PB}} \qquad (4.50)$$

H being the Hamiltonian of the system. Due to both the importance of Poisson brackets in classical mechanics and to formal analogy with commutators, we assume the following:

Quantization Postulate: *the commutators (4.47) are proportional to the corresponding Poisson brackets.*

From (4.49) one has:

$$[q_i, q_j]_{\mathrm{PB}} = 0, \qquad [q_i, p_j]_{\mathrm{PB}} = \delta_{ij}, \qquad [p_i, p_j]_{\mathrm{PB}} = 0$$

so the only nonvanishing commutator between q_i and p_j must be proportional to δ_{ij}. Which is the value of the proportionality constant? Note that owing to 2 of (4.48), such a constant must be pure imaginary and must have the dimensions of an action: we postulate that it is $i\hbar$ (the choice of the sign brings along no physical consequences, since the transformation $i \to -i$ does not touch upon the properties that define the imaginary unit: $i^2 = -1$ and $i^* = -i$).

So, in the end, we have the following commutation rules, or quantization conditions, or Canonical Commutation Relations (usually referred to as CCR):

$$[q_i, q_j] = 0, \qquad [q_i, p_j] = i\hbar\,\delta_{ij}, \qquad [p_i, p_j] = 0. \qquad (4.51)$$

The CCR (4.51) express an important property: observables referring to different ($i \neq j$) degrees of freedom are compatible. There is instead incompatibility between any q_i and its canonically conjugate momentum p_i: for such observables, whose commutator is a multiple of the identity, the uncertainty relation can be written in the form (4.46):

$$\Delta q_i \, \Delta p_i \geq \frac{1}{2}\,\hbar \qquad (4.52)$$

that must be compared with (3.18); in the present case, however, Δx and Δp_x have a precise meaning given by (4.31): they are the root mean squares of the results of measurements of x and p_x in the same state.

Those states for which (4.52) holds with the equality sign (we shall encounter an example in the next chapter) are called **minimum uncertainty states**.

Thanks to the proportionality between (4.51) and the corresponding Poisson brackets, and to the formal properties (4.48) shared by both, it follows that in many cases one has:

$$[f, g] = i\hbar\,[f, g]_{\mathrm{PB}}. \qquad (4.53)$$

The equality (4.53) may break down due to the order of factors that is relevant only in the left hand side. For example:

$$[q^2, p^2] = q\,[q, p^2] + [q, p^2]\,q = 2i\hbar\,(qp + pq)$$

whereas

$$i\hbar\,[q^2, p^2]_{\mathrm{PB}} = 4i\hbar\,qp.$$

In any event, when the problem of the order of factors does not show up, (4.53) holds and may provide a quick way to calculate complicated commutators. For example, in such a way the following important commutators can be calculated:

$$[f(q), g(q)] = 0; \qquad\qquad [f(p), g(p)] = 0;$$
$$[q_i, f(p)] = i\hbar\,\frac{\partial f}{\partial p_i}; \qquad [p_i, f(q)] = -i\hbar\,\frac{\partial f}{\partial q_i}. \qquad (4.54)$$

Chapter 5

The Harmonic Oscillator

5.1 Positivity of the Eigenvalues of Energy

The first truly significant application of the quantization conditions of the previous chapter concerns the determination of the eigenvalues of the energy for a one-dimensional oscillator. In this section we shall limit ourselves to obtain only some qualitative conditions on the energy levels of the oscillator, mainly with the purpose of giving to the reader the occasion to get acquainted with some techniques and concepts of quantum mechanics.

The energy of a linear harmonic oscillator expressed in terms of q and p, i.e. the Hamiltonian, is

$$H = \frac{1}{2m}\,(p^2 + m^2\,\omega^2\,q^2) \qquad (5.1)$$

where q and p are self-adjoint operators that satisfy the commutation relation

$$[\,q\,,\,p\,] = i\,\hbar\ . \qquad (5.2)$$

Let E_0, E_1, $\cdots$ be the eigenvalues of H and $|\,E_0\,\rangle$, $|\,E_1\,\rangle$, $\cdots$ the corresponding eigenvectors. The first result is the following:

1. *The eigenvalues of H are all nonnegative:* $E_n \geq 0$, $n = 0, 1, \cdots$.

Indeed, let $|\,A\,\rangle$ be an arbitrary vector that, for the sake of simplicity, we will assume normalized. If we show that

$$\overline{H} = \langle\,A\,|\,H\,|\,A\,\rangle \geq 0 \qquad \text{for any}\quad |\,A\,\rangle \qquad (5.3)$$

then the desired result follows because the eigenvalues themselves are mean values of H on particular states, the eigenstates of H.

In order to show (5.3) it suffices to observe that $\langle\,A\,|\,p^2\,|\,A\,\rangle$ is the squared norm of the vector $p\,|\,A\,\rangle$ ($p = p^\dagger$!) and, as such, is ≥ 0 – the equality sign holding only if $p\,|\,A\,\rangle = 0$. Likewise $\langle\,A\,|\,q^2\,|\,A\,\rangle \geq 0$, whence (5.3).

Note that the equality sign in (5.3) holds only if both $p\,|\,A\,\rangle = 0$ and $q\,|\,A\,\rangle = 0$, i.e. if there exists a simultaneous eigenstate of p and q: but such

© Springer International Publishing Switzerland 2016
L.E. Picasso, *Lectures in Quantum Mechanics*, UNITEXT for Physics,
DOI 10.1007/978-3-319-22632-3_5

a state cannot exist for it would be in contradiction with the uncertainty relation (4.52) and with the commutation relation (5.2) of which (4.52) is a consequence.

So $\overline{H} > 0$ and one has the stronger result:

2. *All the E_n are positive.*

In classical physics the minimum energy of an oscillator is zero, corresponding to the state in which the oscillator is at rest in the origin ($q = 0$, $p = 0$). According to quantum mechanics, instead, $E_0 > 0$ and this, as just seen, is a consequence of the uncertainty relation (4.52): in other words, due to the incompatibility of q and p, a quantum state like the classical one, in which both position and momentum are well determined, is not possible.

3. *The mean values of q and p in the eigenstates of H are vanishing:*

$$\langle\, E_n \mid p \mid E_n \,\rangle = 0 \qquad\qquad n = 0,\, 1,\, \cdots \qquad\qquad (5.4)$$

$$\langle\, E_n \mid q \mid E_n \,\rangle = 0 \qquad\qquad n = 0,\, 1,\, \cdots \;. \qquad\qquad (5.5)$$

The equality (5.4) holds under more general hypotheses, i.e. it holds for any system whose Hamiltonian has the form:

$$H = \frac{p^2}{2m} + V(q)$$

with arbitrary potential $V(q)$. Not even the fact that the system be one-dimensional is necessary.

Let us show (5.4) in general. Since, thanks to (4.54) $[V(q),\, q] = 0$, one has

$$[\,H\,,\, q\,] = \frac{1}{2m}\,[\,p^2,\, q\,] = \frac{1}{2m}\,\big(p\,[\,p,\, q\,] + [\,p,\, q\,]\,p\big) = -\mathrm{i}\,\frac{\hbar}{m}\,p \qquad (5.6)$$

whence it follows that

$$-\mathrm{i}\,\frac{\hbar}{m}\,\langle\, E_n \mid p \mid E_n \,\rangle = \langle\, E_n \mid [\,H\,,\, q\,] \mid E_n \,\rangle = \langle\, E_n \mid (H\,q - q\,H) \mid E_n \,\rangle$$

but, due to $H = H^{\dagger}$,

$$H \mid E_n \,\rangle = E_n \mid E_n \,\rangle \qquad\Rightarrow\qquad \langle\, E_n \mid H^{\dagger} = \langle\, E_n \mid H = E_n \,\langle\, E_n \mid$$

so that

$$\langle\, E_n \mid (H\,q - q\,H) \mid E_n \,\rangle = E_n \,\langle\, E_n \mid q \mid E_n \,\rangle - \langle\, E_n \mid q \mid E_n \,\rangle\, E_n = 0$$

and in conclusion, thanks to (5.6), $\langle\, E_n \mid p \mid E_n \,\rangle = 0$.

Equation (5.5) can be shown in a similar way: it suffices to observe that

$$[\,H\,,\, p\,] = \frac{1}{2}\,m\,\omega^2\,[\,q^2,\, p\,] = \mathrm{i}\,\hbar\,\omega\,q$$

and one can proceed exactly as above.

4. *The lowest energy level E_0 satisfies $E_0 \geq \frac{1}{2}\hbar\omega$.*

Indeed, if $\langle E_0 \mid E_0 \rangle = 1$, one has:

$$E_0 = \langle E_0 \mid H \mid E_0 \rangle = \frac{1}{2m}(\langle E_0 \mid p^2 \mid E_0 \rangle + m^2\omega^2 \langle E_0 \mid q^2 \mid E_0 \rangle)$$

$$= \frac{1}{2m}(\overline{p^2} + m^2\omega^2\,\overline{q^2})\,.$$

But we have seen that $\overline{p}$ and $\overline{q}$ are vanishing, so $\overline{p^2} = (\Delta p)^2$ and $\overline{q^2} = (\Delta q)^2$ and consequently:

$$E_0 = \frac{1}{2m}\left((\Delta p)^2 + m^2\omega^2\,(\Delta q)^2\right)\,.$$

Now, thanks to the inequality $a^2 + b^2 \geq 2\,a\,b$ with $a = \Delta p$, $b = m\,\omega\,\Delta q$ and to (4.52), one finally has:

$$E_0 \geq \frac{1}{2m}\,2m\,\omega\,\Delta q\,\Delta p \geq \frac{1}{2}\hbar\omega\,.$$

In the next section we will show that $E_0 = \frac{1}{2}\hbar\omega$: this entails that in the above inequalities with the sign $\geq$ (equation (4.52) and $a^2 + b^2 \geq 2\,a\,b$) the equality sign always holds, therefore in the state $\mid E_0 \rangle$ one has (the second equality follows from $a^2 + b^2 = 2ab \Leftrightarrow a = b$):

$$\Delta p\,\Delta q = \frac{1}{2}\hbar,\qquad \Delta p = m\,\omega\,\Delta q\,.$$

So the ground state of the linear harmonic oscillator is a minimum uncertainty state and moreover $\left(\text{as } \overline{p^2} = (\Delta p)^2 \text{ and } \overline{q^2} = (\Delta q)^2\right)$

$$\frac{1}{2}m\,\omega^2\overline{q^2} = \frac{1}{2m}\,\overline{p^2} = \frac{1}{2}\,E_0 \tag{5.7}$$

i.e. the mean value of the kinetic energy equals the mean value of the potential energy: the same result holds for the classical oscillator (virial theorem), although with a different meaning of the term 'mean value': for the classical oscillator the mean is taken over a period, i.e. with respect to time, whereas for the quantum oscillator (in the ground state, but we shall see that the result applies to the other eigenstates of the Hamiltonian as well) one is dealing with the mean of the results of *measurements* respectively of the kinetic and the potential energy.

5.2 The Energy Levels of the Harmonic Oscillator

In order to determine the eigenvalues of the Hamiltonian of the one-dimensional harmonic oscillator, instead of using p and q, we will use the following non-Hermitian operators:

$$\eta = \frac{1}{\sqrt{2m\,\omega\,\hbar}}\,(p - i\,m\,\omega\,q)\,,\qquad \eta^\dagger = \frac{1}{\sqrt{2m\,\omega\,\hbar}}\,(p + i\,m\,\omega\,q)\,. \tag{5.8}$$

In terms of η and $\eta^\dagger$, p and q are given by:

$$q = -i\sqrt{\frac{\hbar}{2m\omega}}\,(\eta^\dagger - \eta)\,, \qquad p = \sqrt{\frac{m\omega\hbar}{2}}\,(\eta^\dagger + \eta)\,. \qquad (5.9)$$

One has:

$$\eta^\dagger\eta = \frac{1}{2m\hbar\omega}\left(p^2 + m^2\omega^2 q^2 + i\,m\omega\,[q,p]\right) = \frac{1}{\hbar\omega}\left(H - \frac{\hbar\omega}{2}\right) \quad \Rightarrow$$

$$\eta^\dagger\eta = \frac{1}{\hbar\omega}\,H - \frac{1}{2}\,. \qquad (5.10)$$

Likewise:

$$\eta\,\eta^\dagger = \frac{1}{\hbar\omega}\,H + \frac{1}{2}$$

therefore, also directly from (4.51),

$$[\eta, \eta^\dagger] = 1\,. \qquad (5.11)$$

Let us now calculate the commutators $[H, \eta]$ and $[H, \eta^\dagger]$. From (5.10) and (5.11) one has

$$[H, \eta] = \hbar\omega\,[\eta^\dagger\eta, \eta] = -\hbar\omega\,\eta\,, \qquad (5.12)$$

whose adjoint is

$$[H, \eta^\dagger] = \hbar\omega\,[\eta^\dagger\eta, \eta^\dagger] = +\hbar\omega\,\eta^\dagger\,. \qquad (5.13)$$

Let now $|E\rangle$ be an eigenvector of H belonging to the eigenvalue E: $H|E\rangle = E|E\rangle$ and consider the vector $\eta|E\rangle$: if $\eta|E\rangle \neq 0$, it is an eigenvector of H belonging to the eigenvalue $E - \hbar\omega$. Indeed, by use of (5.12) one has:

$$H\,\eta\,|E\rangle = \eta\,H\,|E\rangle - \hbar\omega\,\eta\,|E\rangle = (E - \hbar\omega)\,\eta\,|E\rangle\,. \qquad (5.14)$$

Owing to this property, η is called either a **step down** or a **lowering** operator, because when applied to an eigenvector $|E\rangle$ of H – if the result is nonvanishing – it transforms $|E\rangle$ into another eigenvector corresponding to another eigenvalue: the eigenvalue E decreased by the quantity $\hbar\omega$.

Let us now start with the vector $|E\rangle$ and apply η to it; then, if the result is nonvanishing, let us apply η once more ... and so on until this chain possibly stops by giving the null vector. In this way the eigenvalues $E, E - \hbar\omega, E - 2\hbar\omega, \cdots$ are obtained. But this chain must necessarily stop somewhere, otherwise we would find negative eigenvalues when, instead, we know that all the eigenvalues must be $\geq \frac{1}{2}\hbar\omega$. Therefore there must exist a minimum eigenvalue E_0 such that

$$\eta\,|E_0\rangle = 0\,. \qquad (5.15)$$

Then:

$$0 = \langle E_0 \,|\, \eta^\dagger\eta \,|\, E_0 \rangle = \langle E_0 \,|\, \frac{1}{\hbar\omega}\,H - \frac{1}{2} \,|\, E_0 \rangle = \left(\frac{1}{\hbar\omega}\,E_0 - \frac{1}{2}\right)\langle E_0 \,|\, E_0 \rangle$$

whence $E_0 = \frac{1}{2}\hbar\omega$.

It is easy to see that $\eta^\dagger$ is a **step up** or a **raising** operator, i.e. if $|E\rangle$ is an eigenvector of H belonging to the eigenvalue E, then $\eta^\dagger|E\rangle$ is an eigenvector of H belonging to $E + \hbar\omega$. Indeed, by use now of (5.13) one has:

$$H\,\eta^\dagger\,|E\rangle = \eta^\dagger\,H\,|E\rangle + \hbar\omega\,\eta^\dagger\,|E\rangle = (E + \hbar\omega)\,\eta^\dagger\,|E\rangle\,.$$

So, starting with $|E_0\rangle$, we can repeatedly apply $\eta^\dagger$ and obtain a chain of eigenvectors of H belonging to the eigenvalues E_0, $E_0 + \hbar\omega$, $E_0 + 2\hbar\omega$, $\cdots$. Should this chain stop somewhere? In other words, may it happen that there exists a vector $|E\rangle$ such that $\eta^\dagger|E\rangle = 0$? Let us see: if such a vector existed, one should have

$$0 = \langle E\,|\,\eta\,\eta^\dagger\,|\,E\rangle = \langle E\,|\,\frac{1}{\hbar\omega}\,H + \frac{1}{2}\,|\,E\rangle = \left(\frac{1}{\hbar\omega}E + \frac{1}{2}\right)\langle E\,|\,E\rangle$$

that is absurd inasmuch as $E + \frac{1}{2}\hbar\omega$ never vanishes. So the ascending chain never interrupts and all the

$$E_n = \left(n + \frac{1}{2}\right)\hbar\omega \qquad n = 0,\,1,\,2,\,\cdots \qquad (5.16)$$

are eigenvalues of H. Clearly there exist no other eigenvalues different from those given by (5.16): if one existed, the descending chain starting from it, not passing through the value $\frac{1}{2}\hbar\omega$ but overstepping it, would imply the existence of negative eigenvalues. The values given by (5.16) are *all and only* the eigenvalues of H. One should still investigate whether such eigenvalues are degenerate or not: they are nondegenerate, but we will show this property later on, in a more general context.

Also a proof that the $|E_n\rangle$ form a complete set would be essential: this would be a verification of the internal consistency of the theory, because H, being an operator associated with an observable, must have precise properties, among which the completeness of its eigenvectors: should it turn out that H does not possess these properties, it would mean that the quantization postulate (4.51), namely the postulate that determines the properties of the operator H, is not compatible with the previous postulates, in particular the assumption that H is an observable, i.e. a self-adjoint operartor. However, although we will not make a demonstration and not even will we spend a word about what should be demonstrated, we state that the states $|E_n\rangle$ do form a complete set. Therefore, for the time being, everything is all right.

Let us now simplify the notation: we shall denote by $|n\rangle$ ($n = 0,\,1,\,2,\,\cdots$) the eigenvectors of H: $|0\rangle$ represents the ground state, $|1\rangle$ the state corresponding to the first excited level and so on. They are all obtained by repeatedly applying the operator $\eta^\dagger$ to the vector $|0\rangle$: for example $|n\rangle = (\eta^\dagger)^n\,|0\rangle$. However such vectors are not normalized. If we want to normalize them, we must calculate

$$\langle n\,|\,n\rangle = \langle 0\,|\,\eta^n\,(\eta^\dagger)^n\,|\,0\rangle$$

and to this purpose we need the following expression

$$[\eta, (\eta^\dagger)^n] = n\,(\eta^\dagger)^{n-1} \tag{5.17}$$

that we demonstrate by induction. It holds for $n = 1$. We then see that, if (5.17) holds with n replaced by $n - 1$, i.e. if $[\eta, (\eta^\dagger)^{n-1}] = (n-1)\,(\eta^\dagger)^{n-2}$, then also (5.17) holds. Indeed:

$$[\eta, (\eta^\dagger)^n] = [\eta, \eta^\dagger\,(\eta^\dagger)^{n-1}] = \eta^\dagger\,[\eta, (\eta^\dagger)^{n-1}] + [\eta, \eta^\dagger]\,(\eta^\dagger)^{n-1}$$
$$= (n-1)\,\eta^\dagger\,(\eta^\dagger)^{n-2} + (\eta^\dagger)^{n-1} = n\,(\eta^\dagger)^{n-1}\,.$$

Then

$$\langle 0 \mid \eta^n\,(\eta^\dagger)^n \mid 0 \rangle = \langle 0 \mid \eta^{n-1}\eta\,(\eta^\dagger)^n \mid 0 \rangle = \langle 0 \mid \eta^{n-1}\,[\eta, (\eta^\dagger)^n] \mid 0 \rangle$$

(the last step follows from (5.15): $\eta\,|0\rangle = 0$), then, by use of (5.17):

$$\langle 0 \mid \eta^n\,(\eta^\dagger)^n \mid 0 \rangle = n\,\langle 0 \mid \eta^{n-1}\,(\eta^\dagger)^{n-1} \mid 0 \rangle\,.$$

By repeating n times this calculation, that has lowered the exponents of η and $\eta^\dagger$ from n to $n-1$, one obtains:

$$\langle 0 \mid \eta^n\,(\eta^\dagger)^n \mid 0 \rangle = n!\,\langle 0 \mid 0 \rangle\,.$$

We redefine the vectors $|n\rangle$ as

$$|n\rangle \equiv \frac{1}{\sqrt{n!}}\,(\eta^\dagger)^n\,|0\rangle \tag{5.18}$$

so that, if $\langle 0 \mid 0 \rangle = 1$, also $\langle n \mid n \rangle = 1$.

(Operators like $\eta^\dagger$ and η satisfying the commutation relation (5.11) have an important role in the theory of quantum fields, where they are also referred to as **creation** and **destruction** operators since, in that context, they respectively increase and decrease by one unit the number N of particles).

Chapter 6

Representation Theory

6.1 Representations

Let $|e_n\rangle$, $n = 1, 2, \cdots$ be an orthonormal basis of vectors:

$$\begin{cases} \langle e_m \mid e_n \rangle = \delta_{mn} & \text{(orthonormality)} \\ \sum_n |e_n\rangle\langle e_n| = \mathbb{1} & \text{(completeness)}. \end{cases}$$

Then, for any vector $|A\rangle \in \mathcal{H}$ one has

$$|A\rangle = \sum_{n=1}^{\infty} a_n \, |e_n\rangle, \quad a_n = \langle e_n \mid A\rangle, \quad \langle A \mid A\rangle = \sum_n |a_n|^2 < \infty \,.$$

Then any vector determines and is determined by the sequence of complex numbers $\{a_n\}$ such that $\sum |a_n|^2 < \infty$:

$$|A\rangle \to \{a_n\} \,. \tag{6.1}$$

The $\{a_n\}$ are called the **_representatives of_ $|A\rangle$ _in the basis_** $|e_n\rangle$.

Given the two vectors $|A\rangle$ and $|B\rangle$, their scalar product $\langle B \mid A\rangle$ may be expressed by means of the respective representatives $\{a_n\}$ and $\{b_n\}$: indeed, thanks to the completeness relation

$$\langle B \mid A\rangle = \sum_n \langle B \mid e_n\rangle\langle e_n \mid A\rangle = \sum_n b_n^* \, a_n \,.$$

In the above way we have established an isomorphism between the space $\mathcal{H}$ and the space ℓ_2 consisting of the square summable sequences. This is called a **_representation of_ $\mathcal{H}$ _on_ ℓ_2**.

Of course, if the basis $|e_n\rangle$ is changed, the representatives of any vector change, i.e. the isomorphism changes.

Let now ϱ be an operator and $|A\rangle$ any vector in $\mathcal{H}$. Let us put

$$|B\rangle = \varrho |A\rangle \,.$$

© Springer International Publishing Switzerland 2016
L.E. Picasso, *Lectures in Quantum Mechanics*, UNITEXT for Physics,
DOI 10.1007/978-3-319-22632-3_6

How are the representatives $\{a_n\}$ and $\{b_n\}$ of $|A\rangle$ and $|B\rangle$ related to each other? One has

$$b_n = \langle e_n \mid B\rangle = \langle e_n \mid \varrho \mid A\rangle = \sum_m \langle e_n \mid \varrho \mid e_m\rangle \langle e_m \mid A\rangle = \sum_m \varrho_{nm}\, a_m$$

having put

$$\varrho_{nm} \equiv \langle e_n \mid \varrho \mid e_m\rangle .$$

The numbers ϱ_{nm} are called the **matrix elements** of the operator ϱ in the basis $|e_n\rangle$; they are completely determined by the operator (and by the chosen basis) and they completely determine it inasmuch as, by means of them, the action of ϱ on any vector is known:

$$\varrho \to \{\varrho_{nm}\}, \qquad \varrho|A\rangle \to \Big\{ b_n = \sum_m \varrho_{nm}\, a_m \Big\} \qquad (6.2)$$

Let us now examine some properties of the representation of operators.

1. The representatives of the identity operator are, in any basis, δ_{nm}.
2. Let ξ and η be two operators. Which are the representatives $\{(\xi\eta)_{nm}\}$ of the product $\xi\eta$ in terms of the representatives of ξ and η? By use of the completeness relation one has

$$(\xi\eta)_{nm} = \langle e_n \mid \xi\eta \mid e_m\rangle = \sum_k \langle e_n \mid \xi \mid e_k\rangle \langle e_k \mid \eta \mid e_m\rangle = \sum_k \xi_{nk}\, \eta_{km} .$$

Therefore any operator is represented by a matrix of numbers and the matrix representing the product is the product of the matrices, according to the 'row by column' multiplication law (of course one is dealing with infinite-dimensional matrices).

3. Concerning Hermitian conjugation, one has

$$(\xi^\dagger)_{nm} = \langle e_n \mid \xi^\dagger \mid e_m\rangle = \langle e_m \mid \xi \mid e_n\rangle^* = \xi_{mn}^* . \qquad (6.3)$$

Therefore the matrix representing $\xi^\dagger$ is the Hermitian conjugate (= complex conjugate and transposed) of the matrix representing ξ.

Normally the basis $|e_n\rangle$ that identifies the representation consists of the eigenvectors of some observable ξ; however, if ξ is degenerate, we know that there exists some arbitrariness in the choice of an orthonormal basis consisting of eigenvectors of ξ.

If $|\xi_n\rangle$ is an orthonormal basis of eigenvectors of ξ,

$$\xi \to \{\langle \xi_n \mid \xi \mid \xi_m\rangle = \{\xi_n\, \delta_{nm}\}, \qquad \xi|A\rangle \to \{\xi_n\, a_n\}. \qquad (6.4)$$

In general, if the basis consists of simultaneous eigenvectors $|\xi_k, \eta_l, \zeta_m, \cdots\rangle$ of a complete set of compatible observables $\xi, \eta, \zeta, \cdots$, for each of these observables, say ξ, one has:

$$\langle \xi_k, \eta_l, \zeta_m, \cdots \mid \xi \mid \xi_{k'}, \eta_{l'}, \zeta_{m'}, \cdots\rangle = \xi_k\, \delta_{kk'}\, \delta_{ll'}\, \delta_{mm'} \cdots .$$

Therefore, ξ is represented by a *diagonal* matrix, the diagonal matrix elements are the eigenvalues of ξ and each of them appears as many times as its degree of degeneracy. The same applies for η, ζ, $\cdots$. For this reason one says that the considered representation is a representation in which ξ, η, ζ, $\cdots$ are simultaneously diagonal.

Viceversa, as we know that *compatible* observables possess a complete set of simultaneous eigenvectors, it is always possible to choose a basis in which such observables are simultaneously diagonal. As in (6.4), diagonal observables act multiplicatively on the representatives of states.

Let now ξ be an observable and $|\xi_n\rangle$ a basis of eigenvectors of ξ. Let η be an operator that commutes with ξ. How is η represented in the basis $|\xi_n\rangle$? Let us distinguish two cases.

(i) If the vectors $|\xi_n\rangle$ are simultaneous eigenvectors of both ξ and η, then we are in the situation previously described, i.e. η is represented by a diagonal matrix. This certainly happens if ξ is nondegenerate.

(ii) If not all the $|\xi_n\rangle$ are also eigenvectors of η, i.e. ξ is degenerate, then by the lemma of p. 87 (Sect. 4.10) one has that all the matrix elements of η between eigenvectors of ξ belonging to *different* eigenvalues are vanishing; if, viceversa, $|\xi_n\rangle$ and $|\xi_m\rangle$ belong to the same eigenvalue of ξ, then $\langle\xi_n|\eta|\xi_m\rangle$ may be nonvanishing. In the latter case η_{nm} is a "block-diagonal" matrix (see Fig. 6.1b), i.e. a matrix possibly nonvanishing only within square blocks on the principal diagonal, whose dimension coincide with the degree of degeneracy of the eigenvalue of ξ the block refers to (see Fig. 6.1a).

$$\xi \to \begin{pmatrix} \xi_1 & 0 & 0 & 0 & 0 & 0 & \cdots \\ 0 & \xi_1 & 0 & 0 & 0 & 0 & \cdots \\ 0 & 0 & \xi_1 & 0 & 0 & 0 & \cdots \\ 0 & 0 & 0 & \xi_2 & 0 & 0 & \cdots \\ 0 & 0 & 0 & 0 & \xi_3 & 0 & \cdots \\ 0 & 0 & 0 & 0 & 0 & \xi_3 & \cdots \\ \vdots & \vdots & \vdots & \vdots & \vdots & \vdots & \ddots \end{pmatrix} \quad ; \quad \eta \to \begin{pmatrix} \eta_{11} & \eta_{12} & \eta_{13} & 0 & 0 & 0 & \cdots \\ \eta_{21} & \eta_{22} & \eta_{23} & 0 & 0 & 0 & \cdots \\ \eta_{31} & \eta_{32} & \eta_{33} & 0 & 0 & 0 & \cdots \\ 0 & 0 & 0 & \eta_{44} & 0 & 0 & \cdots \\ 0 & 0 & 0 & 0 & \eta_{55} & \eta_{56} & \cdots \\ 0 & 0 & 0 & 0 & \eta_{65} & \eta_{66} & \cdots \\ \vdots & \vdots & \vdots & \vdots & \vdots & \vdots & \ddots \end{pmatrix} .$$

Fig. 6.1a   Fig. 6.1b

Another very important fact has to be remarked: even in the case ξ is a nondegenerate observable (or, more in general, a complete set of commuting observables), the representation is *not* completely specified by saying that the vectors $|\xi_n\rangle$ forming the basis are eigenvectors of ξ: indeed, if the vectors $|\xi_n\rangle$ form an orthonormal basis of ξ, also the vectors

$$|\xi_n\rangle' = e^{i\varphi_n}|\xi_n\rangle \tag{6.5}$$

are such. So the statement 'the representation in which ξ is diagonal' is not correct inasmuch as many such representations do exist; one better says '*a* representation ... '.

Let us examine the differences between two representations in which the nondegenerate observable ξ is diagonal. As the bases are connected by (6.5), for the representatives $\{a_n\}$ and $\{a'_n\}$ of a vector $|A\rangle$ in the bases $|\xi_n\rangle$ and $|\xi_n\rangle'$ one has

$$a'_n = e^{-i\varphi_n}\, a_n$$

and for the representatives of an operator

$$\varrho'_{nm} = e^{i\,(\varphi_m - \varphi_n)}\, \varrho_{nm}\,.$$

So only the diagonal elements of an operator are not changed; in particular the representation of diagonal operators (i.e. represented by diagonal matrices), such as e.g. ξ, is not affected. In order to fully characterize a representation in which ξ is diagonal it is necessary to somehow fix the phases of the eigenvectors of ξ. We shall see in each case how this can be done. Note that we are here interested in the relative phases: indeed the case in which all the φ_n in (6.5) are equal to the same φ is totally irrelevant because the representative of operators are left unchanged, whereas the representatives of any vector $|A\rangle$ become those of the vector $e^{-i\varphi}|A\rangle$, that represents the same state.

We conclude this section with a quite relevant observation about the internal consistency of the theory. More precisely: what warranty are we given about the existence of operators obeying the commutation relations (4.51)? That the question is not obvious is shown by the fact that, if $\mathcal{H}$ had finite dimension, (4.51) could not hold: indeed – calling N the dimension of $\mathcal{H}$, and $|k\rangle,\,(k = 1, 2, \cdots, N)$ any orthonormal basis that defines the representation – taking the trace of both sides of $i\,\hbar = [\,q\,,\,p\,]$ gives rise to a contradiction (remember that $\mathrm{Tr}(A\,B) = \mathrm{Tr}(B\,A)$):

$$i\,N\,\hbar = \mathrm{Tr}(q\,p) - \mathrm{Tr}(p\,q) = 0$$

(if the dimension of $\mathcal{H}$ is infinite, the trace of the commutator is $\infty - \infty$).

As a consequence, operators that fulfill (4.51) (that they do exist will be seen in the next section) necessarily operate on an infinite-dimensional space, as already anticipated in Chap. 4.

6.2 Heisenberg Representation for the Harmonic Oscillator

Any representation in which the Hamiltonian operator is diagonal is given the name of **Heisenberg representation**. In the case of the linear harmonic oscillator the Hamiltonian is nondegenerate. The relative phases of the vectors that form the basis are defined if we take, as we will indeed do, the basis $|n\rangle$ of the eigenvectors of H given by (5.18).

The representation of H is obvious and is not worth any further comment. Let us instead see the representation of the operators η, $\eta^\dagger$, q and p. One has

$$(\eta^\dagger)_{nm} = \langle n\,|\,\eta^\dagger\,|\,m\rangle = \frac{1}{\sqrt{m!}}\,\langle n\,|\,(\eta^\dagger)^{m+1}\,|\,0\rangle$$

$$= \frac{\sqrt{(m+1)!}}{\sqrt{m!}}\,\langle n\,|\,m+1\rangle = \sqrt{m+1}\,\,\delta_{n\,m+1}\,.$$

Since $\eta^\dagger$ is represented by a real matrix, the representation of η is obtained by taking its transpose:

$$\eta^\dagger \rightarrow \begin{pmatrix} 0 & 0 & 0 & 0 & \cdots \\ 1 & 0 & 0 & 0 & \cdots \\ 0 & \sqrt{2} & 0 & 0 & \cdots \\ 0 & 0 & \sqrt{3} & 0 & \cdots \\ \vdots & \vdots & \vdots & \vdots & \ddots \end{pmatrix} \quad ; \quad \eta \rightarrow \begin{pmatrix} 0 & 1 & 0 & 0 & \cdots \\ 0 & 0 & \sqrt{2} & 0 & \cdots \\ 0 & 0 & 0 & \sqrt{3} & \cdots \\ 0 & 0 & 0 & 0 & \cdots \\ \vdots & \vdots & \vdots & \vdots & \ddots \end{pmatrix} .$$

The operators q and p are then expressed in terms of η and $\eta^\dagger$ by (5.9), therefore:

$$q \rightarrow -\mathrm{i}\sqrt{\frac{\hbar}{2m\omega}} \begin{pmatrix} 0 & -1 & 0 & 0 & \cdots \\ 1 & 0 & -\sqrt{2} & 0 & \cdots \\ 0 & \sqrt{2} & 0 & -\sqrt{3} & \cdots \\ 0 & 0 & \sqrt{3} & 0 & \cdots \\ \vdots & \vdots & \vdots & \vdots & \ddots \end{pmatrix} \quad ; \tag{6.6}$$

$$p \rightarrow \sqrt{\frac{m\hbar\omega}{2}} \begin{pmatrix} 0 & 1 & 0 & 0 & \cdots \\ 1 & 0 & \sqrt{2} & 0 & \cdots \\ 0 & \sqrt{2} & 0 & \sqrt{3} & \cdots \\ 0 & 0 & \sqrt{3} & 0 & \cdots \\ \vdots & \vdots & \vdots & \vdots & \ddots \end{pmatrix} . \tag{6.7}$$

It is straightforward to verify that the matrices (6.6) and (6.7) satisfy the commutation relation (5.2), namely:

$$\sum_k (q_{nk}\,p_{km} - p_{nk}\,q_{km}) = \mathrm{i}\,\hbar\,\delta_{nm} .$$

The substitution of (6.6) and (6.7) into (5.1) gives H in diagonal form, with the eigenvalues $E_n = \hbar\omega\left(n + \frac{1}{2}\right)$ on the diagonal.

We have exhibited a 'concrete' representation of the abstract theory and this is important for two reasons:

1. it shows that the set of operators satisfying (5.2) is non-empty;
2. all the discussion of Chap. 5 about the harmonic oscillator is based on the assumption that H has at least one eigenvector: so the existence of a concrete representation, that reproduces all the results found with the above assumption in Chap. 5 in an abstract way, is something more than a mere verification, it rather is a proof of existence.

Note that q is represented by a matrix whose elements are all pure imaginary, whereas p is represented by a real matrix: in several textbooks the situation is reversed, i.e. q is represented by a real matrix, p by a purely imaginary one. This depends on the different choice of the phases of the vectors that form the basis, precisely it corresponds to taking as the basis that characterizes the representation the following vectors:

$$|n\rangle' = (-\mathrm{i})^n \, |n\rangle = \frac{(-\mathrm{i})^n}{\sqrt{n!}} \, (\eta^\dagger)^n \, |n\rangle \tag{6.8}$$

instead of (5.18), which is perfectly legitimate. One must only pay attention not to use results obtained by one representation within a problem where use of the second is being made.

6.3 Unitary Transformations

Let $|e_n\rangle$ and $|\tilde{e}_n\rangle$ two orthonormal bases and U the operator such that

$$U\,|e_n\rangle = |\tilde{e}_n\rangle \,. \tag{6.9}$$

The operator U is defined by linearity on any vector $|A\rangle \in \mathcal{H}$:

$$|A\rangle = \sum_n a_n\,|e_n\rangle \quad \Rightarrow \quad U\,|A\rangle = U\left(\sum_n a_n\,|e_n\rangle\right) = \sum_n a_n\,|\tilde{e}_n\rangle \,.$$

By use of (6.9) and its conjugate $\langle \tilde{e}_n| = \langle e_n|\,U^\dagger$ one has

$$\langle e_n \mid U^\dagger U \mid e_m\rangle = \langle \tilde{e}_n \mid \tilde{e}_m\rangle = \delta_{nm}$$

$$U\,U^\dagger = U\left(\sum_k |e_k\rangle\langle e_k|\right) U^\dagger = \sum_k |\tilde{e}_k\rangle\langle \tilde{e}_k| = \mathbb{1} \,.$$

So both products $U^\dagger U$ and $U\,U^\dagger$ coincide with the identity operator, i.e. $U^\dagger$ is the inverse of the operator U:

$$U\,U^\dagger = U^\dagger U = \mathbb{1} \quad \Rightarrow \quad U^\dagger = U^{-1} \tag{6.10}$$

and in the latter case one says that U is a ***unitary*** operator.

In an infinite dimensional space $UV = \mathbb{1}$ is not sufficient to conclude that $U = V^{-1}$, also $VU = \mathbb{1}$ is needed: for example the operator defined by $V\,|e_n\rangle = |\tilde{e}_n\rangle \equiv |e_{n+1}\rangle$ satisfies $V^\dagger V = \mathbb{1}$ (it transforms orthonormal vectors into orthonormal vectors) but *not* $V\,V^\dagger = \mathbb{1}$ (it does *not* transform a complete set into a complete set).

Viceversa, if U is a unitary operator and $|e_n\rangle$ ($n = 1, 2, \cdots$) an orthonormal basis, then

$$|\tilde{e}_n\rangle \equiv U\,|e_n\rangle, \qquad n = 1, 2, \cdots$$

is an orthonormal basis: the demonstration is straightforward.

Since unitary operators preserve the norm of vectors:

$$\langle A \mid U^\dagger U \mid A\rangle = \langle A \mid A\rangle$$

they are bounded operators, therefore they are defined onto the whole $\mathcal{H}$ and for them no domain problem arises.

If U is a unitary operator, thanks to $U^{-1} = U^\dagger$, the operators

$$\tilde{q}_i \equiv U\,q_i\,U^{-1}, \qquad\qquad \tilde{p}_i \equiv U\,p_i\,U^{-1} \tag{6.11}$$

are self-adjoint and, in addition, satisfy the commutation relations (4.51):

$$[\tilde{q}_i, \tilde{p}_j] = U\, q_i\, U^{-1} U\, p_j\, U^{-1} - U\, p_j\, U^{-1} U\, q_i U^{-1} = U\,[q_i, p_j]\, U^{-1} = \mathrm{i}\,\hbar\,\delta_{ij}$$

and likewise for the others. In classical mechanics the transformations that preserve the Poisson brackets are the canonical transformations: we then see that the *unitary transformations* correspond to canonical transformations.

The transformation $q \to \tilde{q}$, $p \to \tilde{p}$ induces on any observable (more in general, on any operator) $f(q,p)$ the transformation $f(q,p) \to f(\tilde{q},\tilde{p})$ and one has

$$f(\tilde{q},\tilde{p}) \equiv f(U\, q\, U^{-1}, U\, p\, U^{-1}) = U\, f(q,p)\, U^{-1} \qquad (6.12)$$

and, if $\xi \,|\, \xi'\,\rangle = \xi'\,|\,\xi'\,\rangle$ and $\tilde{\xi} = U\,\xi\, U^{-1}$, setting $|\,\tilde{\xi}'\,\rangle \equiv U\,|\,\xi'\,\rangle$ one has

$$\tilde{\xi}\,|\,\tilde{\xi}'\,\rangle = \xi'\,|\,\tilde{\xi}'\,\rangle\,. \qquad (6.13)$$

The last equation is immediate; in order to justify the last step of (6.12) let us start with the case in which $f(q,p)$ is a polynomial:

$$f(q,p) = \sum a_{nm}\, q^n p^m \quad \Rightarrow \quad f(\tilde{q},\tilde{p}) \equiv \sum a_{nm}\, \tilde{q}^n \tilde{p}^m =$$

$$= \sum a_{nm}\, \overbrace{(U\,q\,U^{-1}) \cdots (U\,q\,U^{-1})}^{n\ \text{times}}\, \overbrace{(U\,p\,U^{-1}) \cdots (U\,p\,U^{-1})}^{m\ \text{times}} =$$

$$= U\left(\sum a_{nm}\, q^n p^m\right) U^{-1} = U\, f(q,p)\, U^{-1}\,.$$

Furthermore, unitary transformations preserve algebraic relations, so that if, for example, $f(q,p)$ is a polynomial and $g(q,p) = 1/f(q,p)$, thanks to $(\xi\,\eta)^{-1} = \eta^{-1}\,\xi^{-1}$, one has

$$g(\tilde{q},\tilde{p}) = [f(\tilde{q},\tilde{p})]^{-1} = [U\, f(q,p)\, U^{-1}]^{-1} = U\,[f(q,p)]^{-1} U^{-1} = U\, g(q,p)\, U^{-1}\,.$$

In this way (6.12) can be demonstrated at least for all functions defined by algebraic relations.

Viceversa, for systems with a finite number of degrees of freedom (e.g. a system with a fixed number of particles), if the transformation:

$$q_i \to \tilde{q}_i\,, \quad p_i \to \tilde{p}_i\,; \qquad\qquad \tilde{q}_i^\dagger = \tilde{q}_i\,, \quad \tilde{p}_i^\dagger = \tilde{p}_i \qquad (6.14)$$

preserves the commutation relations (4.51) ($[\tilde{q}_i, \tilde{p}_j] = \mathrm{i}\,\hbar\,\delta_{ij}$ etc.) then there exists a unitary operator U such that (6.11) hold: one says that U *imple-ments* the transformation (6.14). (Since unitary operators do not change the physical dimensions of a variable, it is understood that q and $\tilde{q}$ have the same dimensions, as well as p and $\tilde{p}$).

The above result is a very important one and is known as **von Neumann theorem**.

The basic idea of the proof, if all domain problems (which however are relevant for a correct demonstration) are ignored, is very simple: with q and p as in the harmonic oscillator, let $|\,n\,\rangle_{qp}$ and $|\,\tilde{n}\,\rangle_{\tilde{q}\tilde{p}}$ ($n, \tilde{n} = 0, 1, \cdots$) be respectively the eigenvectors of the operators $p^2 + m\omega^2 q^2$ and $\tilde{p}^2 + m\omega^2 \tilde{q}^2$ (m and ω arbitrary); the operator U defined by

$$U \, | \, n \, \rangle_{qp} = | \, n \, \rangle_{\tilde{q}\tilde{p}}$$

implements the transformation (6.14): obviously it is sufficient to verify that (η and $\tilde{\eta}$ defined as in (5.8))

$$U \, \eta \, U^{-1} \, | \, n \, \rangle_{\tilde{q}\tilde{p}} = \tilde{\eta} \, | \, n \, \rangle_{\tilde{q}\tilde{p}} \,, \quad U \, \eta^\dagger U^{-1} \, | \, n \, \rangle_{\tilde{q}\tilde{p}} = \tilde{\eta}^\dagger | \, n \, \rangle_{\tilde{q}\tilde{p}} \qquad \text{for all } | \, n \, \rangle_{\tilde{q}\tilde{p}} \,.$$

Indeed (and similarly for the second of the above equations):

$$U \, \eta \, U^{-1} \, | \, n \, \rangle_{\tilde{q}\tilde{p}} = U \, \eta \, | \, n \, \rangle_{qp} = \sqrt{n} \, U \, | \, n - 1 \, \rangle_{qp} = \sqrt{n} \, | \, n - 1 \, \rangle_{\tilde{q}\tilde{p}} = \tilde{\eta} \, | \, n \, \rangle_{\tilde{q}\tilde{p}} \,.$$

Clearly U and $e^{i\,\varphi} U$ are equivalent to each other; we now show the converse: two operators U_1 and U_2 that implement the same transformation differ by a phase factor. Indeed, the (unitary) operator $V \equiv U_2^{-1} U_1$ implements the identity transformation $q \to q$, $p \to p$; so V commutes with any $f(q,p)$:

$$V \, f(q,p) \, V^{-1} = f(V \, q \, V^{-1}, V \, p \, V^{-1}) = f(q,p) \quad \Rightarrow \quad V \, f(q,p) = f(q,p) \, V \,.$$

We now exploit the fact that any state $| \, A \, \rangle$ is eigenstate of some observable corresponding to a nondegenerate eigenvalue: from this and the lemma of p. 87 it follows that any state $| \, A \, \rangle$ is eigenstate of the unitary operator V, and this is possible only if V is a multiple of the identity: $V = c \, \mathbb{1}$; since V is unitary ($V^\dagger V = \mathbb{1}$), $|c|^2 = 1$ follows.

Just for completeness we can now exhibit one observable such that the (arbitrary) state $| \, A \, \rangle$ is eigenstate corresponding to a nondegenerate eigenvalue: indeed, $| \, A \, \rangle$ can be taken (in infinitely many ways) to be a member of an orthonormal basis $| \, e_n \, \rangle$ and let W be the unitary operator such that $W \, | \, n \, \rangle = | \, e_n \, \rangle$, where $| \, n \, \rangle$ are the eigenvectors of the operator $p^2 + m^2\omega^2 q^2$ whose eigenvalues we know are nondegenerate; therefore $| \, A \, \rangle$ is an eigenvector corresponding to a nondegenerate eigenvalue of the operator $W \left(p^2 + m^2\omega^2 q^2 \right) W^\dagger$.

What's more, if V were not equivalent to the identity operator: $V \neq e^{i\varphi} \, \mathbb{1}$, the statement made at the end of Sect. 4.8 according to which the collection of all the expectation values univocally determines the state, would not be true: with $| \, B \, \rangle = V | \, A \, \rangle$, $\langle \, A \, | \, f(q,\, p) \, | \, A \, \rangle = \langle \, B \, | \, f(q,\, p) \, | \, B \, \rangle$.

The fact that any operator that commutes with all the observables is a multiple of the identity entails that no subspace of $\mathcal{H}$ is left invariant *by the whole set* of the observables: one then says that the representation of the observables on $\mathcal{H}$ is **irreducible** (the connection between the above statement and Schur's lemma should not have gone unnoticed); thanks to von Neumann's theorem all irreducible representations are (unitarily) equivalent.

It is worth noticing that the irreducibility of the representation of the observables is also implicit in the superposition principle: if $\mathcal{H}_1$ and $\mathcal{H}_2$ are invariant subspaces of $\mathcal{H}$ (i.e. no observable has matrix elements between the vectors of $\mathcal{H}_1$ and $\mathcal{H}_2$) then, with $| \, A_1 \, \rangle \in \mathcal{H}_1$ and $| \, A_2 \, \rangle \in \mathcal{H}_2$), all the vectors

$\alpha_1|A_1\rangle + \alpha_2|A_2\rangle$ with given $|\alpha_1|$ and $|\alpha_2|$ represent the same state, which therefore is the statistical mixture $\{\,|A_1\rangle,\,|\alpha_1|^2\,;\,|A_2\rangle,\,|\alpha_2|^2\,\}$.

An instructive application of the von Neumann theorem is given by the following example: the transformation

$$q \to \tilde{q} = \frac{p}{m\,\omega}\,, \quad p \to \tilde{p} = -m\,\omega\,q \qquad (6.15)$$

is a canonical transformation ($[\tilde{q},\,\tilde{p}] = i\,\hbar$), so there exists a unitary operator that implements it. The Hamiltonian (5.1) of the harmonic oscillator is invariant under the transformation (6.15): $H(\tilde{q},\tilde{p}) = H(q,p)$, therefore

$$U\,H(q,p)\,U^{-1} = H(q,p) \qquad \Rightarrow \qquad U\,H = H\,U\,.$$

Since U commutes with H and H is nondegenerate, thanks to the lemma of p. 87, the eigenvectors of the latter also are eigenvectors for U:

$$U\,|E\rangle = \lambda\,|E\rangle\,; \qquad \langle E\,|\,E\rangle = \langle E\,|\,U^\dagger U\,|\,E\rangle \qquad \Rightarrow \qquad |\lambda|^2 = 1$$

(the eigenvalues of unitary operators are complex numbers of modulus 1). So, since $\langle E\,|\,U^\dagger \cdots U\,|\,E\rangle = |\lambda|^2\langle E\,|\,\cdots\,|\,E\rangle = \langle E\,|\,\cdots\,|\,E\rangle$,

$$\langle E\,|\,\frac{p^2}{2m}\,|\,E\rangle = \langle E\,|\,\frac{\tilde{p}^2}{2m}\,|\,E\rangle = \langle E\,|\,\frac{1}{2}m\,\omega^2 q^2\,|\,E\rangle$$

that extends (5.7) to any eigenstate of the Hamiltonian.

6.4 The Schrödinger Representation: Preliminary Considerations

One of the problems of quantum mechanics is that of determining eigenvalues and eigenvectors of the observables: it would therefore be advisable to have at one's disposal a rather general technique apt to solve this problem. The representation theory discussed in Sect. 6.1 may be useful for this purpose: indeed, if we know a representation of the operators q and p, as e.g. that given by (6.6) and (6.7), then we know the representation of any observable $f(q,p)$ of the theory.

Therefore problem is that of finding eigenvalues and eigenvectors of an infinite dimensional matrix. For example, if we are interested in the Hamiltonian, we shall have to solve the equations:

$$\sum_k H_{nk}\,c_k^E = E\,c_k^E\,, \qquad n = 1,\,2,\,\cdots \qquad (6.16)$$

that translate in terms of representatives the equation $H\,|E\rangle = E\,|E\rangle$.
The unknowns in (6.16) are the eigenvalues E and, in correspondence with each of them, the representatives $\{c_n^E\}$ of the eigenvector $|E\rangle$.

Regarding the $\{c_n^E\}$, the acceptable solutions are those for which the condition $\sum_n |c_n^E|^2 < \infty$ holds, i.e. $\{c_n^E\} \in \ell_2$ (we will come back to this point in the next section, although the context will be slightly different).

So by means of (6.16) (or the analogue for other observables) one can solve, in principle, the problem of finding eigenvalues and eigenvectors of an observable. In practice things go in a different way because the equations (6.16) are a system of infinite linear and homogeneous equations in the unknowns $\{c_n^E\}$ and there is no sufficiently general method of solution at one's disposal. There also is one further problem: (6.6) and (6.7) are just one out of many possible representations for q and p: therefore the problem of understanding, case by case, which representation is the more convenient arises. One is led to think that, in view of the fundamental role q and p have in the whole theory, a representation, in which one of such observables be diagonal, should be particularly meaningful.

Let us then assume we want to find a representation in which the q's are diagonal: in the first place, it is necessary to find eigenvalues and eigenvectors. Here we find the first difficulty: *the q's have no eigenvectors.*

The above statement can be proved is several – more or less rigorous – ways: for example, if $|\,x\,\rangle$ is an eigenvector of q belonging to the eigenvalue x, from $i\,\hbar = [\,q,\,p\,]$ and $q^\dagger = q$ one has the contradiction:

$$ i\hbar \,\langle\,x\,|\,x\,\rangle = \langle\,x\,|\,(q\,p - p\,q)\,|\,x\,\rangle = x\,\langle\,x\,|\,p\,|\,x\,\rangle - \langle\,x\,|\,p\,|\,x\,\rangle\,x = 0 \,. $$

The proof is not rigorous for it assumes that $|\,x\,\rangle$ lies in the domain of definition of the operator p: we have reported it because by means of 'demonstrations' of this type several paradoxes can be invented.

Let us examine a rigorous and more instructive demonstration.

Consider, for any real a, the transformation $q \to \tilde{q} = q - a$, $p \to \tilde{p} = p$. Since commutation relations and self-adjointness are preserved, thanks to the von Neumann theorem there exist a unitary operator $U(a)$ that implements the transformation $q \to \tilde{q}$, $p \to \tilde{p}$:

$$ U(a)\,q\,U(a)^{-1} = q - a \,, \qquad U(a)\,p\,U(a)^{-1} = p \,. \tag{6.17} $$

By multiplying the first of (6.17) from the right by $U(a)$ we get:

$$ [\,q,\,U(a)\,] = a\,U(a) \,. \tag{6.18} $$

Let us now assume (by contradiction) that $|\,x\,\rangle$ be an eigenvector of q belonging to the eigenvalue x: since (6.18) has the same structure as (5.12) and (5.13), also in the present case one has that $U(a)\,|\,x\,\rangle$ is an eigenvector of q belonging to the eigenvalue $x + a$:

$$ q\,U(a)\,|\,x\,\rangle = (x + a)\,U(a)\,|\,x\,\rangle \quad \longleftrightarrow \quad U(a)\,|\,x\,\rangle = |\,x + a\,\rangle \,. \tag{6.19} $$

But a is an arbitrary real number, therefore a continuous set of eigenvalues is obtained: this is the contradiction, because in a *separable* Hilbert space $\mathcal{H}$ the eigenvalues of an operator form a countable (either finite or infinite) set, otherwise there would exist orthonormal bases with the cardinality of the continuum. In conclusion q has neither eigenvectors nor eigenvalues (in $\mathcal{H}$).

The same situation occurs for p: it is sufficient to consider the canonical transformation:

$$V(b)\,q\,V(b)^{-1} = q\,, \qquad V(b)\,p\,V(b)^{-1} = p - b \qquad (6.20)$$

and proceed as above. So also p has neither eigenvectors nor eigenvalues.

The above result is very unpleasant: indeed, if q and p are operators associated with observables, they should possess eigenvectors and eigenvalues. But this does not happen, so q and p are not observables, contrary to all what we have said so far!

Note that all this is a consequence of the commutation rules (4.51): it appears that the quantization postulate is incompatible with the postulate that q and p be observables. But eventually the situation is not so dramatic. Also from a physical point of view it is clear that, for example, q is not an observable in the strict sense: think. e.g, of the Heisenberg microscope as the instrument suitable to measure the position of a particle. It is clear from the discussion of Sect. 3.6 that, in order to make a *precise* measurement of position (i.e. to find an eigenvalue of the position q) radiation of infinite frequency should be at one's disposal, which is clearly impossible: in other words the position of a particle can be measured with an arbitrarily high, but not infinite, accuracy: Δx may be as small as one wishes, but never vanishing. It is the same as saying that no device can measure the position exactly, but that there exist instruments that measure it with an arbitrarily high accuracy (Heisenberg microscopes with smaller and smaller λ). So q is not an observable, but it can be considered as the limit of observables: guided by these considerations of physical character, in Sect. 6.9 we shall extend the name of observables also to operators like q and p, by giving a precise mathematical form to the idea of 'limit of observables'.

We conclude this section by first exhibiting the explicit expression in terms of the operators q and p of the unitary operator $U(a)$, then with a discussion about its physical meaning.

Since $U(a)$ commutes with p, it must exclusively be a function of p, then from (6.18) and (4.54) one has

$$[q,U(a)] = a\,U(a) = \mathrm{i}\,\hbar\,\frac{dU(a)}{dp} \quad \Rightarrow \quad \frac{dU(a)}{dp} = -\frac{\mathrm{i}\,a}{\hbar}\,U(a) \qquad (6.21)$$

and the last equality is just the definition of the exponential, therefore (up to an unessential phase factor):

$$U(a) = e^{-\mathrm{i}\,p\,a/\hbar} \qquad (6.22)$$

$\big($by the way, $V(b) = e^{+\mathrm{i}\,q\,b/\hbar}$: the $+\mathrm{i}$ in the exponent is due to the change of sign of the commutator $[q,p] = \mathrm{i}\,\hbar$ if q and p are interchanged$\big)$.

As far as the physical meaning is concerned, the transformation (6.17) represents a **translation** and if we adopt the active point of view, it corresponds to a translation of $+a$ of the apparata associated with the observables (if instead we adopt the passive point of view, it corresponds to a translation by $-a$ of the Cartesian axes): if in some sense (that we will clarify in Sect. 6.9) q represents an Heisenberg microscope, $\tilde{q}$ represents the same microscope translated by a.

Clearly, as $\tilde{p} = p$, if a device existed for measuring momentum, this should be invariant under translations and, therefore, infinitely extended: this helps understanding that not even p (as we have already seen for q) may be considered an observable in the strict sense.

Let now $|A\rangle$ be a vector (normalized to 1) representing some state of a system. The vector

$$|A_a\rangle \equiv U(a)\,|A\rangle$$

represents the translated state, i.e. the state that is prepared exactly as the state $|A\rangle$, but with the instruments translated by a: indeed, for any observable $f(q,p)$ one has

$$\langle A_a \mid f(\tilde{q},\tilde{p}) \mid A_a \rangle = \langle A \mid U^\dagger \left(U f(q,p)\, U^{-1}\right) U \mid A \rangle = \langle A \mid f(q,p) \mid A \rangle$$

and this is exactly the relation that defines the translated state: all the expectation values do not change if both states and instrument are translated in the same way.

6.5 The Schrödinger Representation

We started with the idea of finding a representation in which the q_i are diagonal and we have immediately found the difficulty that the q_i have no eigenvectors.

So such a representation does not exist, at least according to the meaning we have given to this term in Sect. 6.1.

In order to understand how we should proceed, and with the purpose that our exposition will not sound too abstract and deductive, we start with some heuristic considerations in which the mathematical rigour is provisionally ignored, but that will give us an intuition about the path to be taken.

We are interested in a representation in which the q_i are 'diagonal': for the sake of simplicity, we will focus on a system with just one q and one p (one degree of freedom) and let us reason as if q had a basis of eigenvectors. Moreover we positively admit that *any* real number x is an eigenvalue of q and that, for each of them, q possesses an eigenvector $|x\rangle$. Therefore q has a continuous set of eigenvalues and eigenvectors (once the mistake of admitting that q has an eigenvalue has been made, (6.19) forces us to consider all the real numbers as eigenvalues). If we now proceed formally as in Sect. 6.1, one has that the representatives of any vector $|A\rangle$ in the 'basis' $|x\rangle$ $(-\infty < x < +\infty)$ are $\langle x \mid A\rangle$, i.e. they are functions of x. Therefore:

$$|A\rangle \to \langle x \mid A\rangle \equiv \psi_A(x)\,.$$

The function $\psi_A(x)$ is named **wavefunction** of the state represented by the vector $|A\rangle$. How is the scalar product $\langle B \mid A\rangle$ expressed in terms of the wavefunctions $\psi_A(x)$ and $\psi_B(x)$? Given that x is a continuous variable (that takes the place of the index n of Sect. 6.1), it is natural to write the completeness relation (4.25) as

$$\int |x\rangle\, dx \,\langle x| = \mathbb{1}$$

(the dx is put between the ket and the bra only for aesthetical reasons); in this way one has

$$\langle B \mid A \rangle = \int \langle B \mid x \rangle \langle x \mid A \rangle\, dx = \int \psi_B^*(x)\, \psi_A(x)\, dx$$

that is the scalar product in the space L_2 of the square summable functions. We have therefore obtained a representation of $\mathcal{H}$ on L_2. Let us now examine how the operator q is represented in this representation. One has

$$q\,|A\rangle \to \langle x \mid q \mid A \rangle$$

but, as $q = q^\dagger$, $\langle x | q = x \langle x |$ and in conclusion

$$q\,|A\rangle \to \langle x \mid q \mid A \rangle = x\,\psi_A(x)$$

i.e. q is represented in the space L_2 of wavefunctions by the multiplication by x, much as in a representation on ℓ_2 a *diagonal* operator acts multiplicatively on the representatives of any vector $\big($see (6.4)$\big)$. In this sense we may say that, in our representation, q is diagonal.

We have arrived at an interesting result, even if in a way that is all but rigorous: what can be legitimately considered a representation in which q is diagonal is a representation on L_2 in which q is represented by the multiplication by x. What is not rigorous in the above reasoning is the use of the basis $|x\rangle$, so we will now try to obtain the same results without using it.

Indeed, in Sect. 6.1 a representation was nothing but an isomorphism of the abstract space $\mathcal{H}$ on the 'concrete' space ℓ_2: the use of the basis $|e_n\rangle$ only was a tool to build up a particular isomorphism. From this point of view we can call 'representation' whatever identification (isomorphism) of the space $\mathcal{H}$ with a 'concrete' Hilbert space, as e.g. the space L_2 (all separable Hilbert spaces are isomorphic with one another), regardless of whether this isomorphism is realized either by means of an orthonormal basis (as in Sect. 6.1) or in a different way, as we will now do.

If the system has n degrees of freedom, let us consider the space $L_2^{(n)}$ of the square summable functions of n variables. One must establish an isomorphism between the space $\mathcal{H}$ of the state vectors and the space $L_2^{(n)}$: while in Sect. 6.1 we first established the isomorphism between $\mathcal{H}$ and ℓ_2 (namely the representation of the state vectors) and only after we determined the representation of the operators, now – viceversa – we will first establish the representation of the operators and from this we will determine the representation of the vectors. Of course it is sufficient to state how we represent the q's and the p's to be able to find the representation of any $f(q,p)$. There exist infinite ways to represent the q's and the p's, i.e. infinite representations of $\mathcal{H}$ on $L_2^{(n)}$, but we shall consider only two of them. The first is the

Schrödinger (or Coordinate) Representation

In this representation to any vector $|A\rangle$ of the space $\mathcal{H}$ there corresponds a **wavefunction** $\psi_A(x_1,\cdots,x_n) \in L_2^{(n)}$:

$$|A\rangle \longleftrightarrow \psi_A(x_1,\cdots,x_n) \tag{6.23}$$

and, this correspondence being an isomorphism, one has

$$\langle B\,|\,A\rangle = \int \psi_B^*(x_1,\cdots,x_n)\,\psi_A(x_1,\cdots,x_n)\,\mathrm{d}x_1\cdots\mathrm{d}x_n$$

where the right hand side is the scalar product in $L_2^{(n)}$.

Obviously infinite isomorphisms of $\mathcal{H}$ on $L_2^{(n)}$ exist: we choose the one in which the q_i are represented by the multiplication by x_i:

$$|A'\rangle = q_i\,|A\rangle \longleftrightarrow \psi_{A'}(x_1,\cdots,x_n) = x_i\,\psi_A(x_1,\cdots,x_n)\,. \tag{6.24}$$

The above requirement (6.24) does not yet completely fix the isomorphism (6.23). This can be understood, for example, by referring to the heuristic presentation in the beginning of the present section: indeed there still remains the possibility of changing any vector $|x\rangle$ of the basis by an arbitrary phase factor $\mathrm{e}^{\mathrm{i}\,\varphi(x)}$.

The equivalent of this is, in the presentation we are proposing, a certain arbitrariness in the way of representing the p_i: we have said 'a certain arbitrariness' because the commutation relations (4.51) must be satisfied by the representations of the q_i and the p_i. The Schrödinger representation is that for which the p_i are represented by $-\mathrm{i}\hbar\,\partial/\partial x_i$ (since we will see that in this way the (4.51) are satisfied):

$$|A''\rangle = p_i\,|A\rangle \longleftrightarrow \psi_{A''}(x_1,\cdots,x_n) = -\mathrm{i}\hbar\frac{\partial}{\partial x_i}\psi_A(x_1,\cdots,x_n)\,. \tag{6.25}$$

The Schrödinger representation is therefore defined by

$$|A\rangle \longleftrightarrow \psi_A(x_1,\cdots,x_n) \in L_2^{(n)}, \qquad q_i \to x_i, \quad p_i \to -\mathrm{i}\hbar\frac{\partial}{\partial x_i}\,. \tag{6.26}$$

At this point we must ask ourselves two things:

1. whether the operators x_i and $-\mathrm{i}\hbar\,\partial/\partial x_i$ are self-adjoint operators on $L_2^{(n)}$;
2. whether the representation of the q_i and the p_i is compatible with the commutation relations (4.51).

Let us start with the first problem: as usual we will only show that the operators are Hermitian. As for the x_i, it is necessary to verify that the equation $\langle B \mid q_i \mid A \rangle = \langle A \mid q_i \mid B \rangle^*$ holds also for the representatives, i.e.

$$\int \psi_B^*(x_1, \cdots, x_n)\, x_i\, \psi_A(x_1, \cdots, x_n)\, dx_1 \cdots dx_n$$
$$= \left(\int \psi_A^*(x_1, \cdots, x_n)\, x_i\, \psi_B(x_1, \cdots, x_n)\, dx_1 \cdots dx_n \right)^*$$

that is obviously true. Likewise, in order to show that $-i\hbar\,\partial/\partial x_i$ is Hermitian, one must verify that

$$-i\hbar \int \psi_B^*(x_1, \cdots, x_n) \frac{\partial}{\partial x_i} \psi_A(x_1, \cdots, x_n)\, dx_1 \cdots dx_n =$$
$$= \left(-i\hbar \int \psi_A^*(x_1, \cdots, x_n) \frac{\partial}{\partial x_i} \psi_B(x_1, \cdots, x_n)\, dx_1 \cdots dx_n \right)^* .$$

Indeed, by performing a partial integration in the right hand side one has

$$\left(-i\hbar \int \psi_A^* \frac{\partial}{\partial x_i} \psi_B\, dx \right)^* = i\hbar \int \psi_A \frac{\partial}{\partial x_i} \psi_B^*\, dx$$
$$= -i\hbar \int \psi_B^* \frac{\partial}{\partial x_i} \psi_A\, dx + i\hbar \int \frac{\partial}{\partial x_i}(\psi_B^* \psi_A)\, dx .$$

But the last term is vanishing for functions in L_2 that vanish at infinity, and these are dense in L_2. This shows that the p_i are represented by Hermitian operators.

As for the second problem, we must show that, much as the following equations

$$[q_i, q_j] \mid A \rangle = 0 , \quad [p_i, p_j] \mid A \rangle = 0 , \quad [q_i, p_j] \mid A \rangle = i\hbar\, \delta_{ij} \mid A \rangle \quad (6.27)$$

hold in $\mathcal{H}$, in $L_2^{(n)}$ one has

$$\begin{cases} (x_i x_j - x_j x_i) \psi_A(x_1, \cdots, x_n) = 0 , \\ -\hbar^2 \left(\dfrac{\partial}{\partial x_i} \dfrac{\partial}{\partial x_j} - \dfrac{\partial}{\partial x_j} \dfrac{\partial}{\partial x_i} \right) \psi_A(x_1, \cdots, x_n) = 0 , \\ -i\hbar \left(x_i \dfrac{\partial}{\partial x_j} - \dfrac{\partial}{\partial x_j} x_i \right) \psi_A(x_1, \cdots, x_n) = i\hbar\, \delta_{ij}\, \psi_A(x_1, \cdots, x_n) . \end{cases} \quad (6.28)$$

The first two equations obviously are satisfied. As for the third, one has

$$-i\hbar \left(x_i \frac{\partial}{\partial x_j} - \frac{\partial}{\partial x_j} x_i \right) \psi_A = -i\hbar\, x_i \frac{\partial}{\partial x_j} \psi_A + i\hbar \frac{\partial}{\partial x_j}(x_i \psi_A)$$
$$= -i\hbar\, x_i \frac{\partial \psi_A}{\partial x_j} + i\hbar\, \delta_{ij}\, \psi_A + i\hbar\, x_i \frac{\partial \psi_A}{\partial x_j} = i\hbar\, \delta_{ij}\, \psi_A .$$

It is useful to remark that (6.28) cannot be written for whatever function $\psi_A \in L_2^{(n)}$: for example, in order that the first of (6.28) be meaningful in $L_2^{(n)}$, it is necessary that $x_i\, x_j\, \psi_A(x_1, \cdots, x_n)$ still is square summable; whereas for the second of (6.28) it is necessary that $\psi_A(x_1, \cdots, x_n)$ is twice differentiable and that $\partial^2\, \psi_A/\partial x_i\, \partial x_j$ still is square summable. Likewise for the third of (6.28).

Not all the functions $\psi_A \in L_2^{(n)}$ satisfy the above conditions, they only form a dense set. Indeed we know that unbounded operators are not defined on all the vectors in $\mathcal{H}$, but only on a dense subset of vectors: this is the case for the q_i and p_i, for which equations (6.27) had a meaning not for all the vectors $|\,A\,\rangle \in \mathcal{H}$, but only for vectors $|\,A\,\rangle$ in a suitable dense subset of $\mathcal{H}$. Therefore both (6.27) and (6.28) for the representatives hold on a dense subset, respectively in $\mathcal{H}$ and in $L_2^{(n)}$.

We have come back to domain problems of unbounded operators, like q_i and p_i on $\mathcal{H}$ and x_i and $-\mathrm{i}\,\hbar\,\partial/\partial x_i$ on $L_2^{(n)}$, not only for the sake of correctness: it is indeed an important problem, but any further discussion about it is beyond the scope of this presentation. In any event, and not only to relieve our conscience, we will shortly come back to this point in the end of this section.

The representation given by (6.26) of the q_i and p_i being now known, any observable $f(q_i, p_i)$ is represented by $f(x_i, -\mathrm{i}\,\hbar\,\partial/\partial x_i)$. In particular, an observable $f(q_1, \cdots, q_n)$ is represented by $f(x_1, \cdots, x_n)$ that acts multiplicatively on wavefunctions:

$$f(q_1, \cdots, q_n)\,|\,A\,\rangle \to f(x_1, \cdots, x_n)\,\psi_A(x_1, \cdots, x_n)\,.$$

The idea of seeking a representation in which the q are diagonal had arisen from the necessity of solving the problem of the determination of eigenvector and eigenvalues of the several observables of the system: let us see what has been gained, in this respect, by the introduction of the Schrödinger representation (SR).

Let us, for example, consider the problem of determining eigenvalues and eigenvectors of the energy for a system of particles (n degrees of freedom). The Hamiltonian (in the absence of magnetic fields) has the form:

$$H = \sum_{i=1}^{n} \frac{p_i^2}{2m} + V(q_1, \cdots, q_n)$$

and, since

$$p_i^2 \to \left(-\mathrm{i}\,\hbar\,\frac{\partial}{\partial x_i}\right)\left(-\mathrm{i}\,\hbar\,\frac{\partial}{\partial x_i}\right) = -\hbar^2\,\frac{\partial^2}{\partial x_i^2}$$

(and *not* $p^2\,|\,A\,\rangle \to (\partial\,\psi_A/\partial x)^2$!!), the eigenvalue equation for H in the Schrödinger representation takes the form:

$$\left(\sum_{i=1}^{n} -\frac{\hbar^2}{2m}\,\frac{\partial^2}{\partial x_i^2} + V(x_1, \cdots, x_n)\right)\psi_E(x_1, \cdots, x_n) = E\,\psi_E(x_1, \cdots, x_n) \quad (6.29)$$

that, for historical reasons, is named **Schrödinger equation** (this is due to the particular importance that the determination of the energy levels of a system has always had, for which (6.29) has been written by Schrödinger before quantum mechanics had taken the form we are presenting here).

Equation (6.29) is a second order homogeneous linear differential equation (it contains indeed derivatives at most of the second order): it is true that not always the solution of this equation can be explicitly found (it all depends on the particular form of the potential $V(x_1, \cdots, x_n)$), however for differential equations we do have several techniques for an either exact, or approximate or numeric solutions – much more than we have for equations of the type (6.16). Therefore, from this point of view, the introduction of the Schrödinger representation is a remarkable step forward.

In the equation $H \,|\, E \,\rangle = E \,|\, E \,\rangle$ the unknowns are both the eigenvalues E and the corresponding eigenvectors $|\, E \,\rangle$. In the case of (6.29) the unknowns are the eigenvalues E and the $\psi_E(x_1, \cdots, x_n)$: it is worth spending some words to understand in which sense E is an unknown in (6.29). Indeed, existence theorems for differential equations ensure that (6.29) possesses solutions for *any* value of E; but we should remember that we are interested only in those solutions $\psi_E(x_1, \cdots, x_n)$ that are square summable (and that, therefore, vanish at infinity sufficiently fast). Now not only (6.29) possesses a solution for any value of E, but indeed it possesses so many of them (for it is a partial differential equation); however what happens is that square summable solutions exist only for *particular* values of E that form a discrete – finite or infinite – set. The determination of such values of E for which (6.29) admits solutions in $L_2^{(n)}$ is therefore an essential part of the problem of solving (6.29) itself.

Having introduced the Schrödinger representation in which the q_i 'are diagonal', we can introduce in the same way a second representation in which – with the same meaning, i.e. of being represented by operators that act multiplicatively – the p_i are diagonal: this is the

Momentum Representation

The momentum representation is obtained by interchanging the role of the q_i with that of the p_i in (6.26): any vector $|\, A \,\rangle$ is represented by the function $\varphi_A(k_1, \cdots, k_n) \in L_2^{(n)}$ (we use k instead of x for the independent variables only to distinguish the momentum from the Schrödinger representation), the p_i are represented by the multiplication by k_i and the q_i by $i\,\hbar\,\partial/\partial k_i$ (the change of sign in the latter, with respect to the Schrödinger representation of p_i, is due, as already pointed out, to the fact that interchanging the roles of q and p changes the sign of the commutator: $[\,p_i\,,\,q_j\,] = -i\,\hbar\,\delta_{ij}$).

All what has been said for the Schrödinger representation can be repeated for the momentum representation. Either one of the two representation can be used to determine eigenvalues and eigenvectors of some observable $f(q,p)$, depending on which one provides the simplest equation. If, for example, the observable is the Hamiltonian, normally the Schrödinger representation is the more convenient because usually the potential is a complicated function of the q and it is therefore preferable that it be represented by a multiplicative operator $V(x_1, \cdots, x_n)$, rather than by a complicated differential operator $V(-i\hbar\partial/\partial x_1, \cdots, -i\hbar\partial/\partial x_n)$.

In the sequel we will make use of the following terminology: if $|A\rangle$ is an eigenvector of some observable $f(q, p)$, the wavefunction $\psi_A(x_1, \cdots, x_n)$ (or $\varphi_A(k_1, \cdots, k_n)$) that corresponds to it in the Schrödinger representation (respectively in the momentum representation) will be called **eigenfunction** in the Schrödinger representation (in the momentum representation) **of the observable** $f(q, p)$.

We conclude this section with the following observations: von Neumann theorem ensures us that all the representations of the q and p that obey the commutation rules (4.51) are unitarily equivalent to one another (therefore, for example, if ψ_A and φ_A are the wavefunctions of the state $|A\rangle$ respectively in the Schrödinger and momentum representation, there exists a unitary transformation U such that $U\psi_A = \varphi_A$ for any $|A\rangle$), so the Schrödinger representation has no privileged role: the results found in this representation hold in any representation, therefore in the abstract space $\mathcal{H}$.

In particular, in the Schrödinger representation it is easy to identify a dense (in $L_2^{(n)}$) set of functions on which the x_i and the $\partial/\partial x_i$ and, more in general, all the *polynomials* $P(x_i, -i\hbar\partial/\partial x_i)$ are well defined as operators on $L_2^{(n)}$ (i.e. when applied to the elements of this set give, as a result, square summable functions): we mean the Schwartz space $\mathcal{S}^{(n)}$ of the infinitely differentiable functions (of n variables) that share the property of fast decrease at infinity (i.e. that, for $|x| \to \infty$, vanish with all their derivatives faster that the inverse of any polynomial, as e.g. $x^n e^{-x^2}$): $\mathcal{S}^{(n)}$ is dense in $L_2^{(n)}$.

Therefore, thanks to von Neumann theorem, it is true in any representation, and therefore in the abstract space $\mathcal{H}$, that at least the operators that can be expressed as (or can be approximated by) polynomial functions of the q_i and the p_i are always well defined.

Let us consider, for example, the translation operator $U(a)$ given by (6.22): in the Schrödinger representation one has ($-ip/\hbar \to -d/dx$)

$$U(a)|A\rangle \xrightarrow{\text{SR}} e^{-a\,d/dx}\,\psi_A(x).$$

Then:

$$e^{-a\,d/dx}\,\psi_A(x) = \sum_n \frac{(-a)^n}{n!}\left(\frac{d^n\psi_A(x)}{dx^n}\right) = \psi_A(x - a)$$

for any *entire* function (i.e. a function whose Taylor series converges for any a). Since such functions constitute a set that is dense in L_2 and moreover $\|U(a)\psi_A(x)\| = \|\psi_A(x)\|$, the equation

$$U(a)\,\psi_A(x) = \psi(x - a) \tag{6.30}$$

can be, without any problem, extended by continuity to any $\psi_A \in \mathcal{H}$.

Recalling the problem hinted at in the beginning of Chap. 4, namely whether all the vectors in $\mathcal{H}$ do represent states of the system, it looks natural, at this point, to consider the space $\mathcal{S}$ as a good candidate for a bijective projective (i.e. up to a factor) correspondence with the states of the system.

6.6 Physical Interpretation of Schrödinger and Momentum Representations

Schrödinger and momentum representations not only are a useful mathematical tool to solve the problem of the determination of eigenvalues and eigenvectors of the various observables, but have an important physical meaning. Let us first consider the Schrödinger representation.

Let $\psi_A(x_1, \cdots, x_n)$ be the *normalized* wavefunction of the vector $|A\rangle$:

$$\int |\psi_A(x_1, \cdots, x_n)|^2 \, dx_1 \cdots dx_n = \langle A \,|\, A \rangle = 1 .$$

The mean values of q_i and p_i in the state represented by $|A\rangle$ respectively are

$$\overline{q_i} = \langle A \,|\, q_i \,|\, A \rangle = \int \psi_A^*(x_1, \cdots, x_n) \, x_i \, \psi_A(x_1, \cdots, x_n) \, dx_1 \cdots dx_n ,$$

$$= \int x_i \, |\psi_A(x_1, \cdots, x_n)|^2 \, dx_1 \cdots dx_n$$

$$\overline{p_i} = \langle A \,|\, p_i \,|\, A \rangle = -i\hbar \int \psi_A^*(x_1, \cdots, x_n) \frac{\partial}{\partial x_i} \psi_A(x_1, \cdots, x_n) \, dx_1 \cdots dx_n$$

and the mean value of a generic $f(q_1, \cdots, q_n)$ is

$$\overline{f(q_1, \cdots, q_n)} = \int f(x_1, \cdots, x_n) \, |\psi_A(x_1, \cdots, x_n)|^2 \, dx_1 \cdots dx_n . \qquad (6.31)$$

In the expression of the mean value of q_i and in general of $f(q_1, \cdots, q_n)$ (but *not* of p_i and of any generic $f(q,p)$) the factor $|\psi_A(x_1, \cdots, x_n)|^2$ appears: we want now to establish the physical meaning of this quantity.

Let us first consider, for the sake of simplicity, the case of one degree of freedom; later we will generalize the result. Consider, on the x-axis, a small interval Δ centered around the point x' and let us also consider the operator $E_{\Delta,x'}(q)$ that in the Schrödinger representation is represented by the function $E_{\Delta,x'}(x)$ whose value is 1 inside the interval Δ and 0 outside of it

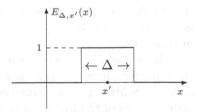

Fig. 6.2

$(E_{\Delta,x'}(x)$ – that acts multiplicatively – is called *characteristic function of the interval* Δ: see Fig. 6.2$)$.

It is evident that the eigenvalues of this self-adjoint operator $E_{\Delta,x'}(q)$ (or $E_{\Delta,x'}(x)$ in the Schrödinger representation) are only 0 and 1. Indeed all and only the functions $\psi(x)$ that are nonvanishing only *inside* the interval

Δ belong to the eigenvalue 1, while all and only those that are nonvanishing only *outside* of it belong to the eigenvalue 0. In addition, there exist no other eigenvalues, because the eigenfunctions of $E_{\Delta,x'}(x)$ corresponding to either 0 or 1 form a complete set: any function $\psi(x)$ can indeed be written as the sum of a function $\psi_0(x)$ nonvanishing only out of the interval and a function $\psi_1(x)$ nonvanishing only inside it. So $E_{\Delta,x'}(x)$ is a (very degenerate) observable. (Note that all the above follows immediately from the fact that the self-adjoint operator $E_{\Delta,x'}(x)$ is a projection operator: $E_{\Delta,x'}(x) E_{\Delta,x'}(x) = E_{\Delta,x'}(x)$) (see (4.26)).

Which is its physical meaning? Let us think of the corresponding classical observable: one is dealing with the observable whose value is 1 if the particle is inside the interval and is 0 if the particle is elsewhere. This being the meaning of $E_{\Delta,x'}(q)$, the probability p_1 that a measurement of $E_{\Delta,x'}(q)$ gives as a result the eigenvalue 1 also is the probability to find the particle inside the interval Δ, whereas the probability p_0 to find the eigenvalue 0 is the probability to find the particle outside of Δ.

The expectation value of $E_{\Delta,x'}(q)$ is given by

$$\overline{E_{\Delta,x'}(q)} = 0 \times p_0 + 1 \times p_1 = p_1$$

i.e. it coincides with p_1. Let now $|A\rangle$ be the state of the particle and $\psi_A(x)$ the corresponding wavefunction. According to (6.31) one has

$$p_1 = \overline{E_{\Delta,x'}(q)} = \int E_{\Delta,x'}(x) |\psi_A(x)|^2 \, dx = \int_\Delta |\psi_A(x)|^2 \, dx \simeq |\psi_A(x')|^2 \times \Delta$$

the last step holding only if Δ is small enough. So $|\psi_A(x)|^2 \, dx$ is the probability of finding the particle with a value of its abscissa between x and $x+dx$ (we have replaced Δ with dx) and, as a consequence, $|\psi_A(x)|^2$ is the **density of probability** relative to the position of the particle. This is a very important result: given a state $|A\rangle$ and consequently a wavefunction $\psi_A(x)$, $|\psi_A(x)|^2$ gives us all the information we want about the probability of finding the particle either in a region or in another one of the x-axis: in particular, if in a region of the x-axis $\psi_A(x)$ is vanishing, the corresponding probability to find the particle in that region is vanishing as well; likewise, if $\psi_A(x)$ is nonvanishing only in a given interval, one has the certainty to find the particle in that interval.

The probabilistic interpretation of the wavefunction was proposed by M. Born in 1926, in a series of works that appeared immediately after the works in which Schrödinger used the equation, that now bears his name, to find the energy levels of the hydrogen atom.

The fact we have given a physical meaning to $|\psi_A(x)|^2$ and not directly to $\psi_A(x)$ should *not* make one think that all the information content of $\psi_A(x)$ is completely contained in $|\psi_A(x)|^2$ or, in other words, that the knowledge of the state be determined by the only knowledge of $|\psi_A(x)|^2$: for example, for the calculation of $\bar{p}$ the knowledge of $\psi_A(x)$ is needed, that of $|\psi_A(x)|^2$ is not sufficient.

More to the point: the two wavefunctions $\psi_1(x)$ and $\psi_2(x) = e^{i\,\varphi(x)}\,\psi_1(x)$, with real $\varphi(x)$, possess the same modulus and are undistinguishable if only position measurements (i.e. measurements of observables $f(q)$) are performed, but being different wavefunctions (if $\varphi(x)$ is not a constant), they represent different states, so they must be distinguishable by measuring some other observables (for example, the expectation value of p is different). Note that this discussion is identical with that made after (4.11) concerning the physical meaning of the representatives $\{a_i\}$ of a vector $|A\rangle$ in the basis $|\xi_i\rangle$: one is indeed dealing with the same problem.

The generalization to the case of many degrees of freedom is immediate: $|\psi_A(x_1, \cdots, x_n)|^2 \, dx_1 \cdots dx_n$ is the probability to find the system (of particles) with coordinates between x_1 and $x_1 + dx_1$, x_2 and $x_2 + dx_2$, $\cdots$ x_n and $x_n + dx_n$.

Consider, as an example, the system consisting of two particles, 1 and 2 (e.g. an electron and a proton), described by the coordinates $\vec{r}_1 = (x_1, y_1, z_1)$ and $\vec{r}_2 = (x_2, y_2, z_2)$. In the present case the wavefunctions of the system are functions of the six variables $\vec{r}_1$ and $\vec{r}_2$: $\psi(\vec{r}_1, \vec{r}_2)$; $|\psi_A(\vec{r}_1, \vec{r}_2)|^2 \, dV_1 \, dV_2$ is the probability to find simultaneously particle 1 in the small volume dV_1 around the point $\vec{r}_1$ and particle 2 in the small volume dV_2 around $\vec{r}_2$; the probability to find particle 1 in dV_1 (we are now uninterested in where particle 2 is) is given by

$$\left(\int |\psi_A(\vec{r}_1, \vec{r}_2)|^2 \, dV_2 \right) dV_1 \ .$$

All what we have said for the Schrödinger representation can be repeated for the momentum representation: if $\varphi_A(k_1, \cdots, k_n)$ is the *normalized* wavefunction corresponding to the vector $|A\rangle$ in the momentum representation, for the mean values of q_i and p_i in the state represented by $|A\rangle$ one has

$$\overline{q_i} = i\,\hbar \int \varphi_A^*(k_1, \cdots, k_n) \, \frac{\partial}{\partial k_1} \, \varphi_A(k_1, \cdots, k_n) \, dk_1 \cdots dk_n$$

$$\overline{p_i} = \int k_i \, |\varphi_A(k_1, \cdots, k_n)|^2 \, dk_1 \cdots dk_n$$

and by the same reasoning as in the case of the Schrödinger representation one arrives at the result that $|\varphi_A(k_1, \cdots, k_n)|^2 \, dk_1 \cdots dk_n$ is the probability to find the system (of particles) with momenta between k_1 and $k_1 + dk_1$, k_2 and $k_2 + dk_2$, $\cdots$ k_n and $k_n + dk_n$. Therefore $|\varphi_A(k_1, \cdots, k_n)|^2$ is the probability density relative to the momenta of the particles.

6.7 The Improper Eigenvectors of q_i and p_i

We have already insisted on the fact that the Schrödinger representation can be considered as a representation in which the q_i are diagonal inasmuch as the q_i are represented by operators that act in a multiplicative way on wavefunctions, but not in the sense that is a representation determined by a basis of eigenvectors of the q_i, because the q_i do not possess eigenvectors. Nonetheless, we want to develop the formalism of the Schrödinger representation in such a way that, at least in a formal way (but the content of the present section has a mathematical dignity largely superior to what we let appear: it is related to what in mathematics is known as 'rigged Hilbert space' or 'Gelfand triple'), there is a more cogent analogy between the Schrödinger (or the momentum) representation and those discussed in Sect. 6.1.

Let $x = \{x_1, \cdots, x_n\}$ and

$$\psi_A(x) = \langle\, x \mid A \,\rangle, \qquad\qquad \psi_A^*(x) = \langle\, A \mid x \,\rangle \qquad (6.32)$$

understanding that, provisionally, (6.32) only stands for a change in notation. Since

$$q_i \mid A \,\rangle \to x_i\, \psi_A(x)$$

with the notation (6.32) one has

$$x_i \langle\, x \mid A \,\rangle = x_i\, \psi_A(x) = \langle\, x \mid q_i \mid A \,\rangle$$

so, taking the complex conjugate,

$$\langle\, A \mid q_i \mid x \,\rangle = x_i \langle\, A \mid x \,\rangle. \qquad (6.33)$$

Since (6.33) holds for arbitrary $\mid A \,\rangle$, we set

$$q_i \mid x \,\rangle = x_i \mid x \,\rangle. \qquad (6.34)$$

The meaning of (6.34) is only formal, because we know that in $\mathcal{H}$ there exists no eigenvector of the q_i: (6.34) is meaningful only if it is multiplied on the left by a vector $\mid A \,\rangle$ in $\mathcal{H}$ (more precisely: by any vector in the domain of q): it is nothing but a shortened way for writing (6.33). These objects $\mid x \,\rangle$, that are *not* vectors in $\mathcal{H}$, are called **improper vectors** (and it would perhaps be advisable to use for them a notation different from that used for the vectors of $\mathcal{H}$). So we say that $\mid x \,\rangle$, i.e. $\mid x_1, \cdots, x_n \,\rangle$, is an improper eigenvector of the q_i, corresponding to improper eigenvalues x_i that form a **continuous spectrum**: $-\infty < x_i < +\infty$.

Let us now consider $\left(\mathrm{d}x \equiv \mathrm{d}x_1, \cdots, \mathrm{d}x_n\right)$

$$\langle\, B \mid A \,\rangle = \int \psi_B^*(x)\, \psi_A(x)\, \mathrm{d}x = \int \langle\, B \mid x \,\rangle \langle\, x \mid A \,\rangle\, \mathrm{d}x \qquad (6.35)$$

and therefore in the same sense in which we have passed from (6.33) to (6.34) we may write

$$\int \mid x \,\rangle\, \mathrm{d}x \, \langle\, x \mid = \mathbb{1} \qquad (6.36)$$

that, as usual, has a meaning only if both sides are multiplied by two arbitrary vectors $\langle\, B \mid$ on the left and $\mid A \,\rangle$ on the right. Equation (6.36) is the analogue of (4.25): it is the completeness relation for the improper basis consisting of the improper eigenvectors of the q_i.

In the case of a basis $|n\rangle$ of eigenvectors of some observable one has the orthogonality relation $\langle n \mid m \rangle = \delta_{nm}$. We will now find the analogue relation for the improper basis, i.e. we want to calculate $\langle x \mid x' \rangle$. Indeed it would not be legitimate to pose this problem since $|x\rangle$ and $|x'\rangle$ are not vectors in $\mathcal{H}$, so the scalar product between them is not defined. We shall therefore pose the problem in the following terms: if we want to give a sense to the expression $\langle x \mid x' \rangle$, what should this expression be?

We begin by considering the one-dimensional case.

Let us start from $\psi_A(x) = \langle x \mid A \rangle$ and use (6.36):

$$\langle x \mid A \rangle = \int \langle x \mid x' \rangle \langle x' \mid A \rangle \, dx'$$

i.e. putting $\langle x \mid x' \rangle = g(x, x')$:

$$\psi_A(x) = \int g(x, x') \, \psi_A(x') \, dx' . \tag{6.37}$$

The problem then is to determine the $g(x, x')$ that satisfies (6.37) for any $\psi_A(x)$. Start with observing that (6.37) is satisfied also when $g(x, x')$ is replaced by $g(x + a, x' + a)$ with any real a. Indeed, putting $\tilde{\psi}_A(x + a) = \psi_A(x')$, one has

$$\int g(x + a, x' + a) \, \psi_A(x') \, dx' = \int g(x + a, x' + a) \, \tilde{\psi}_A(x' + a) \, dx'$$

$$= \int g(x + a, y) \, \tilde{\psi}_A(y) \, dy = \psi_A(x) .$$

This means that g depends only on the difference $x - x'$, so we write

$$g(x, x') = \delta(x - x')$$

i.e.

$$\psi_A(x) = \int \delta(x - x') \, \psi_A(x') \, dx' . \tag{6.38}$$

By a similar reasoning one sees that (6.38) is still satisfied if $\delta(x - x')$ is replaced by $\delta(-x + x')$, i.e. $\delta(x) = \delta(-x)$. Putting $x = 0$ in (6.38) one has

$$\psi_A(0) = \int \delta(x') \, \psi_A(x') \, dx' . \tag{6.39}$$

Let us now see which are the meaning and the properties of the $\delta(x)$ that satisfies (6.39).

1. For $x \neq 0$, $\delta(x) = 0$ (up, at most, to a set of measure 0). It is indeed sufficient to take $\psi_A(x) = 1$ in an interval Δ that does not contain the point $x = 0$, and $\psi_A(x) = 0$ outside of it. Then

$$\int_\Delta \delta(x)\, \mathrm{d}x = \psi_A(0) = 0$$

for any Δ, which proves the statement.

2. If we now take the interval Δ (where $\psi_A(x) = 1$) straddling the point $x = 0$, one has

$$\int_\Delta \delta(x)\, \mathrm{d}x = \psi_A(0) = 1$$

whatever the length of the interval Δ, and therefore also

$$\int_{-\infty}^{+\infty} \delta(x)\, \mathrm{d}x = 1 \,.$$

The above two properties clearly show that $\delta(x)$ is *not* a function: no function is nonvanishing only in a point ($x = 0$) and its integral equals 1! Think of a unitary point mass placed in the origin: obviously in this case the mass distribution (or density) $\rho(x)$ is not defined (as a function), but if one insists in defining it, then clearly $\rho(x) = 0$ for $x \neq 0$ and, in addition, $\int \rho(x)\, \mathrm{d}x = 1$, i.e. $\rho(x)$ has the same properties as the $\delta(x)$.

For this reason one says that $\delta(x)$ is a **distribution**. The $\delta(x)$ is also called **generalized function** or **Dirac delta function**.

We can think of the Dirac δ as the limit, for $\epsilon \to 0$, of the function whose value is $1/\epsilon$ for $-\epsilon/2 \le x \le +\epsilon/2$ and 0 for $|x| > \epsilon/2$. The $\delta(x-x')$ has the same properties as $\delta(x)$: it is vanishing for $x \neq x'$ and $\int \delta(x - x')\, \mathrm{d}x' = 1$.

It s not surprising that $\langle x \mid x' \rangle$ is such an unusual thing as $\delta(x - x')$, i.e. a distribution: this result expresses the fact that $\langle x \mid x' \rangle$ is not a scalar product of vectors in $\mathcal{H}$, therefore for $x = x'$ it is not a number. For the improper eigenvectors one has in conclusion the orthogonality relation:

$$\langle x \mid x' \rangle = \delta(x - x'). \tag{6.40}$$

In the case of many degrees freedom (6.40) becomes:

$$\langle x_1, \cdots, x_n \mid x'_1, \cdots, x'_n \rangle = \delta(x_1 - x'_1)\, \delta(x_2 - x'_2) \cdots \delta(x_n - x'_n) \tag{6.41}$$

where the product $\delta(x_1)\, \delta(x_2) \cdots \delta(x_n)$ is defined by the equation:

$$\int \delta(x_1)\, \delta(x_2) \cdots \delta(x_n)\, \psi(x_1, \cdots, x_n)\, \mathrm{d}x_1 \cdots \mathrm{d}x_n = \psi(0, \cdots, 0) \tag{6.42}$$

analogous to (6.39).

6.8 The Relationship between Coordinate and Momentum Representations

We now face the problem of finding the momentum representation wave function $\varphi(k_1, \cdots, k_n)$ in terms of the coordinate representation wavefunction $\psi(x_1, \cdots, x_n)$, and viceversa. In order to solve this problem, we will use the notation introduced in the previous section. We often will write x and k instead of the n-tuples $\{x_1, \cdots, x_n\}$ and $\{k_1, \cdots, k_n\}$; we will explicitly use the n-tuples when we will want to emphasize the presence of many degrees of freedom.

Let us assume that $\psi_A(x) = \langle x \mid A \rangle$ is known: we want to determine $\varphi_A(k) = \langle k \mid A \rangle$.

By means of the completeness relation (6.36) one has

$$\varphi_A(k) = \langle k \mid A \rangle = \int \langle k \mid x \rangle \langle x \mid A \rangle \, \mathrm{d}x = \int \langle k \mid x \rangle \psi_A(x) \, \mathrm{d}x \ .$$

If viceversa $\varphi_A(k)$ is known and $\psi_A(x)$ is to be found, by means of the completeness relation of the (improper) eigenvectors of momentum, one has

$$\psi_A(x) = \langle x \mid A \rangle = \int \langle x \mid k \rangle \langle k \mid A \rangle \, \mathrm{d}k = \int \langle x \mid k \rangle \varphi_A(k) \, \mathrm{d}k \ .$$

It appears that the passage from $\psi_A(x)$ to $\varphi_A(k)$, as well as its inverse, is possible once $\langle x \mid k \rangle$ is known, because $\langle k \mid x \rangle = \langle x \mid k \rangle^*$.

The 'transformation function' $\langle x \mid k \rangle$ to be found also is the wavefunction in the Schrödinger representation of the improper vector $\mid k \rangle$, i.e. of the improper simultaneous eigenvector of $p_1, \cdots, p_n$ belonging to the eigenvalues $k_1, \cdots, k_n$. It is true that $\mid k \rangle$ (i.e. $\mid k_1, \cdots, k_n \rangle$) does not represent a physical state, but we shall see in the next section that there exist physical states that can be considered approximate eigenvectors of $p_1, \cdots, p_n$ and possess wavefunctions that approximate $\langle x \mid k \rangle$. We then have two good reasons to calculate $\langle x_1, \cdots, x_n \mid k_1, \cdots, k_n \rangle$.

One has

$$p_i \mid k_1, \cdots, k_n \rangle = k_i \mid k_1, \cdots, k_n \rangle , \qquad i = 1, \cdots, n$$

whence, multiplying on the left by $\langle x_1, \cdots, x_n \mid$,

$$\langle x_1, \cdots, x_n \mid p_i \mid k_1, \cdots, k_n \rangle = k_i \langle x_1, \cdots, x_n \mid k_1, \cdots, k_n \rangle , \quad i = 1, \cdots, n$$

But, thanks to (6.25),

$$\langle x_1, \cdots, x_n \mid p_i \mid k_1, \cdots, k_n \rangle = -\mathrm{i}\hbar \frac{\partial}{\partial x_i} \langle x_1, \cdots, x_n \mid k_1, \cdots, k_n \rangle$$

therefore

$$-\mathrm{i}\hbar \frac{\partial}{\partial x_i} \langle x_1, \cdots, x_n \mid k_1, \cdots, k_n \rangle = k_i \langle x_1, \cdots, x_n \mid k_1, \cdots, k_n \rangle \ . \quad (6.43)$$

The transformation function is therefore determined by the differential equations (6.43). The first of them

$$-\mathrm{i}\hbar \frac{\partial}{\partial x_1} \langle x_1 \cdots x_n \mid k_1 \cdots k_n \rangle = k_1 \langle x_1 \cdots x_n \mid k_1 \cdots k_n \rangle$$

has the solution

$$\langle x_1, \cdots, x_n \mid k_1, \cdots, k_n \rangle = f(x_2, \cdots, x_n)\, e^{i\,k_1\,x_1/\hbar}$$

where $f(x_2, \cdots, x_n)$ is the arbitrary constant (with respect to x_1) of integration, so it is an arbitrary function of $x_2, \cdots, x_n$. The second equation (6.43) becomes

$$-i\hbar\frac{\partial}{\partial x_2}\, f(x_2, \cdots, x_n) = k_2\, f(x_2, \cdots, x_n)$$

whose solution is

$$f(x_2, \cdots, x_n) = g(x_3, \cdots, x_n)\, e^{i\,k_2\,x_2/\hbar}$$

so that, at the end of this procedure one has

$$\langle x_1, \cdots, x_n \mid k_1, \cdots, k_n \rangle = c\, e^{i\,(k_1\,x_1 + k_2\,x_2 + \cdots + k_n\,x_n)/\hbar} \tag{6.44}$$

where c is an arbitrary constant, i.e. it does not depend on any of the x. This constant can be determined by the normalization condition

$$\langle k_1, \cdots, k_n \mid k_1', \cdots, k_n' \rangle = \delta(k_1 - k_1') \cdots \delta(k_n - k_n')$$

i.e.

$$\int \langle k_1 \cdots k_n \mid x_1 \cdots x_n \rangle \langle x_1 \cdots x_n \mid k_1' \cdots k_n' \rangle \, dx_1 \cdots dx_n$$
$$= \delta(k_1 - k_1') \cdots \delta(k_n - k_n')$$

whence

$$|c|^2 \int e^{-i\,[(k_1-k_1')x_1 + \cdots + (k_n-k_n')x_n]/\hbar}\, dx_1 \cdots dx_n = \delta(k_1 - k_1') \cdots \delta(k_n - k_n')\,.$$

The integral in the left hand side is the product of n integrals of the type $\int e^{-i(k-k')x/\hbar}\, dx$ and from the theory of Fourier transform one knows that its value is $2\pi\hbar\,\delta(k - k')$. So it must be that

$$|c|^2\, (2\pi\hbar)^n = 1\,.$$

As expected, c is determined only up to a phase factor (because the vectors $|k\rangle$ themselves are determined up to a common phase factor, and likewise for the $|x\rangle$); if we arbitrarily chose a real positive c, one has $c = 1/(2\pi\hbar)^{n/2}$ and, as a consequence,

$$\langle x_1, \cdots, x_n \mid k_1, \cdots, k_n \rangle = \frac{1}{(2\pi\hbar)^{n/2}}\, e^{i\,(k_1\,x_1 + k_2\,x_2 + \cdots + k_n\,x_n)/\hbar} \tag{6.45}$$

so that in the end

$$\varphi_A(k) = \frac{1}{(2\pi\hbar)^{n/2}} \int e^{-ikx/\hbar}\,\psi_A(x)\,dx$$
$$\psi_A(x) = \frac{1}{(2\pi\hbar)^{n/2}} \int e^{ikx/\hbar}\,\varphi_A(k)\,dk \tag{6.46}$$

We thus have two results.

1. In the Schrödinger representation the eigenfunctions of momentum are of the type $e^{ikx/\hbar}$. So the eigenfunctions of the momentum of a particle (three degrees of freedom) are $e^{i\vec{k}\cdot\vec{r}/\hbar}$, i.e. they are plane waves with a wavelength (spatial periodicity) $\lambda = 2\pi\hbar/|\vec{k}| = h/|\vec{k}|$, which precisely is the de Broglie wavelength for a particle of momentum $\vec{k}$.

   Eventually one understands that de Broglie waves associated with a particle are the wavefunctions of the particle itself. They have a quite different nature from the waves known in classical physics, as e.g. either electromagnetic or elastic waves. The latter have a direct physical meaning and can be measured in any point, while wavefunctions are only a way to represent the states of the system (and, just to make it clear, the momentum representation is as good a way as the Schrödinger representation to represent the states of the system), and in any point only $|\psi(x)|^2$ has got a direct physical meaning (even if, as we know, $|\psi(x)|^2$ does not exhaust all the information contained in the wavefunction $\psi(x)$).

2. The Schrödinger and momentum representation wavefunctions corresponding to the same state are (up to the presence of $\hbar$ in $e^{ikx/\hbar}$) the Fourier transform of each other: if we denote by $\widetilde{\psi}$ the Fourier transform of ψ, one has that, up to the factor $\hbar^{n/2}$,

$$\varphi(k) = \widetilde{\psi}(k/\hbar) . \tag{6.47}$$

Note. From now on, in order to comply with current use notation, we will denote by $\varphi(p)$ the wavefunctions in momentum representation: we shall use the same symbol p both for the momentum operator and for the argument of φ. As a consequence, we will denote by p (possibly p', p'', $\cdots$) also the eigenvalues of the momentum operator. The letter k will be reserved to indicate the quantity $p/\hbar$, i.e. $2\pi/\lambda$; so $\vec{k} = \vec{p}/\hbar$ is the *wave vector*.

So, for example, for the eigenfunction of the momentum $\langle x \mid p \rangle$ one has

$$\langle x \mid p \rangle = \frac{e^{ipx/\hbar}}{\sqrt{2\pi\hbar}} = \frac{e^{ikx}}{\sqrt{2\pi\hbar}}, \qquad \langle p \mid p' \rangle = \delta(p-p') = \frac{1}{\hbar}\delta(k-k')$$

(use has been made of the relation $\delta(ax) = |a|^{-1}\delta(x)$) and the first of (6.46) can be written as

$$\varphi(p) = \frac{1}{(2\pi\hbar)^{n/2}} \int e^{-ipx/\hbar}\,\psi(x)\,dx = \frac{1}{(2\pi\hbar)^{n/2}} \int e^{-ikx}\,\psi(x)\,dx \tag{6.48}$$

whereas (6.47) becomes

$$\varphi(p) = \widetilde{\psi}(p/\hbar) = \widetilde{\psi}(k) \; . \tag{6.49}$$

The relationship we have found between Schrödinger and momentum representations, as expressed e.g. by (6.49), gives us the possibility of looking at the uncertainty relations $\Delta p \, \Delta q \geq \frac{1}{2} \hbar$ from a new point of view. It is a consequence of a theorem on Fourier transform, according to which, if a function $\psi(x)$ has a width Δx defined by

$$(\Delta x)^2 = \frac{\int (x^2 - \overline{x}^2) \, |\psi(x)|^2 \, \mathrm{d}x}{\int |\psi(x)|^2 \, \mathrm{d}x}$$

then its Fourier transform $\widetilde{\psi}(k)$ has a width Δk such that

$$\Delta x \, \Delta k \geq \frac{1}{2} \; . \tag{6.50}$$

This is a general theorem that finds applications not only in quantum mechanics, but in all the cases in which one deals with wave-like phenomena: optics, acoustics etc.: the physical meaning of the uncertainty relations that ensue from them depends on the physical meaning of the variables x and k, i.e. on the physical meaning of the Fourier transform: in quantum mechanics $k = p/\hbar$ and therefore from (6.50) the uncertainty relation between position and momentum of a particle follows; in optics $k = 2\pi/\lambda = 2\pi \nu/c$ so that (6.50) entails

$$\Delta x \, \Delta \nu \geq \frac{c}{4\pi} \tag{6.51}$$

that relates the spatial length Δx of an electromagnetic wave with its spectral width $\Delta \nu$: (6.51) then states that the shorter (in space) the wave packet, the less its monochromaticity or, with the current terminology, the larger is the frequency band, and viceversa.

From (6.51) one can also deduce that $\Delta t \, \Delta \nu \geq 1/(4\pi)$ where $\Delta t = \Delta x/c$ is the 'time duration' of the wave packet (or train): the analogue for a particle in quantum mechanics, that is obtained by setting $\nu = E/h$ where E is the energy, namely $\Delta t \, \Delta E \geq \frac{1}{2} \hbar$, even if often cited, is neither obvious (t is not an observable), nor possesses an obvious interpretation (what is the meaning of Δt?).

6.9 The q and p as Observables

We have seen that the q and p, as observables, are 'pathological' in the sense that they possess neither eigenvectors nor eigenvalues. They possess instead a continuous set of improper eigenvalues and a basis of improper eigenvectors.

This type of pathology shows up for many observables $f(q, p)$ (think, e.g., of the hydrogen atom, for which the energy has discrete eigenvalues, but also continuous eigenvalues that correspond to the electron-proton scattering states), so that it is worth examining this problem in some depth in order to understand in which sense these can be considered observables as well. For the sake of simplicity we shall discuss only the case of one single q (one degree of freedom), but what will be said regarding this particular case can be generalized to all other observables (as e.g. p or the Hamiltonian of the hydrogen atom) that share the same type of pathologies.

In Sect. 6.6 we have already introduced the operator $E_{\Delta,x'}(q)$ whose eigenvalues are 0 and 1; as a consequence, the eigenvalues of the operator $x'\,E_{\Delta,x'}(q)$ obviously are 0 and x'. Let us now partition the x-axis into small intervals Δ_n, e.g. all equal to one another and of length Δ, centered around the points x_n (n both a positive and a negative integer) and let us consider the operator q_Δ that, in the Schrödinger representation, is represented by the function (see Fig. 6.3)

$$q_\Delta(x) = \sum_{n=-\infty}^{n=-\infty} x_n\,E_{\Delta,x_n}(x)\ .$$

It is easy to realize (by the same reasoning as in Sect. 6.6 for the operator $E_{\Delta,x'}(q)$) that the eigenvalues of q_Δ are the x_n ($-\infty < x_n < +\infty$), and that the eigenvectors of q_Δ form a complete set. Therefore q_Δ is an observable.

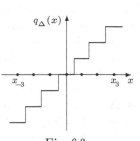

Fig. 6.3

The physical meaning of this observable can be deduced by the discussion we have made in Sect. 6.6 about the physical meaning of $E_{\Delta,x'}(q)$: q_Δ is the observable whose value is x_n when the particle is inside the interval Δ_n around x_n: namely it is an observable that measures the position of the particle with an (experimental) uncertainty $\pm\Delta/2$: think of a measuring rod by which the tenth of millimeter can be estimated; by means of it, the position of the particle can be measured with an uncertainty ±0.05 mm.

In less macroscopic terms we could say that q_Δ is the observable that corresponds to the Heisenberg microscope able to measure the position with an uncertainty $\Delta x = \Delta$. From the physical meaning of q_Δ and from the comparison of Fig. 6.3 and Fig. 6.4 where the Schrödinger representation of q is drawn, one sees that the smaller Δ, the better q_Δ approximates q. This statement can be made precise by noting that $\|q_\Delta - q\| = \Delta$, i.e. that

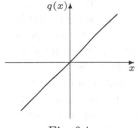

Fig. 6.4

$$\lim_{\Delta\to 0} (q_\Delta - q) = 0$$

in which the "norm convergence" (also called "uniform convergence": it is a very strong one) is to be understood.

So the q_Δ are observables that approximate q as much as we wish, provided Δ is chosen sufficiently small. For this reason, in all the expressions that concern the *observables* (in the strict meaning of the word) q_Δ we can replace q_Δ with q, up to an error of the order of Δ. For example:

$$\overline{q_\Delta} = \langle A \mid q_\Delta \mid A \rangle \simeq \overline{q} = \langle A \mid q \mid A \rangle\,, \qquad \Delta q_\Delta \simeq \Delta q$$

if Δ is smaller then the desired precision.

Having fixed the degree of approximation by which we wish to measure the position of a particle, we will say that a vector $\mid x'_\Delta \rangle$ *approximately* is an eigenvector of q belonging to the eigenvalue x' if in the state represented by $\mid x'_\Delta \rangle$ the conditions $\Delta q \leq \Delta$ and $\overline{q} \simeq x'$ are fulfilled (the $\simeq$ sign means 'up to an error of the order of Δ'). The degree of approximation is measured by the root mean square Δq: indeed, the smaller Δq, the closer the situation one would have if q had eigenvectors, as for them (if they existed!) $\Delta q = 0$.

The eigenvectors of q_Δ (but they are not the only ones) are approximate eigenvectors of q, according to the definition given above: indeed the eigenfunctions of q_Δ are functions whose support is (i.e. they are nonvanishing in) one of the intervals Δ_n, so for them $\Delta q \leq \Delta$.

Note that $\langle x \mid x' \rangle = \delta(x - x')$ is the wavefunction of the improper eigenvector $\mid x' \rangle$ of q: so, in some sense, the wavefunction of $\mid x'_\Delta \rangle$ approximates that of the improper vector $\mid x' \rangle$ – in some sense: indeed, while the functions that approximate Dirac's δ function have an L_2 norm that diverges like Δ^{-1} as $\Delta \to 0$, the L_2 norms of the wavefunctions of the $\mid x'_\Delta \rangle$ stay finite (e.g. equal to 1) whatever the value of Δ.

In conclusion: even if q does not possess eigenvectors and eigenvalues (in the strict sense), we have succeeded in giving both a physical and a mathematical precise meaning to the concepts of approximate eigenvalues and eigenvectors for the operator q. From this point of view q, its improper eigenvectors and eigenvalues are now entitled to have 'citizenship' within the theory, inasmuch as they can be considered as limits of physically meaningful entities: q as limit of the q_Δ, $\mid x \rangle$ as 'limit' (as specified above) of the vectors $\mid x_\Delta \rangle$ that correspond to physically feasible states.

As we have said in the beginning of this discussion, the above conclusions hold also for all the observables that have a continuous spectrum of eigenvalues (in addition to, possibly, a discrete spectrum) and therefore, in particular, for the observable p. Let us now examine which are for this observable the wavefunctions in the Schrödinger representation corresponding to its approximate eigenstates. The improper eigenfunctions are $e^{ipx/\hbar}$: in order to obtain an approximate eigenfunction it will suffice to take a function that: equals $e^{ipx/\hbar}$ for $|x| \leq L$, tends to 0 in a continuous and differentiable way in the intervals $L \leq |x| \leq L'$, and equals 0 for $|x| \geq L'$. This function belongs to L_2 and has $\Delta p \simeq \hbar/(2L)$ so that the higher L, the better it can be considered an eigenfunction of p. If in addition we want to normalize it, we shall have to divide it by a number of the order of $\sqrt{2L}$ (the exact value of the normalization factor depends on the behaviour of the functions in the intervals $L \leq |x| \leq L'$): note that, as L is increased, this function tends to 0 pointwise, so there exists no *exact* eigenfunction of p. In alternative, one can multiply

$e^{i\,p\,x/\hbar}$ by a very wide Gaussian: $e^{i\,p\,x/\hbar}\,e^{-x^2/4L^2}$ and all the considerations made above apply to this function as well.

The previous discussion has emphasized that also the improper eigenvalues of an observable are meaningful from the physical point of view: it is therefore clear that, given an observable, the determination of the improper (or continuous) eigenvalues is just as important as the determination of the proper (or discrete) eigenvalues – the whole spectrum must be found.

Now, in principle, we know how to proceed for the discrete spectrum: one has to solve an eigenvalue equation in $\mathcal{H}$ (or in L_2 if we are in the Schrödinger representation); but, if we want to determine the improper eigenvalues, certainly not in $\mathcal{H}$ (or L_2) the eigenvalue equation must be solved. There arises the problem in which space the solutions of the eigenvalue equation must be searched for. For example, if we are in the Schrödinger representation, the problem is that of knowing in which class of functions one has to look for the improper eigenfunctions of the observable we are interested in. The problem just enunciated is one whose solution is not easy and also is a very technical one, so we shall not discuss it in general: we will face it in a particular case in the next chapter, and from this we will draw some general indications quite sufficient for our purposes.

Chapter 7

Schrödinger Equation for One-Dimensional Systems

7.1 The Hamiltonian for the Free Particle

In this section we will be concerned with the relatively simple problem of determining the eigenvalues of the Hamiltonian of the free particle. We will discuss the one-dimensional case. Our system consists therefore of a particle constrained to move on a straight line. The Hamiltonian is

$$H = \frac{p^2}{2m} \tag{7.1}$$

i.e. only the kinetic energy. Clearly H and p commute with each other: $[H, p] = 0$; therefore they have a complete set of (improper) simultaneous eigenvectors. Furthermore, as p is nondegenerate (i.e. to each eigenvalue p' there corresponds, up to a factor, only one improper eigenvector $|p'\rangle$), each eigenvector of p is an eigenvector of H. Therefore the vectors $|p'\rangle$ are a basis of eigenvectors for H and the corresponding eigenvalues E are obtained by

$$E\,|p'\rangle = H\,|p'\rangle = \frac{p^2}{2m}\,|p'\rangle = \frac{p'^2}{2m}\,|p'\rangle$$

whence:

$$E = \frac{p'^2}{2m}\,.$$

As $-\infty < p' < +\infty$, the eigenvalues of H are continuous and are all the real numbers $E \geq 0$. In the present case H only has improper eigenvalues.

What is the degree of degeneracy of the eigenvalues E of H? Given $E > 0$, there exist two and only two eigenvectors in the basis that correspond to the same eigenvalue of the energy: they are the vectors $|+p'\rangle$ and $|-p'\rangle$ with $p' = \sqrt{2m\,E}$. So all the eigenvalues $E > 0$ of H are twice degenerate (for the $E = 0$ eigenvalue we prefer to say nothing); this means that all and only the vectors of the type:

$$\alpha\,|+p'\rangle + \beta\,|-p'\rangle\,, \qquad \alpha,\,\beta \in \mathbb{C} \tag{7.2}$$

© Springer International Publishing Switzerland 2016
L.E. Picasso, *Lectures in Quantum Mechanics*, UNITEXT for Physics,
DOI 10.1007/978-3-319-22632-3_7

are the eigenvectors of H belonging to E. Note that, H being degenerate, the vectors $|p'\rangle$ are complete, but they are not all the eigenvectors of H: the vectors of the type (7.2) are *all* the eigenvectors of H.

The physical meaning of the twofold degeneracy of the eigenvalues of H is evident: as the energy is only kinetic, provided the velocity keeps its magnitude, the energy is unchanged regardless of whether the particle moves either from the right to the left or from the left to the right.

Observe however the difference between the classical and the quantum situation: classically, for given energy E, the only possible states of motion are those in which the momentum (or the velocity) is either positive or negative; in quantum mechanics, owing to the superposition principle, there exist states of motion (those described by (7.2) with both α and β nonvanishing) in which the direction of the motion of the particle is not defined: it travels both from the left to the right *and* viceversa.

How can such a situation be achieved? As we have to deal with improper vectors, we may only look for physical situations that are described by vectors that approximate those given in (7.2) in a region (of the x-axis) as large as we want, but finite.

Think e.g. that a very far mirror (for example either a crystal or an electric field) is available and that particles are sent on it in a state of quasi-defined momentum: $\Delta p \ll \hbar/L$, where L is the distance between the observation region and the mirror: the particles are described by a wave-

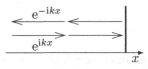

Fig. 7.1

function that, in the Schrödinger representation, is $\mathrm{e}^{i\,k\,x}$ in a region whose length is $\gg L$. This "wave packet" (or "wave train") is reflected by the mirror and returns at the observation region in which, the packet being very long, for a certain time interval both the incident $\mathrm{e}^{i\,k\,x}$ and the reflected $\mathrm{e}^{-i\,k\,x}$ waves are present (Fig. 7.1). In this example, however, $|\alpha| = |\beta|$ since the mirror is totally reflecting.

Once the problem of finding the (improper) eigenvalues, degeneracies and eigenvectors of the Hamiltonian (7.1) has been completely solved, let us re-examine the same problem in the Schrödinger representation: the reason for this repetition is the one put forward at the end of the previous chapter, namely that, taking advantage of already knowing the solution of the problem, we want to understand in which set of functions the improper eigenfunctions of the Hamiltonian are to be searched for. For our system described by the Hamiltonian (7.1), the Schrödinger equation writes:

$$-\frac{\hbar^2}{2m}\,\psi''(x) = E\,\psi(x)\,, \qquad \psi''(x) \equiv \frac{\mathrm{d}^2\psi}{\mathrm{d}x^2} \qquad (7.3)$$

in which the unknowns are both E and $\psi(x)$. Since (7.3) is of the second order, the general solution depends on two arbitrary constants: for any value of E (either real or complex) it has the form

$$\psi(x) = \alpha\,\mathrm{e}^{i\,\sqrt{2m\,E}\,x/\hbar} + \beta\,\mathrm{e}^{-i\,\sqrt{2m\,E}\,x/\hbar}\,, \qquad \alpha,\,\beta \in \mathbb{C} \qquad (7.4)$$

in which $\sqrt{2m\,E}$ is a complex number. It clearly appears that (7.3) possesses solutions for any, even complex, value of E, while in the previous discussion we have found that E must be real and positive. What distinguishes the solutions of (7.4) with $E \geq 0$ from all the other ones? If $E \geq 0$, the exponents in (7.4) are pure imaginary and in this case the solutions (7.4) stay bounded for $x \to \pm\infty$. If either $E \leq 0$, or if E is complex, the exponents in (7.4) contain a real part, so that each one of the two exponentials diverges, either for $x \to +\infty$ or for $x \to -\infty$.

The rule might be the following:

If we want to determine improper eigenvalues and eigenvectors of some observable, we must keep only those solutions of the eigenvalue equations that, in the Schrödinger representation, stay bounded for $x \to \pm\infty$ and reject those that diverge.

Even if this is not the general rule, at least in the cases we are interested in where the observable is the Hamiltonian, this rule works.

Note eventually that (7.4) with $E \geq 0$ is nothing but the Schrödinger representation of (7.2) and that the twofold degeneracy of H comes by as a consequence of the fact that (7.3) is a differential equation of the second order.

7.2 Degeneracy Theorem. Spatial Inversions

In many cases it is not possible to find the eigenvalues of an observable. Nonetheless, there exist general arguments that allow one to obtain information on the degree of degeneracy of the eigenvalues of the observable one is studying. We will have many a time the occasion to appreciate the importance of this information, particularly when we will apply the developed theory to the study of atomic structure and atomic spectroscopy.

The starting point of the general arguments we have hinted at is the following

Theorem: *if two observables η and ζ both commute with some observable ξ, but do not commute with each other:*

$$[\xi, \eta] = 0, \qquad [\xi, \zeta] = 0, \qquad [\eta, \zeta] \neq 0$$

then ξ is degenerate.

The proof is straightforward: if ξ were nondegenerate, *any* of its eigenvectors should be an eigenvector also of η and ζ, therefore η and ζ would have a complete set of simultaneous eigenvectors and, as a consequence, would commute with each other – in contradiction with the assumption.

The above theorem alone does not say which is the degree of degeneracy of the eigenvalues of ξ: this result can be achieved in any single case by means of the detailed knowledge of how many and which observables do commute with ξ and which are the commutation relations they have with one another. We will have the occasion to discuss also this aspect of the problem.

As an application of the degeneracy theorem – and also to understand the mechanism by which ξ must be degenerate – we will re-examine the problem discussed in the previous section, in which we have seen that all the eigenvalues $E > 0$ of the Hamiltonian (7.1) are twice degenerate: we try to find this degeneracy as a consequence of the just shown theorem. As H commutes with p, it is necessary to find another observable that commutes with H and that does not commute with p. The **space-inversion** operator I, that we will now define, just is what we need: it will suffice to put $\xi = H$, $\eta = p$ and $\zeta = I$ in the degeneracy theorem.

Space-Inversion operator I

The space-inversion operator can be defined in two equivalent ways: first, the transformation $q \to -q$, $p \to -p$ being canonical, there exist a unitary operator I such that $I\, q\, I^{-1} = -q$, $I\, p\, I^{-1} = -p$; as an alternative the operator I can be defined by its action on the states of the system; we follow this second possibility.

In the Schrödinger representation the space-inversion operator I is defined on any wavefunction $\psi_A(x)$ in the following way:

$$I\, \psi_A(x) \equiv \psi_A(-x) . \tag{7.5}$$

Properties of I:

1. $I^2 = \mathbb{1}$, namely $I^2\, \psi_A(x) = \psi_A(x)$. It is evident.
2. $I = I^\dagger$, namely $\langle A \mid I \mid B \rangle = \langle B \mid I \mid A \rangle^*$. Indeed

$$
\begin{aligned}
\langle A \mid I \mid B \rangle &= \int_{-\infty}^{+\infty} \psi_A^*(x)\, I\, \psi_B(x)\, \mathrm{d}x \qquad = \int_{-\infty}^{+\infty} \psi_A^*(x)\, \psi_B(-x)\, \mathrm{d}x \\
&= \left(\int_{-\infty}^{+\infty} \psi_B^*(-x)\, \psi_A(x)\, \mathrm{d}x \right)^* = \left(\int_{+\infty}^{-\infty} \psi_B^*(x)\, \psi_A(-x)\, (-\mathrm{d}x) \right)^* \\
&= \left(\int_{-\infty}^{+\infty} \psi_B^*(x)\, I\, \psi_A(x)\, \mathrm{d}x \right)^* = \langle B \mid I \mid A \rangle^*
\end{aligned}
$$

(in the second line of the above equation the change of variable $x \to -x$ has been performed).

3. It follows from 1 and 2 that not only I is self-adjoint, but it also is unitary: $I^\dagger = I^{-1}$.

4. The eigenvalues w of I are $w = +1$ and $w = -1$. Indeed, let $\psi_w(x)$ an eigenfunction of I belonging to the eigenvalue w:

$$I\, \psi_w(x) = w\, \psi_w(x) .$$

By applying I to both sides of the above equation and recalling that $I^2 = \mathbb{1}$ one has

$$\psi_w(x) = w\, I\, \psi_w(x) = w^2\, \psi_w(x)$$

whence $w^2 = 1$, i.e. $w = \pm 1$.

5. The eigenfunctions of I are therefore those functions such that:

$$\psi(-x) = \pm \psi(x) .$$

Those belonging to the eigenvalue $+1$ are the functions *even* with respect to the spatial inversion $x \to -x$: $\psi(x) = \psi(-x)$; those belonging to the eigenvalue -1 are the *odd* functions: $\psi(x) = -\psi(-x)$. The eigenvalue w indicates the parity of the corresponding eigenfunction.

6. The eigenfunctions of I form a complete set. This follows from the fact that any $\psi(x)$ can be expressed as a linear combination of an even and an odd function, thanks to the identity:

$$\psi(x) = \tfrac{1}{2}\left(\psi(x) + \psi(-x)\right) + \tfrac{1}{2}\left(\psi(x) - \psi(-x)\right)$$

in which clearly $\psi(x) + \psi(-x)$ is even and $\psi(x) - \psi(-x)$ is odd.

7. The operator I *anticommutes* with q and p:

$$I\,q\,I^{-1} = -q, \qquad I\,p\,I^{-1} = -p \tag{7.6}$$

or, also, multiplying by I on the right:

$$I\,q = -q\,I, \qquad I\,p = -p\,I. \tag{7.7}$$

Let us demonstrate the first of (7.7). By applying $I\,q$ to a generic vector $|A\rangle$ and going to Schrödinger representation one has:

$$I\,q\,|A\rangle \to I\Big(x\,\psi_A(x)\Big) = -x\,\psi_A(-x) \to -q\,I\,|A\rangle.$$

As for the second one has $(\psi' \equiv d\,\psi/dx)$

$$I\,p\,|A\rangle \to I\Big(-i\,\hbar\,\psi'(x)\Big) = -i\,\hbar\,\psi'_A(-x)$$

$$p\,I\,|A\rangle \to -i\,\hbar\frac{d}{dx}\,\psi_A(-x) = +i\,\hbar\,\psi'_A(-x).$$

Equations (7.6) tell that I is the operator that changes the sign of both the coordinate q and the momentum p, whence the name of space-inversion.

Properties 1–7 hold unchanged for the space-inversion operator I defined, in the case of many degrees of freedom, by

$$I\,\psi(x_1, \cdots, x_n) = \psi(-x_1, \cdots, -x_n). \tag{7.8}$$

In particular:

5′. The eigenfunctions of I are those functions such that:

$$\psi(x_1, \cdots, x_n) = \pm\psi(-x_1, \cdots, -x_n);$$

7′. The action of I on coordinates and momenta obviously is

$$I\,q_i\,I^{-1} = -q_i, \qquad I\,p_i\,I^{-1} = -p_i, \qquad i = 1, \cdots, n. \tag{7.9}$$

It goes without saying that, for example in the case of a particle in three dimensions, it is possible to define the operator that inverts with respect to each of the coordinate planes: $x \to -x$, or $y \to -y$, or $z \to -z$.

The operator I defined by (7.5) commutes with the Hamiltonian (7.1):

$$I \frac{p^2}{2m} I^{-1} = \frac{1}{2m} I p I^{-1} I p I^{-1} = \frac{1}{2m} (-p)^2 = \frac{p^2}{2m} \, .$$

In the same way one sees that I commutes with any observable $f(q,p)$ such that $f(q,p) = f(-q,-p)$. Indeed – see (6.12) – one has that

$$I f(q,p) I^{-1} = f(-q,-p)$$

holds even in the case of many degrees of freedom. So, in particular, I commutes with any Hamiltonian of the type:

$$H = \frac{p^2}{2m} + V(q)$$

provided $V(q) = V(-q)$, i.e. if $V(q)$ is an even function of the q's (e.g. the harmonic oscillator, both isotropic and anisotropic, in whatever number of dimensions).

Going back to the problem of the degeneracy of the Hamiltonian H given by (7.1), one finally has that both I and p are observables that commute with H, but – due to (7.6) – do not commute with each other. This entails, thanks to the degeneracy theorem, that H must be degenerate.

Let us now try to understand the mechanism by which H must be degenerate and, at the same time, let us try to obtain more detailed information about the degree of degeneracy of H.

Let us consider a simultaneous eigenvector of H and p: $|\,p'\,\rangle$, belonging to the eigenvalue E of H; one has

$$I\,|\,p'\,\rangle = |\,-p'\,\rangle \tag{7.10}$$

that is easily shown in the Schrödinger representation: $I \, \mathrm{e}^{\mathrm{i}\,p'x} = \mathrm{e}^{-\mathrm{i}\,p'x}$.

Owing to the lemma of p. 87 (Sect. 4.10), the vector $I\,|\,p'\,\rangle$, i.e. $|\,-p'\,\rangle$, still is an eigenvector belonging to the same eigenvalue E of H, as the vector $|\,p'\,\rangle$ one starts with; but, if $p' \neq 0$, $|\,p'\,\rangle$ and $|\,-p'\,\rangle$ are independent vectors so that the eigenvalue E is degenerate at least twice. If we now apply I to the vector $|\,-p'\,\rangle$, nothing new is obtained (we mean, no eigenvector independent of the previous ones is obtained) because $I^2 = \mathbb{1}$. So the degeneracy theorem guarantees that the degree of degeneracy of each eigenvalue $E > 0$ of H is at least two.

The same strategy can be followed by starting with a simultaneous eigenvector of H and I and applying the 'third' operator p to the latter. Let us check that one arrives at the same result. Which are in the Schrödinger representation the simultaneous eigenfunctions of H and I? As (7.4) display *all* the eigenfunctions of H, the simultaneous eigenfunctions of H and I must be searched for among them. It is easily seen (possibly by means of the symmetrization and antisymmetrization procedure used in point 6 above) that the simultaneous eigenfunctions of H and I are

$$\cos(p'x/\hbar)\,, \quad w = +1\,; \qquad \sin(p'x/\hbar)\,, \quad w = -1\,.$$

Let us consider one of them, e.g. the first and let us apply p, i.e. $-\mathrm{i}\hbar\,\mathrm{d}/\mathrm{d}x$, to it. Up to constant factors, one finds the second and, always as a consequence of the lemma of p. 87, (Sect. 4.10), these functions both correspond to the same eigenvalue of H. In this way the twofold degeneracy is re-obtained.

7.3 General Features of the Solutions of Schrödinger Equation in the One-Dimensional Case

The study of one-dimensional problems is not only academic in character, because – as we shall see – in many cases the solution of a problem in many dimensions is brought back to that of a one-dimensional problem (for example: a particle in a central field and, more in general, any case when the Schrödinger can be solved by 'separation of variables'). It is for this reason that we shall dedicate an exhaustive – although mostly qualitative – discussion to the general features of the eigenvalues and of the eigenfunctions of the Hamiltonian of a particle constrained to move along a straight line, subject to a potential $V(q)$ (not particularly pathological!).

The Hamiltonian is

$$H = \frac{p^2}{2m} + V(q)$$

and the corresponding eigenvalue equation in the Schrödinger representation (Schrödinger's equation) is

$$-\frac{\hbar^2}{2m}\,\psi''(x) + V(x)\,\psi(x) = E\,\psi(x) \tag{7.11}$$

or also

$$\psi''(x) = \frac{2m}{\hbar^2}\left(V(x) - E\right)\psi(x)\,. \tag{7.12}$$

As we have already emphasized, (7.11) is a differential homogeneous linear equation of the second order: as such, it possesses solutions (indeed, two independent solutions) whatever the value of E, either real or complex: we will be interested in those solutions (normally denoted by either $\psi_n(x)$ or $\psi_E(x)$) that either

i) belong to L_2 and, in particular, tend to 0 as $x \to \pm\infty$;

these are the *proper* eigenfunctions of H and correspond to the bound states of the system (indeed the probability to find the particle at $|x| > L$ tends to 0 for $L \to \infty$); the corresponding eigenvalues are the proper eigenvalues, i.e. the discrete energy levels of the system;

or, if the solutions are not in L_2, they

ii) stay bounded for $x \to \pm\infty$;

these are the *improper* eigenfunctions of H and the corresponding eigenvalues are improper eigenvalues that, we shall see, form a *continuous* spectrum.

First of all, let us note that we can limit ourselves to consider only the real solutions of (7.11): in effect, whatever the value of E in (7.11) provided it be real (and there is no reason to consider complex values of E, as the eigenvalues of H must be real), if $\psi(x)$ is a solution also $\psi^*(x)$ is a solution: this is so because the (7.11) is an equation with real coefficients; therefore we can separately study $\Re e\, \psi(x)$ and $\Im m\, \psi(x)$ that also are solutions of (7.11) and are real valued.

Having arbitrarily fixed a value of E in either (7.11) or (7.12) (not necessarily an eigenvalue), the x axis is divided in regions for which:

I. $V(x) - E < 0$ $(E > V)$,   (these will be called *type* I   *regions*);
II. $V(x) - E > 0$ $(E < V)$,   (these will be called *type* II *regions*).

If we think of the corresponding classical problem, the type I regions are those in which a particle endowed with total energy E can move, whereas type II regions are inaccessible to a particle with energy E, because the kinetic energy should be negative. The points that separate the regions of type I and II, i.e. the points where $E = V(x)$, are called **inversion points** as these are the points where the classically moving particle inverts its motion. From (7.12) one has that:

$$\frac{\psi''}{\psi} < 0 \quad \text{in type I regions}, \qquad \frac{\psi''}{\psi} > 0 \quad \text{in type II regions}.$$

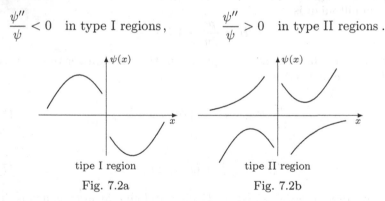

tipe I region   tipe II region

Fig. 7.2a   Fig. 7.2b

Now the value of ψ'' in a point gives the size of the curvature of ψ in that point; in particular, if $\psi'' > 0$, the graph of ψ is concave upwards whereas, if $\psi'' < 0$, the graph of ψ is concave downwards. So in regions of type I (Fig. 7.2a), where $\psi > 0$ the concavity is downwards and where $\psi < 0$ the concavity is upwards. In the type II regions (Fig. 7.2b) the contrary occurs: where $\psi > 0$ the concavity is upwards and where $\psi < 0$ the concavity is downwards. In the inversion points the graph of ψ exhibits an inflection point; inflection points are also all the points where $\psi = 0$: indeed (7.12) shows that wherever $\psi = 0$ also $\psi'' = 0$. The inflection points coinciding with either the inversion points or the points where $\psi = 0$ are easily understood: in all these points the graph of ψ either passes from one type of region to the other type, or crosses the x-axis: in any event the curvature of ψ changes its sign, so the point must be an inflection point.

To summarize: in the regions of type I the behaviour of ψ is oscillating around the value $\psi = 0$, i.e. – as exemplified in Fig. 7.3 – it may cross the x-axis zero or many times, always keeping its concavity towards the x-axis (in Fig. 7.3a the dashed parts of the graph are in regions of type II); the typical behaviours of ψ in regions of type II are exemplified in Fig. 7.4 (the dashed lines are there for future reference).

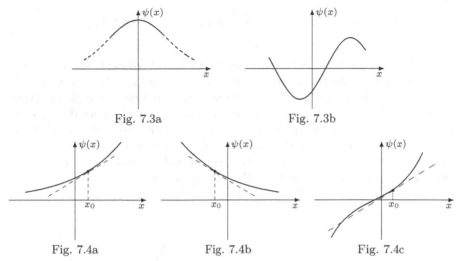

Fig. 7.3a   Fig. 7.3b

Fig. 7.4a   Fig. 7.4b   Fig. 7.4c

7.4 Solutions of Schrödinger Equation: Discrete Eigenvalues

We now take advantage of the obtained results in order to determine, for several types of potentials $V(x)$, the general features of the spectrum of H, the form of its eigenfunctions, the relative degeneracy.

Let us first consider the case in which

$$V(x) \to \infty, \qquad x \to \pm\infty$$

and let us see for which values of E do acceptable solutions of the Schrödinger equation occur. We will distinguish two cases:

1. E smaller than the minimum of $V(x)$: $E < V_{\min}$. In the present case all the x-axis is a type II region; it is easy to see that there exist no acceptable solutions: indeed, let us consider a point x_0 such that $\psi(x_0) \neq 0$ and let us also assume that $\psi(x_0) > 0$ $\big($if this is not the case, one can multiply $\psi(x)$ by -1: this is legitimate, for equation (7.11) is homogeneous$\big)$; let us also consider the tangent to the graph in the point x_0 (dashed lines in Fig. 7.4): it is then evident that, since we always are in a region of type II, if $\psi'(x_0) > 0$ (Figs. 7.4a and 7.4c) at the *right* of x_0 the graph of $\psi(x)$ always is above the tangent in x_0 (as $\psi'' > 0$, ψ' increases), therefore $\psi(x) \to \infty$ for $x \to +\infty$; on the contrary, if $\psi'(x_0) < 0$ (Fig. 7.4b), $\psi(x)$ stays above the tangent at

the *left* of x_0, therefore $\psi(x) \to \infty$ for $x \to -\infty$. In neither case $\psi(x)$ is an acceptable solution.

Alternatively, as we already argued in the case of the harmonic oscillator (Sect. 5.1), in any state $|A\rangle$, and in particular in the eigenstates of the Hamiltonian,

$$\overline{H} = \langle A \mid H \mid A \rangle = \overline{\frac{p^2}{2m}} + \overline{V(q)} \geq V_{\min}$$

since

$$\overline{\frac{p^2}{2m}} \geq 0, \quad \text{and} \quad \overline{V(q)} = \int V(x)\,|\psi(x)|^2\,\mathrm{d}x \geq V_{\min} \ .$$

2. $E > V_{\min}$. In this case, whatever the value of E, there exist two points x_1 and x_2 such that the two regions $-\infty < x < x_1$ and $x_2 < x < +\infty$ are of type II. Between x_1 and x_2 there certainly exists at least one region of type I (Fig. 7.5). In the two type II external regions the $\psi(x)$ either diverges or tends to zero (Fig.7.4): in other words it cannot either oscillate or, more in general, stay bounded without going to zero. This means that only discrete, i.e. proper eigenvalues may occur: they correspond to those functions $\psi_n(x)$ that tend to zero both in the type II region at the right ($x \to +\infty$) and in the type II region at the left ($x \to -\infty$).

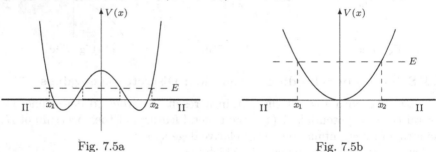

Fig. 7.5a   Fig. 7.5b

It is interesting to try to understand the mechanism by which only for particular values of E – that form a discrete set – there occur acceptable (in the present case L_2) solutions. Let us indeed assume that, given a value of E, we choose a point x_0 in the type II region at the left ($x_0 < x_1$) and fix the value of ψ in x_0 at our will (this is legitimate due, as usual, to the homogeneity of the Schrödinger equation (7.11)) and let, for example, $\psi(x_0) > 0$.

Since the equation is of the second order, it is completely determined if we also give $\psi'(x_0)$. Clearly the behaviour of the solution at the left (i.e. for $x \to -\infty$) depends on the value of $\psi'(x_0)$: if $\psi'(x_0)$ is either negative or positive but not too large, then $\psi(x) \to +\infty$ for $x \to -\infty$ never crossing the x-axis (curves 1 and 2 of Fig. 7.6); if instead $\psi'(x_0)$ is positive and large, the graph of $\psi(x)$ will cross the x-axis and will tend to $-\infty$

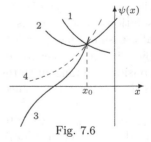

Fig. 7.6

for $x \to -\infty$ (curve 3 of Fig. 7.6). Therefore there will exist an intermediate (positive) value of $\psi'(x_0)$ such that $\psi(x)$ tends to zero for $x \to -\infty$ (curve 4 of Fig. 7.6).

Let us now study the solution $\psi(x)$ determined by the above-mentioned value of $\psi'(x_0)$ and let us examine, in particular, its behaviour for $x \to +\infty$.

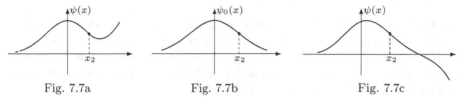

Fig. 7.7a   Fig. 7.7b   Fig. 7.7c

Let us assume, for the sake of simplicity, that the interval $x_1 < x < x_2$ be all a region of type I (as it is the case if $V(x)$ is that of Fig. 7.5b). As a consequence, between x_1 and x_2 $\psi(x)$ has an oscillating behaviour: if we have chosen a value of E slightly higher than $V_{\min}$, $\psi(x)$ never vanishes for x between x_1 and x_2 and one of the three situations plotted in Fig. 7.7 will take place: if the case of Fig. 7.7b occurs, then $\psi(x)$ is acceptable and E is an eigenvalue: $E = E_0$. If the case of Fig. 7.7a occurs, then E is not an eigenvalue and we can say that E is smaller than E_0 (case of Fig. 7.7b): actually, in order to pass from case a to case b, one must increase – in absolute value – the value of $\psi'(x_2)$, i.e. increase the curvature of $\psi(x)$ in the interval $x_1 < x < x_2$. By (7.12) this is achieved by increasing E: note that between x_1 and x_2 $\psi''(x)$ is negative so that increasing E makes $\psi''(x)$ more negative. Likewise: if the case of Fig. 7.7c occurs, then $E > E_0$.

It is clear that, starting from $E = V_{\min}$, and by taking into consideration larger and larger values of E, one shall pass from a situation like that described by Fig. 7.7a to one like that of Fig. 7.7c: between the two there always is a value of E for which $\psi(x)$ is an acceptable solution. Such values of E – the eigenvalues – clearly constitute a discrete set.

From this discussion it emerges that the eigenfunction of H belonging to the lowest eigenvalue (i.e. the lowest energy level) never vanishes for finite x, i.e. it has no nodes.

By increasing the energy, the type I and II regions change (in particular the inversion points x_1 and x_2 respectively move to the left and to the right), but – more important – the curvature of $\psi(x)$ in the region between x_1 and x_2 increases: as a consequence, the graph of $\psi(x)$ may perform more oscillations, i.e. cross once or more than once the x-axis. So the eigenfunctions of H corresponding to the excited levels may have behaviours like, for example, those reported in Fig. 7.8, where one expects that the $\psi(x)$ of Fig. 7.8b corresponds to an eigenvalue higher than that to which the eigenfunction of Fig. 7.8a belongs. As a matter of fact, if the eigenfunction $\psi_0(x)$ of H

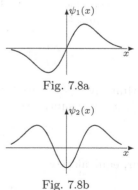

Fig. 7.8a

Fig. 7.8b

belonging to the lowest eigenvalue E_0 never vanishes, certainly the eigenfunctions belonging to the excited levels must possess nodes: this is a consequence of the fact that the eigenfunctions of H corresponding to different eigenvalues are orthogonal:

$$\int \psi_0(x)\,\psi_1(x)\,\mathrm{d}x = 0$$

(we have not written $\psi_0^*(x)$ because we know we may consider real wavefunctions), and since $\psi_0(x)$ has a definite sign, the integral may vanish only if $\psi_1(x)$ does *not* possess a definite sign. The above considerations, that we have illustrated by means of a discussion that is qualitative in character, are confirmed and made precise by the following

Oscillation Theorem: *consider a system with one degree of freedom. Let $E_0 < E_1 < \cdots < E_n \cdots$ be the (proper) eigenvalues of the Hamiltonian in increasing order, and let $\psi_0(x)$, $\psi_1(x)$, $\cdots$, $\psi_n(x)$, $\cdots$ the corresponding eigenfunctions. Then $\psi_0(x)$ has no nodes (i.e. it never vanishes inside its domain of definition), $\psi_1(x)$ has one node, $\cdots$, $\psi_n(x)$ has n nodes.*

(It is implicit in the formulation of the above theorem that the proper eigenvalues of the Hamiltonian are nondegenerate, as we will shortly prove.)

The validity of the above theorem – that we will not demonstrate – is quite general, i.e. it does neither require that $V(x)$ has only one minimum nor – as in the case considered above – that H has only discrete eigenvalues: if H possesses both discrete and continuous eigenvalues, the oscillation theorem only applies to the eigenfunctions that belong to discrete eigenvalues. It should be clear that, in the case we have been considering so far, $\big(V(x) \to +\infty$ for $x \to \pm\infty\big)$ in which only discrete eigenvalues occur, the number of eigenvalues must be infinite, because the space $\mathcal{H}$ of the state vectors has infinite dimension and, as anticipated, the eigenvalues are nondegenerate. Indeed, the problem concerning the degeneracy of the eigenvalues is solved by the following

Nondegeneracy Theorem: *for any one-dimensional system the (proper) eigenvalues of the Hamiltonian are nondegenerate.*

Proof: let $\psi_1(x)$ and $\psi_2(x)$ two eigenfunctions of H belonging to the same eigenvalue E:

$$\psi_1''(x) = \frac{2m}{\hbar^2}\left(V(x) - E\right)\psi_1(x), \qquad \psi_2''(x) = \frac{2m}{\hbar^2}\left(V(x) - E\right)\psi_2(x)\,.$$

Multiplying the first of the above equations by $\psi_2(x)$ and the second by $\psi_1(x)$ and subtracting side by side one has

$$\psi_1''(x)\,\psi_2(x) - \psi_1(x)\,\psi_2''(x) = 0$$

or, equivalently:

$$\frac{\mathrm{d}}{\mathrm{d}x}\left(\psi_1'(x)\,\psi_2(x) - \psi_1(x)\,\psi_2'(x)\right) = 0\,.$$

By integrating with respect to x one arrives at

$$\psi_1'(x)\,\psi_2(x) - \psi_1(x)\,\psi_2'(x) = \text{constant} . \tag{7.13}$$

As by assumption E is a discrete eigenvalue, both ψ_1 and ψ_2 vanish for $x \to \pm\infty$ and, as a consequence, also the constant in the right hand side vanishes (note – this will be useful in the sequel: in order that the constant in the right hand side of (7.13) vanishes, it is sufficient that ψ_1 and ψ_2 vanish either for $x \to +\infty$ or for $x \to -\infty$). In conclusion, from (7.13) one has

$$\frac{\psi_1'}{\psi_1} = \frac{\psi_2'}{\psi_2} \qquad \text{i.e.} \qquad \frac{\mathrm{d}}{\mathrm{d}x}\log\psi_1 = \frac{\mathrm{d}}{\mathrm{d}x}\log\psi_2$$

that, upon integration with respect to x, yields $\psi_1(x) = C\,\psi_2(x)$ or, in other words, ψ_1 and ψ_2 are not independent, i.e. E is nondegenerate.

Let us finally consider the particularly meaningful case in which $V(x)$ is an even function of the coordinate: $V(x) = V(-x)$ (we are now relaxing the hypothesis that $V(x) \to \infty$ for $x \to \pm\infty$). In this case the Hamiltonian H commutes with the space-inversion operator I: $[\,H\,,\,I\,] = 0$. The operators I and H have, as a consequence, a complete set of simultaneous eigenfunctions; but since the (proper) eigenvalues of H are nondegenerate, any eigenfunction of H corresponding to a discrete eigenvalue must also be an eigenfunction of I, i.e. it must have definite parity. Note that even functions either have an even number of nodes or vanish an infinite number of times, whereas odd function have an odd (or infinite) number of zeros. Indeed, if a function with a well defined parity (either $+1$ or -1) vanishes in a point $x \neq 0$, it must vanish also in the point $-x$; in addition, odd functions must vanish also in the origin $\big(f(0) = -f(0) = 0\big)$, while even functions, if they vanish in the origin, the first derivative – being an odd function – must vanish as well, but the solutions of the Schrödinger equation cannot possess zeros of order higher than the first: if in a point both ψ and ψ' vanish then, as these data univocally determine the solution, ψ vanishes everywhere.

By combining the latter result with the oscillation theorem one finds that, if E_0, E_1, $\cdots$, E_n, $\cdots$ are the eigenvalues of H in increasing order, the corresponding eigenfunctions $\psi_0(x)$, $\psi_1(x)$, $\cdots$, $\psi_n(x)$, $\cdots$ alternatively are even and odd and in particular the wavefunction of the ground state is even.

7.5 Solutions of Schrödinger Equation: Continuous Eigenvalues

Let us carry on the discussion on the general properties of the solutions of the Schrödinger equation, by now examining the case of a potential (like e.g. that of Fig. 7.9) that diverges at one side, e.g. for $x \to -\infty$, and tends, for $x \to +\infty$, to a finite value that we can put equal to zero (given that any potential is defined up to a constant) and has the minimum of negative sign:

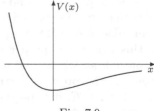

Fig. 7.9

$$\begin{cases} V(x) \to +\infty & x \to -\infty \\ V(x) \to \ \ 0 & x \to +\infty \end{cases} \qquad V_{\min} < 0 \, .$$

Let us distinguish three cases:

1. $E < V_{\min}$: all the x-axis is a type II region and, as discussed in the previous section, there exist no eigenvalues.

2. $V_{\min} < E < 0$: there exist x_1 and x_2 such that the regions $x < x_1$ and $x > x_2$ are type II regions and between x_1 and x_2 there is at least a region of type I. Also this case has been faced in the previous section and we have concluded that only discrete eigenvalues are possible.

Note that, if $V(x) \to 0$ sufficiently fast for $x \to +\infty$, asymptotically, where $|V(x)| \ll |E|$, the *general* solution of (7.11) (with $E < 0$) has the form:

$$\alpha \, e^{\sqrt{2m|E|}\, x/\hbar} + \beta \, e^{-\sqrt{2m|E|}\, x/\hbar} \, , \qquad \alpha, \beta \in \mathbb{C} \, , \qquad x \to +\infty \qquad (7.14)$$

i.e. it is a linear combination of real exponentials, the first of which diverges whereas the second tends to zero. Given a generic value of E between $V_{\min}$ and 0, and having chosen the solution that vanishes for $x \to -\infty$, normally it will happen that for $x \to +\infty$ the wavefunction $\psi(x)$ is of the type (7.14) with both α and β nonvanishing; only if it happens that for $x \to +\infty$ $\psi(x)$ is of the type (7.14) with $\alpha = 0$ (so that $\psi(x) \to 0$), is $\psi(x)$ an acceptable solution and the corresponding E an eigenvalue of H.

The number of discrete eigenvalues depends in this case on the potential: it can be shown that, if $V(x)$ is bounded from below and, for $x \to +\infty$, $V(x) \to 0$ faster than x^{-2}, then there is a *finite* number (possibly vanishing) of discrete eigenvalues.

3. $E > 0$: in this case, for x large and negative, the x axis is a type II region, whereas for x large and positive, it is a type I region. Therefore, given whatever value $E > 0$, it is possible – as discussed in the previous section – to find a solution $\psi(x)$ of the Schrödinger equation that vanishes for $x \to -\infty$; the behaviour of $\psi(x)$ for $x \to +\infty$, i.e. in the type I region, will be oscillatory and $\psi(x)$ stays bounded: in effect in this case the general solution of (7.11) (with $E > 0$) asymptotically has the form:

$$\alpha \, e^{i\sqrt{2m E}\, x/\hbar} + \beta \, e^{-i\sqrt{2m E}\, x/\hbar} \, , \qquad x \to +\infty$$

and is always acceptable. Therefore, contrary to the previously discussed cases, there is no problem to find a 'well behaved' solution also for $x \to +\infty$: no $\psi(x)$ tends to zero, but all of them stay bounded. So any $E > 0$ is an improper eigenvalue of H, i.e. the eigenvalues $E > 0$ constitute a continuous spectrum. In this case the improper eigenvalues are nondegenerate: indeed, as all the acceptable solutions tend to zero for $x \to -\infty$, the non-degeneracy theorem shown in the previous section applies. On the other hand, one understands that, for any E, it is possible to find a solution of (7.11) that tends to zero for $x \to -\infty$, but no more than one, up to a factor.

Let us finally consider a third type of potential $V(x)$ (Fig. 7.10):

$$\begin{cases} V(x) \to V_2 & x \to -\infty \\ V(x) \to V_1 & x \to +\infty \end{cases} \qquad V_{\min} < V_1 < V_2 \,.$$

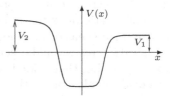

There arise four cases.

1. $E < V_{\min}$: there exist no eigenvalues.

2. $V_{\min} < E < V_1$: for x large, both positive and negative, the x-axis is a type II region, in between there is at least one region of type I.

Fig. 7.10

Therefore, in the present case, only discrete eigenvalues can occur.

3. $V_1 < E < V_2$: for x large and negative the x-axis is a type II region, while for x large and positive it a type I region. All the E are improper, nondegenerate eigenvalues of H. We may explicitly see that for any value of E only one solution can be found that vanishes for $x \to -\infty$: indeed, in this case, asymptotically for $x \to -\infty$ the general solution of (7.11) is

$$\alpha\, e^{\sqrt{2m(V_2-E)}\,x/\hbar} + \beta\, e^{-\sqrt{2m(V_2-E)}\,x/\hbar} \,, \qquad x \to -\infty$$

and it is sufficient to choose the solution with $\beta = 0$.

4. $E > V_2$: for large – both positive and negative – values of x one has type I regions; what happens in the middle depends on the form $V(x)$ (in the case of Fig. 7.10 all the x-axis is a type I region). Then for both $x \to -\infty$ and $x \to +\infty$ any ψ has an oscillatory behaviour and stays bounded:

$$\psi(x) = \alpha\, e^{i\sqrt{2m(E-V_2)}\,x/\hbar} + \beta\, e^{-i\sqrt{2m(E-V_2)}\,x/\hbar} \,, \qquad x \to -\infty$$

$$\psi(x) = \gamma\, e^{i\sqrt{2m(E-V_1)}\,x/\hbar} + \delta\, e^{-i\sqrt{2m(E-V_1)}\,x/\hbar} \,, \qquad x \to +\infty \,.$$

Therefore *any* solution is acceptable and any E is an improper eigenvalue of the Hamiltonian H. Since the Schrödinger equation is of the second order, the *independent* solutions are two (the four coefficients $\alpha, \beta, \gamma, \delta$ are not independent): this means that any $E > V_2$ is a twofold degenerate (improper) eigenvalue.

It appears from the above discussion that, for a given potential $V(x)$ and for a given interval of values of E, the type of eigenvalues and their degeneracy exclusively depends on the type of regions one has for $x \to -\infty$ and for $x \to +\infty$.

We can therefore summarize the results we have found in the following scheme, in which the different cases are distinguished by the type of regions one has for $x \to -\infty$ and for $x \to +\infty$.

1. All the x-axis is a type II region: no eigenvalue.
2. II, II: discrete nondegenerate eigenvalues.
3. II, I or I, II: continuous nondegenerate eigenvalues.
4. I, I: continuous twofold degenerate eigenvalues.

The above are not all the possible cases: it may indeed happen (as e.g in the case of a periodic potential) that for either $x \to -\infty$ or $x \to +\infty$, or both, one as an alternation of regions of the two types. The latter case, of remarkable physical interest, would deserve a separate discussion. We shall limit ourselves by only saying that a kind of 'compromise' between discrete and continuous eigenvalues takes place: the energy levels display a **band** structure, consisting of discrete sets of continuous eigenvalues (the bands may sometimes even partially overlap).

Chapter 8

One-Dimensional Systems

8.1 The One-Dimensional Harmonic Oscillator in the Schrödinger Representation

In Chap. 5 we have found the eigenvalues and the eigenvectors of the Hamiltonian of the one-dimensional harmonic oscillator. We want now to find the eigenfunctions $\psi n(x) \to \psi_n(x)$ of the Hamiltonian in the Schrödinger representation. A way to find such functions is that of solving the Schrödinger equation:

$$-\frac{\hbar^2}{2m}\,\psi_n''(x) + \frac{1}{2}m\,\omega^2 x^2\,\psi_n(x) = E_n\,\psi_n(x)\,.$$

The problem is not difficult and many books on quantum mechanics solve it. We shall instead follow a shorter way, that takes advantage of the fact that we have already solved the problem in abstract way.

We want to calculate $\psi_n(x) = \langle\, x \mid n\,\rangle'$ where the vectors $\mid n\,\rangle'$ are given by (6.8). Then

$$\psi_n(x) = \frac{(-\mathrm{i})^n}{\sqrt{n!}}\,\langle\, x \mid (\eta^\dagger)^n \mid 0\,\rangle\,.$$

Since the Schrödinger representation of $\eta^\dagger$ is

$$\eta^\dagger \to \frac{1}{\sqrt{2m\,\hbar\,\omega}}\left(-\,\mathrm{i}\,\hbar\,\frac{\mathrm{d}}{\mathrm{d}x} + \mathrm{i}\,m\,\omega\,x\right)$$

one has (with the choice of phases given by (6.8) instead of (5.18), the ψ_n are real):

$$\psi_n(x) = \frac{1}{\sqrt{n!}}\,(2m\,\hbar\,\omega)^{-n/2}\left(-\,\hbar\,\frac{\mathrm{d}}{\mathrm{d}x} + m\,\omega\,x\right)^n\,\langle\, x \mid 0\,\rangle\,. \qquad (8.1)$$

So the problem is brought back to that of determining $\psi_0(x) = \langle\, x \mid 0\,\rangle$.

To this end we recall that the vector $\mid 0\,\rangle$ is defined by $\eta\mid 0\,\rangle = 0$, i.e. in the Schrödinger representation:

© Springer International Publishing Switzerland 2016
L.E. Picasso, *Lectures in Quantum Mechanics*, UNITEXT for Physics,
DOI 10.1007/978-3-319-22632-3_8

$$\left(\hbar \frac{\mathrm{d}}{\mathrm{d}x} + m\,\omega\,x \right) \psi_0(x) = 0 \; .$$

This is a first-order differential equation that allows us to determine $\psi_0(x)$ up to a factor. In effect, one has

$$\frac{\psi_0'(x)}{\psi_0(x)} = \frac{\mathrm{d}}{\mathrm{d}x} \log \psi_0(x) = -\frac{m\,\omega}{\hbar}\,x \quad \Rightarrow \quad \psi_0(x) = C\,\mathrm{e}^{-(m\,\omega/2\hbar)\,x^2} \; .$$

The constant C is determined by imposing that the normalization condition $\langle 0 \mid 0 \rangle = 1$ be satisfied, i.e.

$$|C|^2 \int_{-\infty}^{+\infty} \mathrm{e}^{-(m\,\omega/\hbar)\,x^2}\,\mathrm{d}x = 1 \; .$$

As $\int \exp(-\alpha\,x^2)\,\mathrm{d}x = \sqrt{\pi/\alpha}$, one immediately obtains that, choosing it real and positive, $C = (m\,\omega/\pi\,\hbar)^{1/4}$ and, in conclusion, the normalized wavefunction of the ground state of the harmonic oscillator is

$$\psi_0(x) = \left(\frac{m\,\omega}{\pi\,\hbar} \right)^{1/4} \mathrm{e}^{-(m\,\omega/2\hbar)\,x^2} \; . \tag{8.2}$$

The $\psi_0(x)$ is therefore a Gaussian. One can verify that its inflection points occur at $\pm x_0$, with $x_0 = \sqrt{\hbar/(m\,\omega)}$, namely the points that satisfy the equation $V(x) = E_0$, i.e. the inversion points for a classical oscillator with energy E_0.

Note that, just as it must be, for $-x_0 < x < +x_0$ the concavity of $\psi_0(x)$ is downwards (ψ_0 is positive), whereas for $|x| \geq x_0$ the concavity is upwards. Furthermore $\psi_0(x)$ has no nodes and is an even function of x.

We take here the opportunity to discuss a typical and interesting aspect of quantum mechanics: we know that, according to classical mechanics, the motion of a particle of energy E can take place *only* in type I regions. Now we have seen in general, and receive direct confirmation in the case of the harmonic oscillator, that the eigenfunctions of the Hamiltonian are nonvanishing even in the regions of type II, those classically forbidden. So there exists a nonvanishing probability of finding a particle in a type II region, where the potential energy alone exceeds E, and there is in addition the kinetic energy that never is negative! It would seem that one is facing a situation in which energy is not conserved.

To solve this apparent paradox one must keep in mind that the statement 'there exists a nonvanishing probability to find a particle in a type II region' implies that we make measurements on the system suitable to tell us where the particle is: for instance, we measure some q_Δ. After the measurement the particle is in an eigenstate of q_Δ: this means that its wavefunction is localized in an interval of length Δ, for example, in a 'classically forbidden' zone. In this state, however, the particle does not possess a definite energy, because it is not in an eigenstate of the energy. However, the fact that, after finding the particle in a classically forbidden zone, we are not entitled to say it has no definite energy does not fully solve the problem. This is so because, if we find the particle in a forbidden zone, it is straightforward to check that in this state the mean energy $\overline{E}$ exceeds the energy it initially had. Now $\overline{E} > E$ means that, if measurements of energy are made in this state, sometimes results greater than E will be found, so a real increase of energy has taken place and the question comes up again.

The explanation of the paradox is the following: the measurement of position (i.e. of some q_Δ), that allows us to find sometimes the particle in the forbidden zone, perturbs the state of the system: it makes the system pass from an eigenstate of the energy to an eigenstate of q_Δ. This perturbation is due to the interaction between the system (the particle) and the instrument that measures q_Δ, and during this interaction the instrument transfers an uncontrollable quantity of energy to the system. Think that the instrument can be for example an Heisenberg microscope: the light the microscope uses interacts with the particle and transfers energy to it. There is no violation of the principle of conservation of energy: the particle is not an isolated system for it interacts with the measuring apparatus and its energy may be not conserved at the expense of the energy of the instrument. The energy of the isolated system 'particle plus instrument' is instead conserved.

Does it make any sense to ask whether the particle already was in the forbidden region, before making the position measurement? Or should one say that, until the position measurement is not effected, certainly the particle is in an allowed region? To take either one of the above points of view is a matter of taste: from the standpoint of the physicist both questions are irrelevant, because the purpose of the physicist is to be able to predict the results of specific measurements, not that of making statements that cannot be subjected to a verification (like 'until I do not make the measurement, certainly the particle is in the allowed region').

Let us go back to the problem of determining the $\psi_n(x)$. Once $\psi_0(x)$ is known, (8.1) allows for the determination of all the others: in order to find $\psi_n(x)$ it suffices to apply n times the operator $-\hbar \, d/dx + m\,\omega\,x$ to $\psi_0(x)$. Given the (Gaussian) form of $\psi_0(x)$, the result of this operation is, up to the factor $(m\,\hbar\omega)^{n/2}$, a polynomial $H_n(\xi)$ of degree n in the dimensionless variable $\xi \equiv \sqrt{m\,\omega/\hbar}\,x$, multiplied by $\psi_0(x)$.

In conclusion, from (8.1) and from the definition of H_n – that we shall give in (8.4) – the normalized ψ_n are:

$$\psi_n(x) = \frac{1}{\sqrt{2^n\,n!}} \left(\frac{m\,\omega}{\pi\,\hbar}\right)^{1/4} H_n(\sqrt{m\,\omega/\hbar}\,x)\, \mathrm{e}^{-(m\,\omega/2\hbar)\,x^2} . \qquad (8.3)$$

The polynomials $H_n(\xi)$ bear the name of **Hermite polynomials** and it is easy to see that, thanks to the way they are obtained, they have parity $(-1)^n$, i.e. those with even n are even, those with odd n are odd. Since $\mathrm{e}^{-(m\,\omega/2\hbar)\,x^2}$ is even we find again what we already knew, i.e. that the $\psi_n(x)$ given by (8.3) have parity $(-1)^n$. So $H_1(\xi)$, being of the first degree and odd, is proportional to ξ; $H_2(\xi)$ is of the form $\xi^2 + a$ and, since by the oscillation theorem $\psi_2(x)$ must possess two nodes, a is negative:

$$\begin{cases} H_n(\xi) \equiv e^{+\xi^2/2}\left(-\dfrac{d}{d\xi}+\xi\right)^n e^{-\xi^2/2} \\ H_0(\xi) = 1, \quad H_1(\xi) = 2\xi, \quad H_2(\xi) = 4\xi^2 - 2, \quad \cdots . \end{cases} \tag{8.4}$$

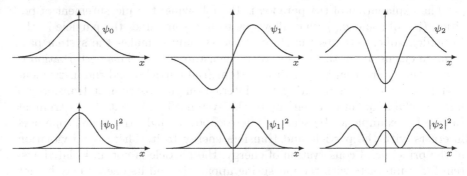

Fig. 8.1

In Fig. 8.1 the first three (equally normalized) eigenfunctions of the Hamiltonian H are reported together with the corresponding probability densities $|\psi_n(x)|^2$.

8.2 Potential Well

In several physically interesting situations (see Sect. 2.5) it may happen one has to deal with potentials $V(x)$ that in a very short interval (of length ϵ) go from a value V_1 to another value V_2, as shown in Fig. 8.2. Think e.g. of an electron close to the surface of a metal: over an interval of a few ångström the potential goes

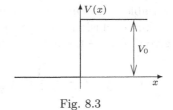

Fig. 8.2

from the value it has inside the metal to that it assumes outside of it, with a jump V_0 that equals the work function W.

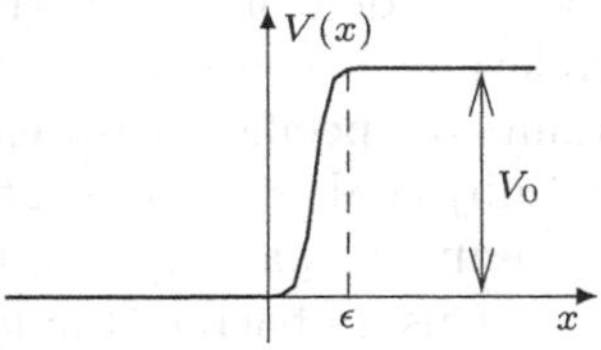

In situations of this type, both for the lacking knowledge we have of the actual behaviour of the potential in the zone where it grows from the value V_1 to the value V_2, and to make computations easier, one replaces the 'true' potential with a 'step potential', like that shown in Fig. 8.3, that displays a discontinuity. Note that one passes from the potential of Fig. 8.2

Fig. 8.3

to that of Fig. 8.3 by taking the limit $\epsilon \to 0$. In addition, in other cases, the intervening energies are so small with respect to V_0 that one is led to the schematization in which also the limit $V_0 \to \infty$ is taken.

In order to solve the Schrödinger equation in all these cases in which the potential displays a (either finite or infinite) discontinuity, it is necessary to

know which conditions of continuity must be enforced on the eigenfunctions of
the Hamiltonian H in the points of discontinuity for $V(x)$. This problem can
be faced from a mathematical point of view by studying the domain (i.e. the
set of functions) that allows for the definition of H as a self-adjoint operator.
We instead will face this problem by recalling that, from the physical point of
view, a discontinuous potential is a schematization, i.e. a limit of continuous
potentials: we will therefore investigate which properties of continuity of the
eigenfunctions of H survive, and which do not, this limiting procedure.

Let us assume that $V(x)$ is like in Fig. 8.2. The Schrödinger equation is

$$\psi''(x) = \frac{2m}{\hbar}\left(V(x) - E\right)\psi(x) .$$

By integrating both sides between $x = 0$ and $x = \epsilon$ one has

$$\psi'(\epsilon) - \psi'(0) = \frac{2m}{\hbar}\int_0^\epsilon \left(V(x) - E\right)\psi(x)\,\mathrm{d}x . \qquad (8.5)$$

Since both $V(x)$ and $\psi(x)$ are finite in the interval $(0, \epsilon)$, the right hand
side of (8.5) tends to zero for $\epsilon \to 0$. Therefore $\psi'(\epsilon) \to \psi'(0)$.

We have shown that $\psi'(x)$ remains continuous in the limit of step potential. Likewise one shows that also $\psi(x)$ remains continuous. So for the step
potential, i.e. with a finite discontinuity, me must require that both ψ and ψ'
are continuous functions. Note that, instead, the Schrödinger equation entails

$$\lim_{x \to 0^+}\psi''(x) \neq \lim_{x \to 0^-}\psi''(x) .$$

As an application we now study the problem of
a particle subject to the 'square well potential',
as in Fig. 8.4:

$$V(x) = \begin{cases} 0 & |x| < a \\ V_0 & |x| > a . \end{cases}$$

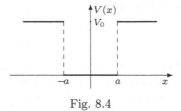

Fig. 8.4

The Schrödinger equation can be written as

$$\begin{cases} \psi''(x) = \dfrac{2m\,E}{\hbar^2}\,\psi(x) & |x| < a \\[2mm] \psi''(x) = \dfrac{2m(V_0 - E)}{\hbar^2}\,\psi(x) & |x| > a . \end{cases} \qquad (8.6)$$

If we want to find the bound states, then $0 < E < V_0$ and in this case the
general solution of (8.6) is

$$\psi(x) = A\,\cos k\,x + B\,\sin k\,x , \qquad\qquad -a < x < +a \qquad (8.7)$$

where

$$k = \frac{\sqrt{2m\,E}}{\hbar} \tag{8.8}$$

and

$$\psi(x) = B\,\mathrm{e}^{-\kappa x} + B'\,\mathrm{e}^{+\kappa x}\,, \qquad\qquad x > +a \tag{8.9}$$

$$\psi(x) = C\,\mathrm{e}^{+\kappa x} + C'\,\mathrm{e}^{-\kappa x}\,, \qquad\qquad x < -a \tag{8.10}$$

where

$$\kappa = \frac{\sqrt{2m\,(V_0 - E)}}{\hbar}\,. \tag{8.11}$$

In order for the solution to be acceptable, it must be L_2 and continuous together with its first derivative in the points $x = \pm a$. The square-integrability condition demands that $B' = 0$ in (8.9) and $C' = 0$ in (8.10).

Before enforcing the continuity conditions, let us observe that $V(x)$ is even in x: $V(x) = V(-x)$. Therefore, since the discrete eigenvalues are nondegenerate, the bound states have definite parity. Let us start with the study of the even states, among which there is the ground state. The wavefunctions of such states are those for which $A' = 0$ in (8.7) and $B = C$ in (8.9) and (8.10) (where, we recall, already $B' = C' = 0$ has been imposed).

So the *even* solutions of the Schrödinger equation (8.6) have the form:

$$\psi(x) = \begin{cases} B\,\mathrm{e}^{+\kappa x} & x < -a \\ A\,\cos k\,x & -a < x < +a \\ B\,\mathrm{e}^{-\kappa x} & x > +a\,. \end{cases} \tag{8.12}$$

Now it is sufficient to impose the continuity conditions in the point $x = +a$ because, thanks to the fact that $\psi(x)$ has a well defined parity, the latter conditions will be automatically satisfied in the point $x = -a$. The continuity conditions are:

$$\begin{cases} A\,\cos k\,a = B\,\mathrm{e}^{-\kappa a} \\ k\,A\,\sin k\,a = \kappa\,B\,\mathrm{e}^{-\kappa a}\,. \end{cases} \tag{8.13}$$

One is dealing with a system of two linear homogeneous equations in the unknowns A and B, so the solution is $A = B = 0$, i.e. $\psi(x)$ is identically vanishing – unless the two equations are linearly dependent, i.e. the determinant of the coefficients of (8.13) vanishes, namely:

$$k\,\tan k\,a = \kappa\,. \tag{8.14}$$

In other words the equations (8.13) possess a nontrivial solution only for those values of E $\big(k$ and κ are functions of E, given by (8.8) and (8.11)$\big)$ such that (8.14) is satisfied. Such values of E are the eigenvalues of H for the eigenstates with parity $+1$.

The *odd* solutions of (8.6) have the form:

$$\psi(x) = \begin{cases} -B\,\mathrm{e}^{+\kappa x} & x < -a \\ A\,\sin k\,x & -a < x < +a \\ B\,\mathrm{e}^{-\kappa x} & x > +a \end{cases} \tag{8.15}$$

and the continuity conditions in $x = +a$ are

$$
\begin{cases}
A \sin k a = & B\,\mathrm{e}^{-\kappa a} \\
k A \cos k a = & -\kappa\, B\,\mathrm{e}^{-\kappa a}\,.
\end{cases}
$$

so that for the odd solutions:

$$
k\,/\tan k a = -\kappa\,. \tag{8.16}
$$

Usually a graphical method is used to solve (8.14) and (8.16). Let us put

$$
k a = \xi\,, \qquad\qquad \kappa a = \eta
$$

so that (8.14) and (8.16) become:

$$
\eta = \xi \tan \xi\,, \qquad\qquad \eta = -\xi\,/\tan \xi\,. \tag{8.17}
$$

Moreover, from (8.8) and (8.11) one has

$$
\xi^2 + \eta^2 = \frac{2m\,V_0\,a^2}{\hbar^2}\,. \tag{8.18}
$$

In Fig. 8.5 the circumference (8.18) as well as the functions $\eta = \xi \tan \xi$ (thicker line) and $\eta = -\xi\,/\tan \xi$ (thinner line) are drawn: the abscissae ξ of the intersection points between these two curves and the circumference in the first quadrant (k and κ, i.e. ξ and η, must be positive) provide us with the values of k that satisfy (8.14) and (8.16) and, via either (8.8) or (8.11), the values of E we are looking for. Fig. 8.5 shows that the number of bound

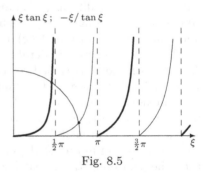

Fig. 8.5

states is finite and that, whatever the value of V_0 and a, there always exists at least one bound state: for $2m\,V_0\,a^2/\hbar^2 < \pi^2/4$ the system has only one (even) bound state; the existence of at least one odd bound state requires $2m\,V_0\,a^2/\hbar^2 > \pi^2/4$.

Let us now investigate what happens for $V_0 \to \infty$, limiting ourselves to the ground-state wavefunction $\psi_0(x)$ (for the other $\psi_n(x)$ the same conclusions will apply).

Always looking at Fig. 8.5 one realizes that, if $V_0 \to \infty$, the abscissa of the first solution tends to $\pi/2$. So one has $k a \to \pi/2$, i.e. $E_0 \to \pi^2 \hbar^2/(8m\,a^2)$. We want also to investigate which continuity properties of $\psi_0(x)$ in the point $x = a$ survive the limit $V_0 \to \infty$, and which do not. Starting from (8.12) and solving the first of (8.13) for B one has

$$
\psi_0(x) = \begin{cases}
A \cos k a\ \mathrm{e}^{+\kappa\,(x+a)} & x < -a \\
A \cos k x & -a < x < +a \\
A \cos k a\ \mathrm{e}^{-\kappa\,(x-a)} & x > +a\,.
\end{cases} \tag{8.19}
$$

and, since $k\,a \to \pi/2$ and therefore $\cos k\,a \to 0$ in the limit $V_0 \to \infty$, one eventually has:

$$\lim_{V_0 \to \infty} \psi_0(x) = \begin{cases} A \cos \dfrac{\pi\,x}{2a} & -a < x < +a \\ 0 & |x| > a \,. \end{cases} \tag{8.20}$$

In conclusion, in the limiting case $V_0 = \infty$ (well of infinite depth), $\psi_0(x)$ is nonvanishing only inside the well and vanishing outside of it. In addition $\psi_0(x)$ stays continuous, therefore vanishing, in the points $x = \pm a$; its first derivative instead develops a discontinuity. Since this conclusion applies to all the $\psi_n(x)$, and the latter constitute a basis, it follows that *any* wavefunction $\psi(x)$ vanishes for $|x| > a$ (not necessarily for $x = \pm a$): for this reason one says that the particle is constrained to move in the segment $(-a, +a)$. As a consequence, one can disregard the regions $|x| > a$ and solve the Schrödinger equation only in the interval $(-a, +a)$, enforcing on the eigenfunctions $\psi_n(x)$ of H the conditions:

$$\psi_n(+a) = \psi_n(-a) = 0\,, \qquad (V_0 = \infty \ \text{for} \ |x| > a)\,. \tag{8.21}$$

The boundary conditions (8.21) are quite general: they must be enforced any time one is searching for the eigenfunctions of H when $V_0 = \infty$ for $|x| > a$ and whatever the form of $V(x)$ for $|x| < a$.

As an example we will determine all the eigenfunctions and eigenvalues of H for a particle constrained to move in a segment, with the potential that vanishes inside the segment (infinite potential well). We could start from the case already discussed, in which V_0 is finite, and take the limit $V_0 \to \infty$, but it is more instructive to face the problem from the scratch. In order to give a unique treatment for both even and odd solutions, it is convenient to choose the origin in one of the end-points of the segment and let $x = a$ be the other end-point (note that now the length of the segment is a, not $2a$ as it was before). The Schrödinger equation is therefore:

$$\psi''(x) = \frac{2m\,E}{\hbar^2}\,\psi(x)\,, \qquad 0 \le x \le a$$

with boundary conditions

$$\psi(0) = \psi(a) = 0\,. \tag{8.22}$$

The general solution of the Schrödinger equation is

$$\psi(x) = A \sin k\,x + B \cos k\,x\,, \qquad k = \frac{\sqrt{2m\,E}}{\hbar}\,, \qquad 0 \le x \le a\,.$$

The condition $\psi(0) = 0$ requires $B = 0$; then the condition $\psi(a) = 0$ demands:

$$A \sin k\,a = 0 \quad \Rightarrow \quad k\,a = n\,\pi\,, \qquad n = 1, 2, \cdots\,. \tag{8.23}$$

Note that $n = 0$ is excluded for it would imply $k = 0$ and $\psi_0(x) = 0$; also the negative integers $n = -1, -2, \cdots$ are excluded because $\sin(-k\,x) = -\sin k\,x$: they give raise to wavefunctions dependent on those obtained by limiting oneself to positive values of n. From (8.23) and the relationship between k and E one obtains:

$$ E_n = \frac{\hbar^2 \pi^2}{2m\,a^2}\, n^2, \qquad \psi_n(x) = A\,\sin\left(n\,\pi\,\frac{x}{a}\right), \qquad n = 1, 2, \cdots. \quad (8.24) $$

Note that the energy levels (8.24) are the same as those obtained in Chap. 2 (see (2.18)) by means of either Bohr or de Broglie theory.

8.3 Tunnel Effect

Let us consider a potential $V(x)$ vanishing for both $x < 0$ and $x > a$ and satisfying $0 \le V(x) \le V_0$ in the region $0 \le x \le a$ (Fig. 8.6). Consider, from the classical point of view, the situation in which a particle is sent, e.g. from the region $x < 0$, against this 'potential barrier': if the energy E of the particle exceeds V_0, then the particle will cross the barrier and will continue its motion in the region $x > a$; if instead $E < V_0$, the particle will be reflected by the barrier and will go back.

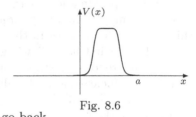

Fig. 8.6

According to quantum mechanics, on the contrary, even if $E < V_0$, there is a nonvanishing probability that the particle crosses the barrier: this typically quantum effect is called **tunnel effect**.

In order to discuss this problem, let us study the eigenfunctions $\psi_E(x)$ of the Hamiltonian. Since the asymptotic regions are type I regions, the eigenvalues are continuous and doubly degenerate: in the regions $x < 0$ and $x > a$ the general solution has the form

$$ \psi_E(x) = \begin{cases} A\,e^{i\,k\,x} + A'\,e^{-i\,k\,x} & x \le 0 \\ C\,e^{i\,k\,x} + C'\,e^{-i\,k\,x} & x \ge a \end{cases} $$

where, since the Schrödinger equation is of the second order, only two from among the four coefficients A, A', C, C' are independent and k is given by (8.8): $k = \pm\sqrt{2mE}/\hbar$.

If we want to describe the situation of a particle incoming from the left (and that can possibly cross the barrier) we search for those solutions with $k > 0$ such that $A \ne 0$ and $C' = 0$; moreover, since the Schrödinger equation is homogeneous, we can set $A = 1$; redefining A' one has (Fig. 8.7):

$$ \psi_E(x) = \begin{cases} e^{i\,k\,x} + A\,e^{-i\,k\,x} & x \le 0 \\ C\,e^{i\,k\,x} & x \ge a. \end{cases} \quad (8.25) $$

The states whose wavefunction has (in the one-dimensional case) the form (8.25) are called **scattering states**.

It is obvious that, whatever the form of
$V(x)$ and for any value of the energy, it
must happen that $C \neq 0$: indeed the solutions of the Schrödinger equation cannot
vanish in a point together with the first
derivative in that point and, a fortiori, cannot vanish in an interval. Therefore the
transmission of the particle through the

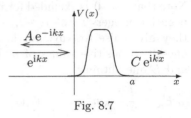

Fig. 8.7

barrier actually takes place. For this reason, having set to 1 the coefficient
of the 'incoming wave' $e^{i k x}$ in (8.25), $|C|^2$ is named the **transmission co-
efficient** of the barrier, and $|A|^2$ the **reflection coefficient**.

It is true that $\psi_E(x)$ is an improper eigenfunction of the Hamiltonian and
does not represent a physical state, however we know that there exist physical
states whose wavefunction approximate $\psi_E(x)$. So, for instance, if we start
with a an approximation to $\psi_E(x)$ consisting of a state whose wavefunction
is of the form $e^{i k x}$ in a very large but bounded region of the negative x-axis,
this state (as we will see in the next chapter) evolve in time and, as we will
explicitly show in Chap. 17, will give rise to both a reflected and a transmitted
wave; therefore for such states it is meaningful to talk about transmission and
reflection coefficients, as probabilities of finding the particle beyond the barrier
or reflected backward.

Of course the values of A and C depend on the potential $V(x)$ but, whatever the form of $V(x)$, the following relationship holds:

$$|A|^2 + |C|^2 = 1 \qquad (8.26)$$

namely, the sum of the transmission and the reflection coefficient equals 1:
this means that if we send many particles against the barrier and detect them
by means of counters, the numbers of the transmitted and of the reflected
particles equals the number of the incident ones.

Equation (8.26) is a consequence of the **continuity equation** that, for
further reference we prove in the three-dimensional case. In three dimensions
the Schrödinger equation is ($\Delta \equiv \partial^2/\partial x^2 + \partial^2/\partial y^2 + \partial^2/\partial z^2$ is the Laplace
operator):

$$-\frac{\hbar^2}{2m} \Delta \psi_E + V \psi_E = E \psi_E \qquad (8.27)$$

Let us multiply (8.27) by ψ_E^* and its complex conjugate by ψ_E; subtracting
side by side we obtain:

$$\psi_E^* \Delta \psi_E - \psi_E \Delta \psi_E^* = 0 \qquad \Rightarrow \qquad \nabla \cdot \left(\psi_E^* \nabla \psi_E - \psi_E \nabla \psi_E^* \right) = 0$$

so that, defining the **probability current density** by

$$\vec{J} = -\mathrm{i}\,\frac{\hbar}{2m} \left(\psi_E^* \nabla \psi_E - \psi_E \nabla \psi_E^* \right) \qquad (8.28)$$

(the prefactor $-\mathrm{i}\,\hbar/2m$ is put in for reasons to be clarified in the next chapter), if ψ_E is a solution of the Schrödinger equation (8.27), one has the **continuity equation:**

$$\nabla \cdot \vec{J}(x, y, z) = 0 \, . \tag{8.29}$$

In the one-dimensional case – the one we are presently interested in – (8.29) becomes:

$$\frac{\mathrm{d}}{\mathrm{d}x}\left(\psi_E^* \, \psi_E' - \psi_E \, \psi'^*_E\right) = 0$$

that, upon integration between x_1 and x_2, gives:

$$\Im m\left(\psi_E^*(x_1)\,\psi_E'(x_1)\right) = \Im m\left(\psi_E^*(x_2)\,\psi_E'(x_2)\right) \, .$$

If $x_1 < 0$ and $x_2 > a$ in the $\psi_E(x)$ given by (8.25), then (8.26) follows.

It is interesting to remark that *independently* of the form of the potential, the transmission and reflection coefficient are the same, regardless of whether the particle hits the barrier on either the left or the right side. Indeed, since the Schrödinger equation is real, if ψ_E is a solution of the type (8.25), ψ_E^* is a solution independent of ψ_E :

$$\psi_E^*(x) = \begin{cases} \mathrm{e}^{-\mathrm{i}\,k\,x} + A^*\,\mathrm{e}^{\mathrm{i}\,k\,x} & x \leq 0 \\ C^*\,\mathrm{e}^{-\mathrm{i}\,k\,x} & x \geq a \, . \end{cases} \tag{8.30}$$

As a consequence, the solution $\widetilde{\psi}_E$ corresponding to a particle incoming from the right (i.e. the region $x > 0$) is a linear combination $\alpha\,\psi_E + \beta\,\psi_E^*$:

$$\widetilde{\psi}_E = \begin{cases} (\alpha + \beta\,A^*)\,\mathrm{e}^{\mathrm{i}\,k\,x} + (\alpha\,A + \beta)\,\mathrm{e}^{-\mathrm{i}\,k\,x} & x \leq 0 \\ \alpha\,C\,\mathrm{e}^{\mathrm{i}\,k\,x} + \beta\,C^*\,\mathrm{e}^{-\mathrm{i}\,k\,x} & x \geq a \end{cases} \tag{8.31}$$

such that $\beta\,C^* = 1$ and $\alpha + \beta\,A^* = 0$: from these conditions it follows that $\beta = 1/C^*$, $\alpha = -A^*/C^*$ and therefore the reflection coefficient is $|\widetilde{A}|^2 = |\alpha\,C|^2 = |A|^2$ and likewise the transmission coefficient $|\widetilde{C}|^2 = 1 - |\widetilde{A}|^2 = |C|^2$.

We now explicitly calculate the transmission and reflection coefficients for the 'rectangular barrier'

$$V(x) = \begin{cases} 0 & x \leq 0 \\ V_0 > 0 & 0 \leq x \leq a \\ 0 & x \geq a \end{cases} \tag{8.32}$$

considering, to begin with, the case $0 < E < V_0$. If in (8.25) we redefine C: $C \to \mathrm{e}^{-\mathrm{i}\,k\,a}\,C$, one has

$$\psi_E(x) = \begin{cases} \mathrm{e}^{\mathrm{i}\,k\,x} + A\,\mathrm{e}^{-\mathrm{i}\,k\,x} & x \leq 0 \\ B\,\mathrm{e}^{\kappa\,x} + B'\,\mathrm{e}^{-\kappa\,x} & 0 \leq x \leq a \\ C\,\mathrm{e}^{\mathrm{i}\,k\,(x-a)} & x \geq a \end{cases} \tag{8.33}$$

where κ is given by (8.11): $\kappa = \sqrt{2m\,(V_0 - E)}/\hbar$. The coefficients A, B, B', C are determined by the requirement of the continuity conditions of ψ_E and ψ_E' in $x = 0$ and $x = a$:

$$
\begin{cases}
1 + A = B + B' \\
i\,(k/\kappa)\,(1 - A) = B - B' \,;
\end{cases}
\qquad
\begin{cases}
C = B\,e^{\kappa a} + B'\,e^{-\kappa a} \\
C = -i\,(\kappa/k)\,(B\,e^{\kappa a} - B'\,e^{-\kappa a})
\end{cases}
\Rightarrow
$$

$$
\begin{cases}
C = (B + B')\,\cosh \kappa a + (B - B')\,\sinh \kappa a \\
C = -i\,(\kappa/k)\,[(B - B')\,\cosh \kappa a + (B + B')\,\sinh \kappa a]
\end{cases}
\Rightarrow
$$

$$
\begin{cases}
C = (1 + A)\,\cosh \kappa a + i\,(k/\kappa)\,(1 - A)\,\sinh \kappa a \\
C = (1 - A)\,\cosh \kappa a - i\,(\kappa/k)\,(1 + A)\,\sinh \kappa a
\end{cases}
\Rightarrow
$$

$$
\begin{cases}
A\,(\cosh \kappa a - i\,(k/\kappa)\,\sinh \kappa a) - C = -(\cosh \kappa a + i\,(k/\kappa)\,\sinh \kappa a) \\
A\,(\cosh \kappa a + i\,(\kappa/k)\,\sinh \kappa a) + C = \cosh \kappa a - i\,(\kappa/k)\,\sinh \kappa a
\end{cases}
\Rightarrow
$$

$$
C = \frac{2}{2\cosh \kappa a + i((\kappa/k) - (k/\kappa))\,\sinh \kappa a} \qquad \Rightarrow
$$

$$
\begin{cases}
|C|^2 = \dfrac{4E\,(V_0 - E)}{4E\,(V_0 - E) + V_0^2\,\sinh^2 \kappa a} \\[2mm]
|A|^2 = 1 - |C|^2 = \dfrac{V_0^2\,\sinh^2 \kappa a}{4E\,(V_0 - E) + V_0^2\,\sinh^2 \kappa a}
\end{cases}
\qquad E \le V_0\,. \quad (8.34)
$$

The case $E > V_0$ is obtained by means of the substitution $\kappa \to i\,k_1 = i\,\sqrt{2m\,(E - V_0)}/\hbar$:

$$
\begin{cases}
|C|^2 = \dfrac{4E\,(E - V_0)}{4E\,(E - V_0) + V_0^2\,\sin^2 k_1 a} \\[2mm]
|A|^2 = 1 - |C|^2 = \dfrac{V_0^2\,\sin^2 k_1 a}{4E\,(E - V_0) + V_0^2\,\sin^2 k_1 a}
\end{cases}
\qquad E \ge V_0\,. \quad (8.35)
$$

We report in Fig. 8.8 the transmission coefficient $|C|^2|$ as a function of E/V_0 in the numerical case $2mV_0a^2/\hbar^2 = 9$ that, for an electron and $V_0 = 1\,\mathrm{eV}$ corresponds to $a \simeq 6\,\mathrm{\AA}$.

The tunnel effect has very many technological applications (scanning electron microscope, electronic devices like the tunnel diode,

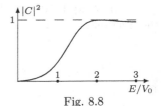

Fig. 8.8

the metal-oxide-semiconductor (MOS), Josephson junction, flash memories, etc.).

Chapter 9

Time Evolution

9.1 Time Evolution in the Schrödinger Picture

So far we have mainly been concerned with the determination of eigenvalues and eigenvectors of observables and we have not yet faced what in quantum mechanics is a fundamental problem, i.e. the problem of determining how the states of a system evolve in time – which is equivalent, in classical mechanics, to solving the equations of motion.

In other words the discussion made so far regards the states of a system at a *given instant* and we have not yet investigated how the states change as time goes by.

The problem of time evolution in quantum mechanics is to be faced in the same way as in classical mechanics: given the state of a system at a certain instant and the forces acting on the system being known, the task is to determine the state at any other instant. (In quantum mechanics the terms 'interaction' and 'potential' are preferred to the term 'force').

In order to solve this problem, we shall have to introduce a certain number of assumptions (or postulates) that will be fully justified in an alternative approach to be described in the next section.

We will use the following notation: if the system initially ($t = 0$, to be concrete) is in the state described by the vector $|A\rangle$, we will denote this vector by $|A,0\rangle$ to emphasize the instant in which the system is in the state $|A\rangle$. By $|A,t\rangle$ we will denote the vector that describes the state of the system at the generic instant t, if the system initially was in the state $|A,0\rangle$. Namely, due to time evolution:

$$|A,0\rangle \xrightarrow{\text{after a time } t} |A,t\rangle \,.$$

We have implicitly assumed that the state of the system at time t is univocally determined (the 'forces' being known) by the state at time 0: this is legitimate if in the interval $(0,t)$ no measurement is made on the system because, the state of the system being only statistically determined after a measurement, any causal relation between $|A,0\rangle$ and $|A,t\rangle$ would fail. Let us assume that

© Springer International Publishing Switzerland 2016
L.E. Picasso, *Lectures in Quantum Mechanics*, UNITEXT for Physics,
DOI 10.1007/978-3-319-22632-3_9

$$|A,t\rangle = U(t,0)\,|A,0\rangle \tag{9.1}$$

where $U(t,0)$ a *linear* and *unitary* operator. The linearity of $U(t,0)$ guarantees that superposition relations are preserved over time: if

$$|C,0\rangle = \alpha\,|A,0\rangle + \beta\,|B,0\rangle$$

then, thanks to the linearity of $U(t,0)$,

$$|C,t\rangle = U(t,0)\,|C,0\rangle = U(t,0)\,(\alpha\,|A,0\rangle + \beta\,|B,0\rangle)$$
$$= \alpha\,|A,t\rangle + \beta\,|B,t\rangle\,.$$

The unitarity brings along the consequence that the scalar products (and therefore the probability transitions) do not depend on time:

$$\langle A,t\,|\,B,t\rangle = \langle A,0\,|\,U(t,0)^{\dagger}\,U(t,0)\,|\,A,0\rangle = \langle A,0\,|\,B,0\rangle\,.$$

In particular, the norm of $|A,t\rangle$ is the same as that of $|A,0\rangle$.

In general, if $U(t,t_0)$ is the time evolution operator between the (initial) instant t_0 and the instant t, one has

$$U(t_2,t_1)\,U(t_1,t_0) = U(t_2,t_0)\,. \tag{9.2}$$

Indeed:

$$|A,t_2\rangle = U(t_2,t_0)\,|A,t_0\rangle = U(t_2,t_1)\,|A,t_1\rangle$$
$$= U(t_2,t_1)\,U(t_1,t_0)\,|A,t_0\rangle\,.$$

If we limit ourselves to consider forces that *do not depend on time*, then $U(t,t_0)$ depends only on the interval $t - t_0$ and *not* on the initial instant t_0: in this case we shall write

$$U(t) \equiv U(t,0)$$

and (9.2) takes the form:

$$U(t_1)\,U(t_2) = U(t_2)\,U(t_1) = U(t_1 + t_2)\,. \tag{9.3}$$

If (9.3) holds (i.e. if the forces do not depend on time), one can show (Stone's theorem) that

$$U(t) = \mathrm{e}^{-\mathrm{i}Kt} \tag{9.4}$$

in which K is a self-adjoint operator. Note the likeness between the time evolution operator in the form (9.4) and the space translation operator $U(a)$ given by (6.22).

It goes without saying that $U(t)$, and therefore K, depend on the type of forces acting on the system; in particular K is an operator that determines the dynamics of the system. The analogy with the space translation operator suggests to postulate that K is proportional to the Hamiltonian H of the system. In effect, from the (classical) theory of canonical transformations we know that, much as the linear momentum p is the generator of space translations, H is the generator of time translations. The proportionality constant must have the dimensions of the reciprocal of an action, so we will set

$$K = \hbar^{-1} H \quad \Rightarrow \quad U(t) = e^{-i H t/\hbar} \tag{9.5}$$

and the choice of the sign of the proportionality constant between H and K will be justified by the requirement that, under suitable circumstances, quantum mechanics reproduces the results of classical mechanics.

In conclusion, for systems subject to forces independent of time:

$$|A, t\rangle = e^{-i H t/\hbar} |A, 0\rangle . \tag{9.6}$$

If we take the derivative of both sides of (9.6) with respect to t (the derivative of a vector $|A, t\rangle$ is defined, as usual, as the limit of the ratio of the increments; as far as the derivative of the operator $e^{-i H t/\hbar}$ is concerned, the ordinary differentiation rules can be followed, as if H were a number: in the expression $e^{-i H t/\hbar}$ there is nothing that does not commute with H), we obtain

$$\frac{d}{dt} |A, t\rangle = -\frac{i}{\hbar} H |A, 0\rangle$$

i.e.

$$i\hbar \frac{d}{dt} |A, t\rangle = H |A, t\rangle . \tag{9.7}$$

Equation (9.7) is the differential form of (9.6) and is known as the **time-dependent Schrödinger equation**.

Equation (9.7) has been deduced from (9.6), that holds only if the forces do not depend on time. However, contrary to (9.6), (9.7) keep its validity even in the case of time dependent forces (e.g. an atom under the action of an electromagnetic wave); in the latter case H *explicitly* depends on the (numeric) parameter t.

Equation (9.7) is a differential equation of the first order: this expresses the fact that the solution $|A, t\rangle$ is completely determined when the state at time $t = 0$ is assigned – as it must be, inasmuch as (9.7) is equivalent to (9.6).

In the Schrödinger representation (9.7) (in the general case of time dependent forces) writes:

$$i\hbar \frac{\partial}{\partial t} \psi(x, t) = H(x, -i\hbar \, \partial/\partial x ; t) \, \psi(x, t) , \qquad x \equiv (x_1, \cdots, x_n) \tag{9.8}$$

where

$$\psi(x, t) = \langle x \mid A, t\rangle .$$

If the system is subject to position-dependent and possibly time-dependent (but not velocity-dependent) forces, in the same way as in Sect. 8.3 we have derived the continuity equation (8.29), one now obtains:

$$\nabla \cdot \vec{J}(x,t) + \frac{\partial \rho(x,t)}{\partial t} = 0 \qquad (9.9)$$

where $\vec{J}$ is defined by (8.28) and $\rho \equiv |\psi|^2$ is the probability density; (9.9) is a conservation equation, analogous to the conservation equation for the electric charge, and expresses the **conservation of probability**.

Equation (9.7), that holds even when H explicitly depends on t, can be written in the form

$$i\hbar \frac{d}{dt} U(t) \, | \, A,0 \, \rangle = H(t) \, U(t) \, | \, A,0 \, \rangle$$

and, since it holds for any vector $| \, A,0 \, \rangle$ (ignoring, as usual, the domain problems), it is equivalent to the operator equation:

$$i\hbar \frac{d}{dt} U(t) = H(t) \, U(t) \qquad (9.10)$$

that will be useful in the sequel. In general, with the exception of the case in which $H(t_1)$ and $H(t_2)$ commute for arbitrary values of t_1 and t_2 (this is certainly so if H does not depend on time), the solution $U(t)$ of (9.10) does not share the form of the solution (9.5).

Let us go back to the case of forces independent of time and let us ask ourselves whether there exist states that do not evolve in time: we will call such states **stationary states**.

A state is a stationary state if, for any t, the vectors $| \, A,t \, \rangle = U(t) \, | \, A,0 \, \rangle$ and $| \, A,0 \, \rangle$ represent the same state: this happens if and only if $| \, A,t \, \rangle$ and $| \, A,0 \, \rangle$ are proportional to each other:

$$| \, A,t \, \rangle = C(t) \, | \, A,0 \, \rangle \qquad (9.11)$$

where the constant $C(t)$ may depend on t.

We will show that, for forces independent of time:

The stationary states of the system are all and only the eigenstates of H.

Indeed, if $H \, | \, E \, \rangle = E \, | \, E \, \rangle$, then

$$U(t) \, | \, E \, \rangle = e^{-iHt/\hbar} \, | \, E \, \rangle = e^{-iEt/\hbar} \, | \, E \, \rangle \qquad (9.12)$$

i.e. the eigenstates of H are stationary states. Viceversa, if a state is stationary, (9.11) applies to it, therefore the Schrödinger equation (9.7) yields:

$$i\hbar \dot{C}(t) \, | \, A,0 \, \rangle = H \, | \, A,0 \, \rangle \, C(t)$$

whence

$$i\hbar \frac{\dot{C}(t)}{C(t)} \, | \, A,0 \, \rangle = H \, | \, A,0 \, \rangle \, . \qquad (9.13)$$

Since the right hand side does not depend on t, the same must happen for the left hand side, i.e. $i\hbar \, \dot{C}(t)/C(t)$ is constant with respect to t. If we call E this constant, (9.13) becomes $H \mid A,0 \rangle = E \mid A,0 \rangle$, i.e. $\mid A,0 \rangle$ is an eigenvector of H corresponding to the eigenvalue E – the statement made above. Moreover, note that from the equation

$$i\hbar \frac{\dot{C}(t)}{C(t)} = E, \qquad C(0) = 1$$

one obtains $C(t) = \mathrm{e}^{-\mathrm{i}\,E t/\hbar}$, in agreement with (9.12).

If one wants to determine explicitly the evolution of a state over time – i.e. to determine $\mid A,t \rangle$ in terms of $\mid A,0 \rangle$ – according to (9.6) one must apply the operator $\mathrm{e}^{-\mathrm{i}\,H t/\hbar}$ to the vector $\mid A,0 \rangle$; usually this is not an easy task, so it is convenient to operate in the following way: one first writes $\mid A,0 \rangle$ as a linear combination of eigenvectors of H, that form a basis (we will provisionally assume that H only has discrete eigenvalues):

$$\mid A,0 \rangle = \sum_n a_n \mid E_n \rangle, \qquad a_n = \langle E_n \mid A,0 \rangle$$

then applies the time evolution operator to both sides of the above equation:

$$\mathrm{e}^{-\mathrm{i}\,H t/\hbar} \mid A,0 \rangle = \mathrm{e}^{-\mathrm{i}\,H t/\hbar} \sum_n a_n \mid E_n \rangle$$
$$= \sum_n a_n \,\mathrm{e}^{-\mathrm{i}\,H t/\hbar} \mid E_n \rangle = \sum_n a_n \,\mathrm{e}^{-\mathrm{i}\,E_n\, t/\hbar} \mid E_n \rangle$$

having used (9.12). In conclusion:

$$\mid A,t \rangle = \sum_n a_n \,\mathrm{e}^{-\mathrm{i}\,E_n\, t/\hbar} \mid E_n \rangle, \qquad a_n = \langle E_n \mid A,0 \rangle. \qquad (9.14)$$

Equation (9.14) shows that the problem of determining the time evolution of a state is completely solved if we know the eigenvectors and eigenvalues of the Hamiltonian H. It is therefore clear that, from among all the observables of a system, the Hamiltonian has a privileged role. It is also for this reason that one of the most important problems in quantum mechanics is that of finding the eigenvectors and eigenvalues of H.

In the case H has continuous eigenvalues, instead of (9.14) one has

$$\mid A,t \rangle = \int a(E)\,\mathrm{e}^{-\mathrm{i}\,E t/\hbar} \mid E \rangle \,\mathrm{d}E, \qquad a(E) = \langle E \mid A,0 \rangle. \qquad (9.15)$$

Very often, particularly when H is degenerate, the eigenstates of H are denoted by the eigenvalues of some (nondegenerate) observable that commutes with H: in this case (9.14) and (9.15) are expressed in terms of the eigenvalues of such an observable. For example, for the free particle

$$\mid A,t \rangle = \int \varphi(p)\,\mathrm{e}^{-\mathrm{i}\,E(p)\, t/\hbar} \mid p \rangle \,\mathrm{d}p, \qquad E(p) = \frac{p^2}{2m}.$$

Formulae similar to (9.14) and (9.15) apply to the case in which H has both discrete and continuous eigenvalues.

In some sense one can say that in quantum mechanics the problem of time evolution does not possess the primary importance it has in classical mechanics: indeed we have seen that the problem of determining the eigenvectors and eigenvalues of H is preliminary to that of time evolution; furthermore the eigenstates of energy are stationary states so that, given the importance (from the physical point of view) of the eigenstates of energy, stationarity is almost more important than time evolution. Of course this statement should be cautiously taken: for example, if H only has continuous eigenvalues, there exist no proper eigenstates so that no state is rigorously stationary. But, even if H only has discrete eigenvalues, as is the case for the harmonic oscillator, not only the stationary states exist: the oscillators, after all, do oscillate! Let us indeed assume we have a one-dimensional harmonic oscillator, initially in the state $|A,0\rangle$ represented by the wavefunction $\psi_A(x,0)$. One has

$$|A,0\rangle = \sum_n a_n |n\rangle$$

$$|A,t\rangle = \sum_n a_n \, e^{-i(n+\frac{1}{2})\omega t}|n\rangle = e^{-\frac{1}{2}i\omega t} \sum_n a_n \, e^{-in\omega t}|n\rangle . \qquad (9.16)$$

If $T = 2\pi/\omega$, since $e^{-in\omega(t+T)} = e^{-in\omega t}$ and $e^{-\frac{1}{2}i\omega T} = -1$, one has

$$|A,t+T\rangle = -|A,t\rangle$$

so after a period, although the vector has changed sign, the *state* has come back the same: indeed, the factor $e^{-\frac{1}{2}i\omega t}$ is totally irrelevant. So in general, and as expected, the time evolution of the nonstationary states of the harmonic oscillator is periodic and the period is the same as in classical physics. However, if in (9.16) only the a_n with even (odd) n are nonvanishing (correspondingly the $\psi(x,t)$ is an even (odd) function of x), the period of the state is $T/2$ and this fact, as well as the existence of stationary states, has no classical analogue. In general, after half a period, one has

$$|A,t+T/2\rangle = e^{-i\pi/2} \sum_n a_n (-1)^n |n\rangle$$

and, since the space-inversion operator I acts on the eigenvectors of H according to $I|n\rangle = (-1)^n |n\rangle$, one has

$$|A,t+T/2\rangle = -i\,I\,|A,t\rangle \qquad \Rightarrow \qquad \psi_A(x,t+T/2) = -i\,\psi_A(-x,t)$$

i.e. after half a period, up to the factor $-i$, we find the wavefunction reflected with respect to the origin. During the interval $(t,\, t+T/2)$ the wavefunction does not preserve its form; indeed, if for example at time t the wavefunction is real, between the instants t and $t+T/2$ the wavefunction is complex (and not only because of a constant phase factor): the time evolution of ψ_A is *not* simply a shift of its graph.

As a byproduct we discover that, independently of the Hamiltonian of the particle, a possible way of representing the space-inversion operator is given by

$$I = \mathrm{i}\, \mathrm{e}^{-\frac{1}{2}\mathrm{i}\pi\left(p^2/(m\hbar\omega)+m\omega q^2/\hbar\right)} \qquad \text{for any real } m,\, \omega\,. \tag{9.17}$$

The time evolution of a given state being known, we can establish how the mean value of whatever observable ξ of the system in that state changes over time: the mean value of ξ in the state at time t, that we will denote by $\overline{\xi}_t$, is

$$\overline{\xi}_t = \langle\, A\,,t\mid \xi \mid A\,,t\,\rangle = \langle\, A\,,0 \mid U(t)^\dagger \xi\, U(t) \mid A\,,0\,\rangle\,. \tag{9.18}$$

Note that if $\mid A\,,t\,\rangle$ is a stationary state (i.e. if it is an eigenstate of H) the mean value of *any* observable is independent of time: indeed, if the state does not change, no mean value can change. Equation (9.18) provides us with the same result: if $\mid A\,,t\,\rangle = \mathrm{e}^{-\mathrm{i}\,E t/\hbar}\mid A\,,0\,\rangle$, then

$$\langle\, A\,,t\mid \xi \mid A\,,t\,\rangle = \mathrm{e}^{\mathrm{i}\,E t/\hbar}\langle\, A\,,0\mid \xi \mid A\,,0\,\rangle\, \mathrm{e}^{-\mathrm{i}\,E t/\hbar} = \langle\, A\,,0\mid \xi \mid A\,,0\,\rangle\,.$$

We now ask ourselves whether there exist observables such that, *for any* state of the system and *for any* t fulfill $\overline{\xi}_t = \overline{\xi}_0$ (i.e. $\overline{\xi}_t$ is independent of t). From (9.18) it follows that one must have

$$\langle\, A\,,0\mid U(t)^\dagger \xi\, U(t) \mid A\,,0\,\rangle = \langle\, A\,,0\mid \xi \mid A\,,0\,\rangle\,. \tag{9.19}$$

Since (9.19) must hold for any $\mid A\,,0\,\rangle$, it follows that

$$U(t)^\dagger \xi\, U(t) = \xi\,. \tag{9.20}$$

In addition, the latter having to be satisfied for any t, we can differentiate both sides with respect to t: recalling that $U(t) = \mathrm{e}^{-\mathrm{i}\,H t/\hbar}$,

$$\frac{\mathrm{i}}{\hbar}\, H\, \mathrm{e}^{\mathrm{i}\,H t/\hbar}\, \xi\, \mathrm{e}^{-\mathrm{i}\,H t/\hbar} - \frac{\mathrm{i}}{\hbar}\, \mathrm{e}^{\mathrm{i}\,H t/\hbar}\, \xi\, H\, \mathrm{e}^{-\mathrm{i}\,H t/\hbar)} = 0$$

and putting $t = 0$ one finally obtains

$$[\,H\,,\xi\,] = 0\,. \tag{9.21}$$

Viceversa: if ξ commutes with H, it commutes also with $U(t)$ and (9.20) is satisfied. Therefore all and only the observables that commute with the Hamiltonian H have mean values that, in any state, do not depend on time. Such observables are called **constants of motion**.

Indeed, if $[\,H\,,\xi\,] = 0$ also $[\,H\,,\xi\,]_{\mathrm{PB}} = 0$ and, due to (4.50), ξ is a quantum observable that corresponds to a classical constant of motion. Much as in classical mechanics, also in quantum mechanics the knowledge of constants of motion allows for a simplification of the problem of time evolution, thanks to the following

Theorem: if ξ is a constant of motion and if $\mid A\,,0\,\rangle$ is an eigenvector of ξ corresponding to the eigenvalue ξ', also $\mid A\,,t\,\rangle$ is such.

Indeed, if ξ is a constant of motion, ξ and $U(t)$ commute with each other and thanks again to the lemma of p. 87 (Sect. 4.10) the thesis follow.

It then appears that the vector $\mid A\,,t\,\rangle$ remains, for any t, in the subspace of $\mathcal{H}$ consisting of the eigenvectors of ξ corresponding to the eigenvalue ξ'.

9.2 Time Evolution in the Heisenberg Picture

As we know, the states of a system are 'described' or represented by the vectors of a Hilbert space $\mathcal{H}$; in any event it is clear that a vector in $\mathcal{H}$ completely describes a state only inasmuch as, by means of rules that should by now be familiar, from the knowledge of the vector we can extract all the pieces of information that characterize the state of the system: transition probabilities, mean values of the observables, the observables for which the state is an eigenstate and the corresponding eigenvalues, etc. We have seen that all the above information can be deduced from the mean values of suitable observables so that we can state that the knowledge of a state is equivalent to the knowledge of the mean values of all the observables in that state.

Also the problem of time evolution can be considered from this point of view: the knowledge of how the states evolve in time is equivalent to the knowledge of how the mean values of the observables evolve in time. This can be obtained, as has been made in the previous section by means of (9.18), from the knowledge of $|\,A\,,t\,\rangle$, given $|\,A\,,0\,\rangle$. We emphasize that, in the above scheme, time dependence is assigned to the vectors that represent the states: such a scheme is known as the ***Schrödinger picture*** for time evolution. But any other way of determining how the mean values of observables evolve in time is acceptable as well, indeed equivalent.

It is clear from (9.18) that, if the state of the system at a given instant is known – e.g. $t = 0$ – and let $|\,A\,,0\,\rangle$ a vector representative of such a state – then for any observable ξ we know $\overline{\xi}_t$ if we know

$$\xi(t) \overset{\text{def}}{=} U(t)^{\dagger}\,\xi\,U(t)\,. \tag{9.22}$$

In other word, as all one is interested in is

$$\overline{\xi}_t = \langle\,A\,,0\,|\,U(t)^{\dagger}\,\xi\,U(t)\,|\,A\,,0\,\rangle$$

we have the two equivalent possibilities of 'sticking' the time evolution operators either to the vectors, i.e.

$$\overline{\xi}_t = \langle\,A\,,t\,|\,\xi\,|\,A\,,t\,\rangle$$

or of sticking the $U(t)$ and $U(t)^{\dagger}$ to the observable, as in (9.22), i.e.

$$\overline{\xi}_t = \langle\,A\,,0\,|\,\xi(t)\,|\,A\,,0\,\rangle\,.$$

The latter possibility takes the name of ***Heisenberg picture*** for time evolution. Therefore in the Heisenberg picture the observable do carry the time dependence:

$$\xi \xrightarrow{\text{after a time } t} \xi(t)$$

whereas the vectors stay fixed. Equation (9.22) for the evolution of the observables in the Heisenberg picture therefore takes over (9.1) for the evolution of the vectors in the Schrödinger picture.

Time evolution in the Heisenberg picture appears in a way that is analogous to that of classical mechanics: in classical mechanics indeed the q and p, and in general the $f(q,p)$, do depend on time according to a law that is determined by the equations of motion (4.50).

The classical equations are in differential form: let us then find which is the differential form of the quantum equations (9.22).

Taking $U(t) = \mathrm{e}^{-\mathrm{i}\,H\,t/\hbar}$ and differentiating with respect to t both sides of (9.22) (pay attention to the order of the operators!) one has

$$\dot{\xi}(t) = \frac{\mathrm{i}}{\hbar} U(t)^\dagger\, H\, \xi\, U(t) - \frac{\mathrm{i}}{\hbar} U(t)^\dagger\, \xi\, H\, U(t)$$

therefore

$$\dot{\xi}(t) = \frac{\mathrm{i}}{\hbar} U(t)^\dagger\, [\,H\,,\,\xi\,]\, U(t) \qquad (9.23)$$

or also, having in mind that H commutes with both $U(t)$ and $U(t)^\dagger$,

$$\dot{\xi}(t) = \frac{\mathrm{i}}{\hbar}\, [\,H\,,\,\xi(t)\,]\,. \qquad (9.24)$$

Equations (9.24) are known as **Heisenberg equations**. If we have in mind the quantization postulate in the form (4.53), we see that (9.24) is identical with (4.50), in other words:

Heisenberg equations are formally identical with the classical equations of motion.

Of course the classical equations are equations for the *number valued* functions $q(t)$, $p(t)$, $f\big(q(t),p(t)\big)$, whereas (9.24) are equations for the corresponding *operator valued* functions. But, apart from this – substantial – difference and the – less substantial – question of the ordering of factors (see the discussion following (4.53)), they are identical with each other. This identity (that, we shall see in the next section, plays a fundamental role in finding classical mechanics as a suitable limit of quantum mechanics) is a consequence of the quantization postulate and of the postulates introduced in order to solve the problem of time evolution: see in particular the postulated proportionality between K and H expressed by (9.5).

As an exercise, let us write the Heisenberg equation for a particle whose Hamiltonian is

$$H = \frac{p^2}{2m} + V(q)\,.$$

One has

$$\dot{q}(t) = \frac{\mathrm{i}}{\hbar}\, [\,H\,,\,q(t)\,]\,, \qquad \dot{p}(t) = \frac{\mathrm{i}}{\hbar}\, [\,H\,,\,p(t)\,]\,. \qquad (9.25)$$

How can we calculate the commutators between $q(t)$, $p(t)$ and H if we do not yet know how $q(t)$ and $p(t)$ are expressed in terms of q and p (i.e. we have not yet solved the equations of motion)? To this end one takes advantage of the form (9.23) of the Heisenberg equations, namely

$$[H, q(t)] = U(t)^\dagger [H, q] U(t) = U(t)^\dagger \left[\frac{p^2}{2m}, q \right] U(t)$$

$$= -i \hbar U(t)^\dagger \frac{p}{m} U(t) = -i \frac{\hbar}{m} p(t); \qquad (9.26)$$

$$[H, p(t)] = U(t)^\dagger [H, p] U(t)$$

$$= U(t)^\dagger [V(q), p] U(t) = i \hbar \frac{\partial V(q(t))}{\partial q}. \qquad (9.27)$$

By inserting (9.26) and (9.27) into (9.25) one finally obtains

$$\begin{cases} \dot{q}(t) = \dfrac{p(t)}{m} \\ \dot{p}(t) = -\dfrac{\partial V(q(t))}{\partial q} \end{cases} \qquad (9.28)$$

that are formally identical with the classical Hamilton equations.

For example, in the case of the harmonic oscillator, the equations of motion being linear, the integration is made as in the classical case:

$$q(t) = A \cos \omega t + B \sin \omega t$$

where A and B are operators determined by the initial conditions $q(0) = q$ and $\dot{q}(0) = p/m$, then

$$q(t) = q \cos \omega t + \frac{p}{m \omega} \sin \omega t, \quad p(t) = p \cos \omega t - m \omega q \sin \omega t$$

from which the result expressed by (9.17) is found again: $q(\pi/\omega) = -q$, $p(\pi/\omega) = -p$.

In the previous section we have defined constant of motion those observables that commute with H: we then see from either (9.23) or (9.22) that in the Heisenberg picture the constant of motion are those observables that do not depend on time: $[\xi, H] = 0$ entails $\xi(t) = \xi$; one finds again that the mean values (and, more in general, the matrix elements) of the constants of motion do not depend on time.

9.3 The Classical Limit of Quantum Mechanics

The purpose of this section is to establish the conditions under which the results of classical mechanics for a system consisting of one particle are a good approximation of the results provided by quantum mechanics.

The problems we must face are essentially two.

1. Which quantum states of the system admit a classical description.
2. In which conditions the classical scheme and the quantum scheme provide, to a good approximation, the same results for the time evolution.

As far as point 1 is concerned, according to classical mechanics the state of a particle is determined by giving position and momentum of the particle; according to quantum mechanics, instead, the state is determined by giving, e.g. in the Schrödinger representation, the wavefunction $\psi(x)$. The transition from the quantum description to the classical one is realized by attributing a position q_{cl} and a momentum p_{cl} that equal the mean values of q and p in the quantum state of the particle:

$$q_{cl} = \bar{q}\,, \qquad\qquad p_{cl} = \bar{p}\,.$$

If the above position is to make any sense, it is necessary that the fluctuations Δq and Δp of the values that q and p may take in the state of the particle around their mean values be negligible. What determines whether Δq and Δp are either small or large is, in essence, the resolution power of the instruments employed to measure q and p and the intrinsic dimensions of the problem (e.g. the dimensions of the holes of an accelerating grid in a cathode ray tube, ...).

It appears that a classical description is possible only for particular quantum states: i.e. those states that are not very far from being *minimum uncertainty* states. In particular $\psi(x)$ must be nonvanishing only in a small region of space and its Fourier transform $\tilde{\psi}(k)$ must be itself appreciably nonvanishing only in a small interval around the value $\bar{k} = \bar{p}/\hbar$: the $\psi(x)$ must be what is called a 'wave packet'; indeed $\psi(x)$ is obtained by superposing – or packing – plane waves $e^{i\,k\,x}$ with k in a small interval around $\bar{k}$; for example:

$$\psi(x) = \int_{\bar{k}-\Delta}^{\bar{k}+\Delta} e^{i\,k\,x}\,\tilde{\psi}(k)\,dk\ .$$

Once one has established in which way and for which quantum states a classical description is possible, the problem of time evolution must be now faced: the state of the system evolves in time, so one must verify, from the one hand, that the conditions allowing for a classical description of the state carry on being satisfied over time (i.e. that Δq and Δp stay small) and, on the other hand, that $q_{cl}(t)$ and $p_{cl}(t)$, obtained by solving the classical equations of motion, do not differ too much from $\bar{q}(t)$ and $\bar{p}(t)$ obtained by solving the quantum equations of motion: in other words the center-of-mass $\bar{q}(t)$ of the wave packet must evolve according to the classical laws of motion.

The equation that determines $q_{cl}(t)$ is

$$m\,\ddot{q}_{cl}(t) = F(q_{cl})\ . \tag{9.29}$$

($F(q)$ is the force), whereas the equation for $\bar{q}(t)$ is obtained by the operator equation

$$m\,\ddot{q}(t) = F(q)$$

that follows from the Heisenberg equations (9.25) or (9.28). Taking the mean value of both sides in the state of the system

$$m\,\ddot{\overline{q}}(t) = \overline{F(q)}\,. \tag{9.30}$$

Comparison of (9.29) and (9.30) shows that, if we want that at any time $q_{cl} = \overline{q}$, since $F(q_{cl}) = F(\overline{q})$, it is necessary that

$$F(\overline{q}) = \overline{F(q)}\,. \tag{9.31}$$

Equation (9.31) is satisfied only if F is a linear function of q: $F(q) = a\,q + b$, so that the center-of-mass moves as a classical particle if the particle is:

i) a free particle;
ii) a particle subject to a constant force;
iii) a particle subject to a harmonic force.

With the exception of these particular cases, let us investigate under which circumstances (9.31) is at least approximately satisfied.

Let us expand $F(q)$ in a powers series of $q - \overline{q}$:

$$F(q) = F(\overline{q}) + F'(\overline{q})\,(q - \overline{q}) + R(q) \tag{9.32}$$

with $R(q)$ including all the powers of $q - \overline{q}$ higher than the first.

By taking the mean value of both sides of (9.32) one obtains

$$\overline{F(q)} = F(\overline{q}) + \overline{R(q)}\,.$$

The smaller $\overline{R(q)}$ with respect to $F(\overline{q})$, the better (9.31) will be satisfied. From

$$\overline{R(q)} = \int |\psi(x)|^2\,R(x)\,dx$$

it appears that $R(x)$ must be small (i.e. $F(q)$ must differ not too much from a linear function) in the region where $|\psi(x)|^2$ is appreciably different from zero. In other words: the force must be 'good' and the wave packet small. If one considers, for example, the first term in $R(q)$, i.e. $\frac{1}{2}F''(\overline{q})\,(q - \overline{q})^2$ then, in order that its mean value $\frac{1}{2}F''(\overline{q})\,(\Delta q)^2$ be small with respect to $F(\overline{q})$, both $F''(\overline{q})$ and Δq must be small; more precisely: it is the variation of $F'(x)$ over an interval of the order of Δq around $\overline{q}$ that must be small.

One could be tempted to conclude that the smaller Δq is, the better the classical approximation holds. Indeed things are not in this way because a too small Δq entails a too large Δp and, therefore, the state does not admit a classical description. But there is more to the point: in general a too large Δp takes along a spreading of the packet as time goes by, so equation (9.31) is more and more poorly satisfied.

It appears that the conditions for the validity of the classical approximation are realized when a compromise situation is reached: Δq must be small enough for (i) the wave packet be well described by the knowledge of its center-of-mass, (ii) equation (9.31) be satisfied to a good approximation; on the other hand Δq cannot be too small otherwise Δp is big, with the consequence that the state does not admit a classical description and, in addition, the wave packet spreads too much.

The following discussion will, however, emphasize that for macroscopic systems (a grain of size $\simeq 10^{-4}$ cm and mass $\simeq 10^{-12}$ g can be considered a macroscopic system from this point of view) there exists a large window in which this compromise situation can be fulfilled.

Let us explicitly examine, to this end, how the width of a wave packet changes in time for a free particle. This is the same as determining how Δq depends on time: let us solve this problem in the Heisenberg picture for time evolution.

By integrating (9.28) in the case $V(q) = 0$ one has

$$p(t) = p, \qquad q(t) = q + \frac{p}{m} t$$

whence:

$$\left(\Delta q(t)\right)^2 = \overline{q(t)^2} - \overline{q(t)}^2$$
$$= (\Delta q)^2 + \frac{1}{m}\left(\overline{qp} + \overline{pq} - 2\overline{q}\,\overline{p}\right) t + \frac{(\Delta p)^2}{m^2} t^2 . \qquad (9.33)$$

This shows that $\left(\Delta q(t)\right)^2$ is a positive definite quadratic function of t (see Fig. 9.1).

It appears that until some instant t_0 the wave packet shrinks, then it spreads. So inevitably for large times the packet will spread: this is an important observation for it sets a limit to the time span during which the packet can be treated as a classical particle. For sufficiently large times, assuming that for $t = 0$ the packet approximately corresponds to a minimum uncertainty state ($\Delta p\,\Delta q \simeq \hbar$), by (9.33) one has

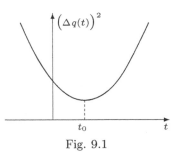

Fig. 9.1

$$\Delta q(t) \simeq \frac{\Delta p}{m} t \simeq \frac{\hbar}{m\,\Delta q} t . \qquad (9.34)$$

Equation (9.34) emphasizes the relevant parameters on which the rate of spreading depends: $\hbar$, m and Δq. It is worth evaluating this rate of spreading in some typical cases.

1. A free electron with $\Delta q \simeq 1$ Å: $\hbar/(m\,\Delta q) \simeq 10^9$ cm s^{-1}. This result speaks for itself.
2. An electron with $\Delta q \simeq 10^{-2}$ cm (for example an electron that has been focalized either in a cathode ray tube or in an electron microscope): $\hbar/(m\,\Delta q) \simeq 10^3$ cm s^{-1}. This rate of spreading may seem large; note however that an electron endowed with energy 10^3 eV should go a path of about 10^4 cm in order that the spreading would result of the order of magnitude of the initial Δq.

3. A dust grain of mass 10^{-12} g and with $\Delta q \simeq 10^{-5}$ cm: $h/(m\,\Delta q) \simeq$ 10^{-9} cm s^{-1}.

The phenomenon of the spreading of wave packets we have just discussed is nothing but the well known phenomenon of diffraction that manifests itself in any problem involving wave propagation. We may better emphasize this fact by considering electrons of momentum $\simeq p$ that cross a slit of width d and are detected on a screen a distance D away from the slit: initially the transversal dimension of the packet associated with each electron is $\Delta q \simeq d$; the electrons take a time $\tau \simeq m\,D/p$ to reach the screen, so by (9.34) the transversal dimension of the packet on the screen (i.e. the dimension of the diffraction pattern produced by the electrons) can be computed:

$$\Delta q(\tau) \simeq \frac{h\,D}{p\,d} \simeq \frac{\lambda}{d}\,D$$

that is a well known formula in diffraction theory.

Chapter 10

Angular Momentum

10.1 Angular Momentum: Commutation Rules

Given the importance angular momentum has in classical physics, it will be no surprise to ascertain the fundamental role it also plays in quantum mechanics. This is mainly due to the fact that, in many cases of physical interest (isolated systems, particles in a central force field, etc.), the components of angular momentum are constants of motion. We will furthermore see that the different components of angular momentum do not commute with one another, so that the assumptions of the degeneracy theorem discussed in Sect. 7.2 are verified and this will enable us to find extremely relevant information on the degree of degeneracy of energy levels.

Angular momentum is defined as in classical mechanics: for a particle with coordinates $\vec{q} = (q_1, q_2, q_3)$ and momentum $\vec{p} = (p_1, p_2, p_3)$, the angular momentum $\vec{M}$ is defined by

$$\vec{M} = \vec{q} \wedge \vec{p} \tag{10.1}$$

and it is straightforward to verify that $\vec{M} = \vec{M}^\dagger$.

The total angular momentum of a system consisting of n particles is

$$\vec{M} = \sum_{\alpha=1}^{n} \vec{q}_\alpha \wedge \vec{p}_\alpha = \sum_{\alpha=1}^{n} \vec{M}_\alpha . \tag{10.2}$$

In the sequel several symbols will be used to represent angular momenta: $\vec{L}$, $\vec{s}$, $\vec{S}$, $\vec{J}$, $\cdots$: each of them stands for a particular type of angular momentum. In the general discussion of the present and the next section the angular momentum will be denoted by $\vec{M}$.

We now wish to determine the commutation rules of the components of the angular momentum given by (10.1) with the components of $\vec{q}$ and $\vec{p}$, and among the different components of $\vec{M}$ itself. Let us start with the calculation of $[\,M_i\,,\,q_j\,]$. There are several ways to arrive at the result:

1. direct calculation;

© Springer International Publishing Switzerland 2016
L.E. Picasso, *Lectures in Quantum Mechanics*, UNITEXT for Physics,
DOI 10.1007/978-3-319-22632-3_10

2. calculation of the Poisson brackets and of the commutators as a consequence of (4.53);
3. to exploit the relationship between angular momentum and rotations.

We will follow the third way (but also the (4.53)), which certainly is the most meaningful.

It is known from analytical mechanics that, if $(q, p) \rightarrow (q', p')$ is an infinitesimal **canonical transformation** of coordinates, then there exists a function $G(q, p)$, called the **generating function** of the transformation, such that

$$\begin{cases} q'_j = q_j + \epsilon \, [\, G \, , \, q_j \,]_{\text{PB}} \\ p'_j = p_j + \epsilon \, [\, G \, , \, p_j \,]_{\text{PB}} \end{cases} \tag{10.3}$$

in which ϵ is the infinitesimal parameter of the transformation.

If the transformation is a **rotation** around some given axis, then G coincides with the component of $\vec{M}$ parallel to the rotation axis. So, for example, for a rotation by an angle ϕ around the 3 (or z) axis one has

$$\begin{cases} q'_1 = q_1 \, \cos\phi + q_2 \, \sin\phi \\ q'_2 = -q_1 \, \sin\phi + q_2 \, \cos\phi \\ q'_3 = q_3 \end{cases} \tag{10.4}$$

and if $\phi \rightarrow \epsilon$ (infinitesimal):

$$\begin{cases} q'_1 = q_1 + \epsilon \, q_2 \\ q'_2 = q_2 - \epsilon \, q_1 \\ q'_3 = q_3 \, . \end{cases} \tag{10.5}$$

For the transformation (10.5) $G = M_3$, so that comparison of (10.3) with (10.5) yields

$$[\, M_3 \, , \, q_1 \,]_{\text{PB}} = q_2 \, , \qquad [\, M_3 \, , \, q_2 \,]_{\text{PB}} = -q_1 \, , \qquad [\, M_3 \, , \, q_3 \,]_{\text{PB}} = 0$$

whence:

$$[\, M_3 \, , \, q_1 \,] = i \, \hbar \, q_2 \, , \qquad [\, M_3 \, , \, q_2 \,] = -i \, \hbar \, q_1 \, , \qquad [\, M_3 \, , \, q_3 \,] = 0 \, . \tag{10.6}$$

The commutation rules of M_1 and M_2 with the components of $\vec{q}$ are found likewise:

$$\begin{aligned} [\, M_1 \, , \, q_1 \,] &= 0 \, , & [\, M_1 \, , \, q_2 \,] &= i \, \hbar \, q_3 \, , & [\, M_1 \, , \, q_3 \,] &= -i \, \hbar \, q_2 \\ [\, M_2 \, , \, q_1 \,] &= -i \, \hbar \, q_3 \, , & [\, M_2 \, , \, q_2 \,] &= 0 \, , & [\, M_2 \, , \, q_3 \,] &= i \, \hbar \, q_1 \, . \end{aligned} \tag{10.7}$$

Equations (10.6) and (10.7) are summarized in the formula:

$$[\, M_i \, , \, q_j \,] = i \, \hbar \, \epsilon_{ijk} \, q_k \tag{10.8}$$

where in the right hand side, as usual, the sum over the repeated index k is understood; ϵ_{ijk} is the Ricci symbol, nonvanishing only if the indices $i \, j \, k$ are all different from one another, equals $+1$ if ijk are a cyclic permutation of $1, 2, 3$ and equals -1 otherwise.

We have established (10.8) by means of a certainly longer way than the direct calculation: we now know, however, that (10.8) are the consequence of the fact that the components of $\vec{q}$ are the components of a vector, i.e. they transform under rotations according to (10.4) and its analogues for rotations around axes different from the z-axis. We can therefore conclude that we will find the same commutation rules between $\vec{M}$ and the components of *any* vector $\vec{V}$, i.e.

$$[\, M_i \,,\, V_j \,] = i\,\hbar\,\epsilon_{ijk}\,V_k \; . \tag{10.9}$$

In particular, for $\vec{V} = \vec{p}$:

$$[\, M_i \,,\, p_j \,] = i\,\hbar\,\epsilon_{ijk}\,p_k \tag{10.10}$$

and for $\vec{V} = \vec{M}$:

$$[\, M_i \,,\, M_j \,] = i\,\hbar\,\epsilon_{ijk}\,M_k \; . \tag{10.11}$$

The above are the commutation rules we had resolved to determine.

The advantages of the method we have followed for the determination of (10.8), (10.10) and (10.11) appear also in another aspect of the problem of determining the commutation rules between the components of $\vec{M}$ and the observables $f(q, p)$ of the theory. Indeed, for an infinitesimal transformation $(q, p) \rightarrow (q', p')$, from (10.3) one has (to the first order in ϵ):

$$\begin{aligned}
f(q,p) \rightarrow f(q',p') &= f\big(q + \epsilon\,[\,G\,,\,q\,]_{\mathrm{PB}}\,,\; p + \epsilon\,[\,G\,,\,p\,]_{\mathrm{PB}}\big) \\
&= f(q,p) + \epsilon \sum_i \Big(\frac{\partial f}{\partial q_i}\,[\,G\,,\,q_i\,]_{\mathrm{PB}} + \frac{\partial f}{\partial p_i}\,[\,G\,,\,p_i\,]_{\mathrm{PB}} \Big) \\
&= f(q,p) + \epsilon \sum_i \Big(-\frac{\partial f}{\partial q_i}\frac{\partial G}{\partial p_i} + \frac{\partial f}{\partial p_i}\frac{\partial G}{\partial q_i} \Big) = f(q,p) + \epsilon\,[\,G\,,\,f\,]_{\mathrm{PB}}
\end{aligned}$$

i.e., in general, $\epsilon\,[\,G\,,\,f\,]_{\mathrm{PB}}$ exactly equals the variation $\delta f \equiv f(q',p') - f(q,p)$ of $f(q,p)$ under the infinitesimal transformation generated by G. In this way one has that, in the case of rotations, the Poisson brackets – and the commutators – between the components of $\vec{M}$ and those observables $f(q,p)$ that are invariant under rotations do vanish. The latter observables are called **scalars**: for example, in the case of a single particle, $\vec{q}^{\,2}$, $\vec{p}^{\,2}$, $\vec{q}\cdot\vec{p}$ and their functions are scalars.

We thus have the general result, analogous to (10.9) that

the components of $\vec{M}$ commute with all the scalar operators S:

$$[\, M_i \,,\, S \,] = 0 \; . \tag{10.12}$$

In particular, for $S = \vec{M}^{\,2} = M_1^2 + M_2^2 + M_3^2$:

$$[\, M_i \,,\, \vec{M}^{\,2} \,] = 0 \; . \tag{10.13}$$

Equation (10.11) expresses the fact that the different components of $\vec{M}$ are not compatible observables whereas (10.13) ensures that $\vec{M}^2$ is compatible separately with each component of $\vec{M}$: as a consequence – with the exception of a particular case we will discuss later – there exist no simultaneous eigenstates of M_1, M_2 and M_3, i.e. states with well defined angular momentum $\vec{M}$; there exists however a complete set of states in which both $\vec{M}^2$ and a single component M_i have a well defined value.

As a consequence of (10.12) we find the well known result that for a particle in a central force field all the components M_i are constants of motion: indeed the potential for a central force field is a function only of the distance $r = \sqrt{q_1^2 + q_2^2 + q_3^2}$ of the particle from the center, therefore

$$H = \frac{\vec{p}^{\,2}}{2m} + V(r)$$

is a scalar, i.e. it commutes with any M_i.

Concerning (10.9) and (10.12), note that they hold only if $\vec{V}$ and S respectively are vectors and scalars built with the variables of the system ($\vec{q}$ and $\vec{p}$ in the case at hand): for example, the potential energy of an electron (charge $-e$) in an external uniform electric field $\vec{E}$ is written as $e\,\vec{E}\cdot\vec{q}$ and the latter, despite its appearance, is not a scalar because $\vec{E}$, despite its appearance, is not a vector: indeed $\vec{E}$ is an *external* field, i.e. it does not depend on the dynamical variables of the system ($\vec{q}$ and $\vec{p}$) and therefore commutes with the M_i: one says (see Sect. 4.11) that $\vec{E}$ is a 'c-number' (in order to distinguish it from the 'q-numbers', i.e. the operators).

Another way to see that $\vec{E}\cdot\vec{q}$ is not a scalar is the following: $e\,\vec{E}\cdot\vec{q}$ is unchanged (i.e. it is a scalar) if both the electron and the capacitor that generates the field are rotated, so it commutes with the angular momentum of the whole system consisting of both the electron and the capacitor; but if, as in the case we are considering, the system consists of the electron alone, then, as the result of a rotation *of the system* (the electron), the components of $\vec{q}$ change, the components of $\vec{E}$ do not, therefore $\vec{E}\cdot\vec{q}$ changes, i.e. it is not a scalar.

10.2 Angular Momentum: Eigenvalues

In this section we set out to determine the eigenvalues of the components M_i of the angular momentum, and of $\vec{M}^2$. First of all, it should be clear that all the M_i have the same eigenvalues (but not the same eigenvectors!), so it suffices to find the eigenvalues of one component of $\vec{M}$: by tradition M_3 is chosen, but the same procedure would be used for M_1 or M_2 and the same results would be obtained.

Let $\mu^2\hbar^2$ stand for the eigenvalues of $\vec{M}^2$ and $m\,\hbar$ for those of M_3: in this way μ^2 and m are dimensionless numbers. Indeed, often μ^2 and m will be referred to (at the expense of precision) as the eigenvalues of $\vec{M}^2$ and M_3: one may consider the latter as a convenient terminology, or in alternative one may think of measuring the angular momenta in units of $\hbar$.

According to (10.13) the operators $\vec{M}^2$ and M_3 commute with each other, so they possess a complete set of simultaneous eigenvectors that we will denote by $|\mu^2, m\rangle$. One must however expect that $\vec{M}^2$ and M_3 alone do not form a complete set of compatible observables, so the state is not completely determined by the knowledge of the eigenvalues μ^2 and m. Let us then add to $\vec{M}^2$ and M_3 a certain number of observables $\sigma_1, \sigma_2, \cdots$, globally denoted by Σ, such that $\Sigma, \vec{M}^2, M_3$ form a complete set of compatible observables. To this end it is necessary that Σ commutes with both $\vec{M}^2$ and M_3: this is certainly true if Σ (i.e. $\sigma_1, \sigma_2, \cdots$) commutes with all the M_i, i.e. if Σ is a scalar – which we will assume from now on.

We can now consider the complete set of vectors $|\Sigma', \mu^2, m\rangle$, Σ' – i.e. $\sigma'_1, \sigma'_2, \cdots$ – standing for the generic eigenvalue of Σ.

The following calculation, with the aim of finding the eigenvalues of $\vec{M}^2$ and M_3, will automatically provide us with some information on the degeneracy of the eigenvalues Σ' of Σ: since normally in applications Σ (better: one of the σ_i) is the Hamiltonian, the latter are also information about the degeneracy of the energy levels.

The procedure to be followed to determine the eigenvalues of $\vec{M}^2$ and M_3 strongly resembles that followed in Chap. 5 to determine the eigenvalues of the Hamiltonian of the harmonic oscillator: let us introduce the operators

$$M_+ = M_1 + i\,M_2, \qquad\qquad M_- = M_1 - i\,M_2 = (M_+)^\dagger. \qquad (10.14)$$

From (10.11) one obtains

$$[M_3, M_+] = \hbar\,M_+, \qquad\qquad [M_3, M_-] = -\hbar\,M_- \qquad (10.15)$$

whereas from (10.12) and (10.13) one has

$$[M_+, \Sigma] = [M_-, \Sigma] = 0, \qquad [M_+, \vec{M}^2] = [M_-, \vec{M}^2] = 0. \qquad (10.16)$$

Equations (10.15), analogous to (5.12) and (5.13), express the fact that M_+ and M_- respectively behave as 'step up' and 'step down' operators by the amount $\hbar$ for the eigenvalues of M_3: from this, and from the lemma of p. 87 (Sect. 4.10), one has

$$M_+|\Sigma', \mu^2, m\rangle = \begin{cases} 0 & \text{or} \\ |\Sigma', \mu^2, m+1\rangle \end{cases} \qquad (10.17)$$

$$M_-|\Sigma', \mu^2, m\rangle = \begin{cases} 0 & \text{or} \\ |\Sigma', \mu^2, m-1\rangle \end{cases} \qquad (10.18)$$

i.e. M_+ and M_- (if they do not give a vanishing result) make m change by one unit, but *neither* Σ' *nor* μ^2 are changed (if $|\Sigma', \mu^2, m\rangle$ is normalized, the vectors $|\Sigma', \mu^2, m\pm1\rangle$ are not such).

Then, starting from a vector $|\Sigma', \mu^2, m\rangle$ and repeatedly applying to it the operator M_+ one obtains the ascending chain of vectors

$$| \Sigma', \mu^2, m+1 \rangle, | \Sigma', \mu^2, m+2 \rangle, \cdots$$

whereas if one applies the operator M_- to it one obtains the descending chain

$$| \Sigma', \mu^2, m-1 \rangle, | \Sigma', \mu^2, m-2 \rangle, \cdots .$$

Can these chains stop? In other words one is asking whether there exists a maximum value for m – that we will denote by j – and a minimum – that we will denote by $\bar{\jmath}$ – such that

$$M_+ | \Sigma', \mu^2, j \rangle = 0, \qquad\qquad M_- | \Sigma', \mu^2, \bar{\jmath} \rangle = 0 . \qquad (10.19)$$

Both chains must interrupt: indeed

$$M_3^2 = \vec{M}^{\,2} - M_1^2 - M_2^2$$

and taking the mean value of this equation in the state $| \Sigma', \mu^2, m \rangle$ (that is an eigenstate of both M_3 – therefore of M_3^2 – and of $\vec{M}^{\,2}$) one has

$$m^2 \le \mu^2 \qquad\qquad\qquad (10.20)$$

because the mean value of $M_1^2 + M_2^2$ is nonnegative. Therefore both the ascending and the descending chain must interrupt, otherwise (10.20) would be violated. When the first of (10.19) is satisfied, one has (we recall that $M_- = (M_+)^\dagger$)

$$\langle \Sigma', \mu^2, j \,|\, M_- M_+ \,|\, \Sigma', \mu^2, j \rangle = 0 . \qquad (10.21)$$

But

$$\begin{aligned} M_- M_+ &= (M_1 - i\, M_2)\,(M_1 + i\, M_2) = M_1^2 + M_2^2 + i\,[M_1\,,\,M_2] \\ &= \vec{M}^{\,2} - M_3^2 - \hbar\, M_3 \end{aligned} \qquad (10.22)$$

that, inserted into (10.21), gives:

$$\mu^2 = j\,(j+1) . \qquad\qquad\qquad (10.23)$$

This relation provides us with the maximum value m may have, once μ has been fixed.

Likewise, when the second of (10.19) is satisfied, one has

$$\langle \Sigma', \mu^2, \bar{\jmath} \,|\, M_+ M_- \,|\, \Sigma', \mu^2, \bar{\jmath} \rangle = 0 . \qquad (10.24)$$

but

$$M_+ M_+ - (M_1 + i\, M_2)\,(M_1 - i\, M_2) = \vec{M}^{\,2} - M_3^2 + \hbar\, M_3 \qquad (10.25)$$

that, inserted into (10.24), gives:

$$\mu^2 = \bar{\jmath}\,(\bar{\jmath}-1) . \qquad\qquad\qquad (10.26)$$

From (10.23) and (10.26) one has

$$j(j+1) = \bar{j}(\bar{j}-1).$$

The above equation has two solutions: $\bar{j} = -j$ and $\bar{j} = j+1$ of which only the first is acceptable, because $j \geq \bar{j}$. Therefore

$$m = -j, -j+1, -j+2, \cdots, j-1, j$$

and, as a consequence $2j$ must be an integer, either positive or vanishing. In conclusion:

1. the possible eigenvalues $\mu^2 \hbar^2$ of $\vec{M}^2$ are $j(j+1)\hbar^2$ with $j = 0, \frac{1}{2}, 1, \frac{3}{2}, 2, \cdots$;
2. the possible eigenvalues of any component M_i of $\vec{M}$ are $m\hbar$ with m either integer or half-integer according to whether j is either integer or half-integer;
3. given Σ' and μ^2 (i.e. j), m may only take the values $-j, -j+1, -j+2, \cdots, j-1, j$, i.e. $2j+1$ values: therefore there exist $2j+1$ independent states in correspondence with any pair of eigenvalues of Σ and $\vec{M}^2$.

This means that both $\vec{M}^2$ and Σ (i.e. any scalar operator) are degenerate observables. Note that this degeneracy is a consequence of the degeneracy theorem of Sect. 7.2: it is due to the fact that M_+, M_- and M_3, i.e. all the components M_i, commute with Σ (as well as with $\vec{M}^2$), but no two of them commute with each other.

For an (isolated) atomic system, if Σ is the Hamiltonian H, normally it happens (the hydrogen atom is an exception to this rule) that for a given value E of H, i.e. for any energy level, only the states corresponding to one value of $\vec{M}^2$ are possible: in this case the knowledge of μ^2, i.e. the knowledge of the angular momentum of the level, automatically brings along the knowledge of the degeneracy of that level. Most times, instead, the value of μ^2 is obtained by the laboratory measurement (we shall see how) of the degeneracy of the level.

Since we have seen that giving μ^2 is the same as giving j, from now on instead of writing $|\,\Sigma', \mu^2, m\,\rangle$ we will write $|\,\Sigma', j, m\,\rangle$ and will briefly say that the latter 'is a state of angular momentum j', understanding that it is an eigenstate of $\vec{M}^2$ with eigenvalue $j(j+1)\hbar^2$.

The results found in the present section are quite general inasmuch as depending only on the commutation rules (10.15) and (10.16) (namely on (10.11) and (10.12)), but not on the fact that $\vec{M}$ is defined as in (10.1), or (10.2) or even in a different way: for this reason one can't take for granted that for a particular system, i.e. for a particular type of angular momentum, j may take all the values we have found as the possible ones: $0, \frac{1}{2}, 1, \cdots$; we will indeed see in the next section that, for any given system, j may take only either the integer values $0, 1, 2, \cdots$, or the half-integer values $\frac{1}{2}, \frac{3}{2}, \frac{5}{2}, \cdots$. In Sect. 10.4 we will show that the first possibility (integer j) takes place if $\vec{M}$

is defined as in (10.1) or (10.2) (*orbital* angular momentum), while we will later have the occasion to see that the second possibility (half-integer j) takes place for example for the electron when one takes into account that, besides the orbital angular momentum, it also possesses – as many other particles – an additional *intrinsic* or *spin angular momentum*.

10.3 Rotation Operators

Let us consider a rotation by an angle ϕ around an *arbitrary* axis, that we will call the 3 (or z) axis. The rotation induces the transformation (10.4) on the q_i and the analogue on the p_i, thus being a canonical transformation. Therefore, by the von Neumann theorem of Sect. 6.3, there exists a unitary operator $U(\phi)$ (unique up to a phase factor) that implements the transformation. Since the composition of two rotations by angles ϕ_1 and ϕ_2 is a rotation by the angle $\phi_1 + \phi_2$, one has that the operators $U(\phi_1)\,U(\phi_2)$ and $U(\phi_1 + \phi_2)$ may at most differ by a phase factor (possibly depending on ϕ_1 and ϕ_2). It is not obvious that it be possible to choose the phase factors, up to which the operators $U(\phi)$ are defined, in such a way that

$$U(\phi_1)\,U(\phi_2) = U(\phi_1 + \phi_2) \qquad \text{for any } \phi_1,\,\phi_2 \qquad (10.27)$$

but for the group of rotations in $\mathbb{R}^3$ this is possible, thanks to a theorem by V. Bargmann (*Bargmann theorem*).

Equation (10.27) is similar to (9.3) and also in the present case one can show (Stone's theorem) that

$$U(\phi) = \mathrm{e}^{-\mathrm{i}\,G\,\phi}$$

where $G\ (=G^\dagger)$ is the generator of the rotations around the considered axis (the 3-axis). The term 'generator' forewarns that G is proportional to M_3: indeed, for infinitesimal rotations ($\phi \to \epsilon$) one has

$$U(\epsilon)\,q_i\,U^{-1}(\epsilon) \simeq (1 - \mathrm{i}\,\epsilon\,G)\,q_i\,(1 + \mathrm{i}\,\epsilon\,G) \simeq q_i - \mathrm{i}\,\epsilon\,[G,\,q_i]$$

and, thanks to (10.5) and (10.6) (and the analogues for p_i) one has that $G = M_3/\hbar$:

$$U(\phi) = \mathrm{e}^{-\mathrm{i}\,M_3\,\phi/\hbar} \qquad (10.28)$$

and in general, for a rotation around an axis parallel to the unit vector $\vec{n}$,

$$U(\vec{n}, \phi) = \mathrm{e}^{-\mathrm{i}\,\vec{M}\cdot\vec{n}\,\phi/\hbar} \ . \qquad (10.29)$$

Let us now assume that for a given system both integer values $j\,'$ and half-integer values $j\,''$ are possible, and let $|\,A\,\rangle$ be whatever state. One has

$$|\,A\,\rangle = \sum |\cdots j\,'\,m'\,\rangle + \sum |\cdots j\,''\,m''\,\rangle \equiv |\,A'\,\rangle + |\,A''\,\rangle \qquad (10.30)$$

in which the first sum is performed over the states with integer j and the second over those with half-integer j. Let us make a rotation by 2π around the z-axis: one has

$$U(2\pi)\,|\,A'\,\rangle = \sum e^{-2\pi\mathrm{i}\,M_3/\hbar}\,|\cdots j'\,m'\,\rangle = \sum e^{-2\pi\mathrm{i}\,m'}\,|\cdots j'\,m'\,\rangle$$
$$= |\,A'\,\rangle \tag{10.31}$$

given that, for integer m', $e^{-2\pi\mathrm{i}\,m'} = 1$. Instead, since for half-integer m'' one has $e^{-2\pi\mathrm{i}\,m''} = -1$,

$$U(2\pi)\,|\,A''\,\rangle = \sum e^{-2\pi\mathrm{i}\,M_3/\hbar}\,|\cdots j''\,m''\,\rangle = -|\,A''\,\rangle\,. \tag{10.32}$$

Of course $|\,A''\,\rangle$ and $-|\,A''\,\rangle$ represent the same state – as it must be, a rotation of 2π being the identity transformation on the q_i and p_i and, more in general, on all the observables (i.e. $U(2\pi)$ is a multiple of the identity); but

$$U(2\pi)\,|\,A\,\rangle = U(2\pi)\left(|\,A'\,\rangle + |\,A''\,\rangle\right) = |\,A'\,\rangle - |\,A''\,\rangle \neq |\,A\,\rangle\,. \tag{10.33}$$

For this reason integer and half-integer angular momenta are not simultaneously possible: the rotations by 2π would not be equivalent to the identity transformation.

10.4 Orbital Angular Momentum

The orbital angular momentum for a particle is defined by (10.1) and for it the notation $\vec{L}$ will be used:

$$\vec{L} = \vec{q} \wedge \vec{p}\,. \tag{10.34}$$

The eigenvalues of any component L_i of $\vec{L}$ are still denoted by $m\,\hbar$ whereas those of $\vec{L}^2$ are denoted by $\hbar^2\,l\,(l+1)$: instead of the letter j of the previous section, the letter l is used here.

For a system of several particles (10.2) defines the *total* angular momentum: this is still denoted by $\vec{L}$; in the latter case, however, upper case letters are used to denote the eigenvalues of L_i and $\vec{L}^2$: eigenvalues of $L_i \to M\,\hbar$; eigenvalues of $\vec{L}^2 \to L\,(L+1)\,\hbar^2$.

We will now show that m, M – therefore l, L – are integer numbers; we want also to find the Schrödinger representation of the states $|\,\Sigma'\!,\,l,\,m\,\rangle$.

Let us first consider the case of a single particle. The Schrödinger representation of L_z (i.e. of L_3) is

$$L_z \xrightarrow{\text{SR}} -\mathrm{i}\,\hbar\left(x\,\frac{\partial}{\partial y} - y\,\frac{\partial}{\partial x}\right)\,.$$

It is convenient to use spherical coordinates with the z-axis as the polar axis (should we consider either L_x or L_y, we would respectively take either x or y as the polar axis):

$$\begin{cases} x = r\,\sin\theta\,\cos\phi \\ y = r\,\sin\theta\,\sin\phi \\ z = r\,\cos\theta \end{cases}$$

whence

$$L_z \xrightarrow{\text{SR}} -i\hbar \left[\left(x\frac{\partial r}{\partial y} - y\frac{\partial r}{\partial x} \right) \frac{\partial}{\partial r} + \left(x\frac{\partial \theta}{\partial y} - y\frac{\partial \theta}{\partial x} \right) \frac{\partial}{\partial \theta} + \left(x\frac{\partial \phi}{\partial y} - y\frac{\partial \phi}{\partial x} \right) \frac{\partial}{\partial \phi} \right]$$

but, as $\partial r/\partial y = y/r$ and $\partial r/\partial x = x/r$, the coefficient of $\partial/\partial r$ vanishes. The same happens for the coefficient of $\partial/\partial \theta$. There remains only the coefficient of $\partial/\partial \phi$ that a direct calculation shows to be 1, so in conclusion:

$$L_z \xrightarrow{\text{SR}} -i\hbar \frac{\partial}{\partial \phi} . \tag{10.35}$$

Equation (10.35) is analogous to $p \to -i\hbar\partial/\partial x$ and it is understandable that it must be so, because ϕ and L_z are canonically conjugate variables, much as x and p are (however the commutation rule $[L_z, \phi] = -i\hbar$ presents some delicate domain problems because, since ϕ varies between 0 and 2π, the multiplication by ϕ introduces a discontinuity).

The eigenvalue equation for L_z (using spherical coordinates) is therefore:

$$-i\hbar \frac{\partial}{\partial \phi} \psi_m(r,\theta,\phi) = m\hbar\, \psi_m(r,\theta,\phi) . \tag{10.36}$$

This equation is similar to (6.42) and its general solution is

$$\psi_m(r,\theta,\phi) = f(r,\theta)\, e^{i m \phi} \tag{10.37}$$

in which $f(r,\theta)$ is an arbitrary function of r and θ. Since the azimuth angle ϕ is defined modulo 2π, (r,θ,ϕ) and $(r,\theta,\phi+2\pi)$ represent the same point. So it must be that

$$\psi_m(r,\theta,\phi) = \psi_m(r,\theta,\phi + 2\pi)$$

i.e.

$$e^{i m \phi} = e^{i m (\phi + 2\pi)}$$

that holds only if m is an integer number.

We have thus shown the orbital angular momentum of a particle only has integer eigenvalues: m, and therefore l are integer numbers.

Equation (10.37) displays the dependence of the eigenfunctions of L_z on the variable ϕ; if we now want the eigenfunctions of $\vec{L}^2$, the expression of $\vec{L}^2$ in the Schrödinger representation is needed; in spherical coordinates one has (but we will skip proving it):

$$\vec{L}^2 \xrightarrow{\text{SR}} -\hbar^2 \left[\frac{1}{\sin\theta} \frac{\partial}{\partial \theta} \left(\sin\theta \frac{\partial}{\partial \theta} \right) + \frac{1}{\sin^2\theta} \frac{\partial^2}{\partial \phi^2} \right] . \tag{10.38}$$

In order to determine the simultaneous eigenfunctions of $\vec{L}^2$ and L_z, one must solve the system of equations:

$$\begin{cases} -\left[\dfrac{1}{\sin\theta} \dfrac{\partial}{\partial \theta} \left(\sin\theta \dfrac{\partial}{\partial \theta} \right) + \dfrac{1}{\sin^2\theta} \dfrac{\partial^2}{\partial \phi^2} \right] \psi_{lm}(r,\theta,\phi) = l(l+1)\, \psi_{lm}(r,\theta,\phi) \\[4mm] -i\dfrac{\partial}{\partial \phi} \psi_{l\,m}(r,\theta,\phi) = m\, \psi_{lm}(r,\theta,\phi) . \end{cases} \tag{10.39}$$

We will not make this calculation. One can however note that the operators $\vec{L}^2$ and L_z do not contain the variable r, so the $\psi_{lm}(r,\theta,\phi)$ must have the form:

$$\psi_{lm}(r,\theta,\phi) = f(r)\, Y_{lm}(\theta,\phi) \qquad (10.40)$$

where $f(r)$ is an arbitrary function of r and the $Y_{lm}(\theta,\phi)$ are solutions of (10.39). They are named **spherical harmonics**. Owing to (10.36) and (10.37), the $Y_{lm}(\theta,\phi)$ must depend on ϕ through the factor $e^{im\phi}$.

We report the spherical harmonics with $l=0$ and $l=1$: the numerical prefactors are chosen in such a way that they are normalized according to

$$\int Y_{l'm'}(\theta,\phi)^* \, Y_{lm}(\theta,\phi)\, \mathrm{d}\Omega = \delta_{ll'}\, \delta_{mm'} . \qquad (10.41)$$

$$Y_{00}(\theta,\phi) = \frac{1}{\sqrt{4\pi}} ; \qquad
\begin{cases}
Y_{11}(\theta,\phi) = \sqrt{\dfrac{3}{8\pi}}\, \sin\theta\, e^{i\phi} \\[2ex]
Y_{10}(\theta,\phi) = \sqrt{\dfrac{3}{4\pi}}\, \cos\theta \\[2ex]
Y_{1\,-1}(\theta,\phi) = \sqrt{\dfrac{3}{8\pi}}\, \sin\theta\, e^{-i\phi} .
\end{cases} \qquad (10.42)$$

As far as Y_{00} is concerned, it was foreseeable it should depend neither on θ nor on ϕ: indeed, if $l=0$, *any* component of $\vec{L}$ (not only the z component) may only have the eigenvalue 0 ($-l \le m \le l$): the states with angular momentum $l=0$ are simultaneous eigenstates of all the components of $\vec{L}$ belonging to the eigenvalue 0. This means that, regardless of how the polar axis is oriented, Y_{00} must not depend on the corresponding azimuthal angle, i.e. Y_{00} possesses spherical symmetry or, in different words, it depends neither on θ nor on ϕ. The fact that the components of $\vec{L}$ have simultaneous eigenstates (those with vanishing angular momentum) is not in contradiction with the fact that the components of the angular momentum do not commute with one another: the states with vanishing angular momentum do not form a complete set of states.

We leave to the reader to verify that Y_{11}, Y_{10}, $Y_{1\,-1}$ satisfy (10.39) with $l=1$ and m respectively equal to 1, 0, -1. If we rewrite the spherical harmonics with $l=1$ by means of the Cartesian coordinates, up to a common normalization factor, we obtain:

$$Y_{11} = \frac{x+iy}{\sqrt{2}\,r}, \qquad Y_{10} = \frac{z}{r}, \qquad Y_{1\,-1} = \frac{x-iy}{\sqrt{2}\,r} . \qquad (10.43)$$

In general the spherical harmonics of order l are the product of r^{-l} with a homogeneous polynomial of degree l in the variables x, y and z.

Equation (7.9) immediately shows that the space-inversion operator I commutes with the components of $\vec{L}$:

$$I\, L_i\, I^{-1} = L_i .$$

Therefore I is a scalar. As a consequence the operators I, $\vec{L}^2$ and L_z have a complete set of simultaneous eigenvectors whose wavefunctions, as all the eigenfunctions of $\vec{L}^2$ and L_z, have the form (10.40): since $f(r)$ is even under $x \to -x$, $y \to -y$, $z \to -z$, their parity is that of Y_{lm}, i.e. it is function of l and m. But I is a scalar, so all the states with the same l and m between $-l$ and l have the same parity. This result is in agreement with what has been stated before, i.e. that the Y_{lm} are, up to the factor r^{-l}, homogeneous polynomials of degree l and this also tells us that the parity of any state $|\, \Sigma', l, m \,\rangle$ coincides with the parity of l, i.e. it is $(-1)^l$.

Let us now consider the case of n particles: the *total* orbital angular momentum is

$$\vec{L} = \sum_{\alpha=1}^{n} \vec{L}_\alpha \,.$$

It is clear that also in this case the eigenvalues L and M are integer numbers. Indeed, since all the $L_{\alpha z}$ commute with one another, they possess a complete set of simultaneous eigenstates: $|\, m_1, m_2 \cdots \rangle$; these also are a complete set of eigenvectors of L_z belonging to the eigenvalues $M = m_1 + m_2 + \cdots + m_n$ and, since each m_α is integer, also M is integer.

For historical reasons, whose origin will be told about in the sequel, the states of a particle with $l = 0$ are denoted by the letter s, those with $l = 1$ by p, those with $l = 2$ by d, those with $l = 3$ by f. In the case of a system consisting of several particles, the upper case letters S, P, D, F are used to indicate the eigenstates of the total angular momentum of the system, respectively with $L = 0, 1, 2, 3$.

Chapter 11

Particle in a Central Force Field

11.1 The Schrödinger Equation in a Central Force Field

If a particle is subject to a central force field, the potential depends only on
the distance r of the particle from the center of the force, that we will take
as the origin of our frame: $V = V(r)$. In this case all the components of the
angular momentum $\vec{L}$ commute with the Hamiltonian and we will see that
this fact allows for a remarkable simplification of the Schrödinger equation.

We will need the expression of $\vec{L}^2$ in terms of $\vec{q}$ and $\vec{p}$: we will not make
this calculation, that we leave to the reader with the recommendation of not
interchanging the operators that do not commute with each other and of using
(4.54). The result is

$$\vec{L}^2 = r^2\,\vec{p}^2 - r\,p_r^2\,r \tag{11.1}$$

where

$$r^2 \equiv \vec{q}^2, \qquad p_r \equiv r^{-1}\,(\vec{q}\cdot\vec{p})\,. \tag{11.2}$$

The expression in the right hand side of (11.1) differs (as it must do) only for
the order of factors from the classical expression calculated without worrying
about the order of factors: indeed the 'classical' calculation gives

$$\vec{L}^2 = |\,\vec{q}\wedge\vec{p}\,|^2 = \vec{q}^2\,\vec{p}^2\,\sin^2\theta = \vec{q}^2\,\vec{p}^2\,(1-\cos^2\theta) = \vec{q}^2\,\vec{p}^2\left[1 - \frac{(\vec{q}\cdot\vec{p})^2}{\vec{q}^2\,\vec{p}^2}\right]$$

$$= \vec{q}^2\,\vec{p}^2 - \vec{q}^2\,p_r^2\,.$$

The operator p_r defined in (11.2) is the **radial momentum**, i.e. the projection of the momentum $\vec{p}$ along the radial direction $\vec{q}$. Observe that p_r is not
a Hermitian operator: $p_r^\dagger = (\vec{q}\cdot\vec{p})\,r^{-1} \neq p_r$, therefore it is not an observable;
however we shall need just the p_r defined in (11.2). Let us indeed examine
the Schrödinger representation of p_r. One has

$$p_r \equiv r^{-1}\,(\vec{q}\cdot\vec{p}) \xrightarrow{\text{SR}} -\mathrm{i}\,\hbar\sum_{i=1}^{3}\frac{x_i}{r}\frac{\partial}{\partial x_i} = -\mathrm{i}\,\hbar\sum_{i=1}^{3}\frac{\partial x_i}{\partial r}\frac{\partial}{\partial x_i} = -\mathrm{i}\,\hbar\frac{\partial}{\partial r}\,.$$

Therefore

© Springer International Publishing Switzerland 2016
L.E. Picasso, *Lectures in Quantum Mechanics*, UNITEXT for Physics,
DOI 10.1007/978-3-319-22632-3_11

$$p_r \xrightarrow{\text{SR}} -i\hbar \frac{\partial}{\partial r} \qquad (11.3)$$

so, even if it is not Hermitian, p_r has the advantage of having a simple Schrödinger representation.

The reason why the operator $-i\partial/\partial x$ is Hermitian (Sect. 6.5), whereas $-i\partial/\partial r$ is not such, is in the factor r^2 appearing in the volume element in polar coordinates intervening in the integral that defines the scalar product between two wavefunctions.

Let us go back to (11.1): if both its sides are multiplied on the left by r^{-2}, one obtains

$$\vec{p}^{\,2} = \frac{\vec{L}^2}{r^2} + \frac{1}{r} p_r^2 r \qquad (11.4)$$

(we have written $\vec{L}^2/r^2$ and not $r^{-2}\vec{L}^2$ because $\vec{L}^2$ commutes with r^2 that is a scalar and therefore the notation $\vec{L}^2/r^2$ is not ambiguous).

If we now divide both sides of (11.4) by $2m$, the following expression for the kinetic energy is obtained

$$\frac{\vec{p}^{\,2}}{2m} = \frac{1}{2m} \frac{1}{r} p_r^2 r + \frac{\vec{L}^2}{2m\,r^2} \qquad (11.5)$$

whose physical meaning (much as in the classical case) is evident: the first term in the right hand side of (11.5) is the contribution to kinetic energy due to the component of the velocity of the particle parallel to $\vec{r}$, whereas the second term is the contribution due to the component orthogonal to $\vec{r}$: it is called **centrifugal potential** at fixed angular momentum. Indeed the centrifugal force is

$$F_c = m\omega^2 r = \frac{m^2 \omega^2 r^4}{m\,r^3} = \frac{\vec{L}^2}{m\,r^3}$$

whose potential at fixed $\vec{L}^2$ is $\vec{L}^2/2m\,r^2$ (just the opposite of the centrifugal potential at fixed angular velocity).

Once (11.5) is inserted into the expression for the Hamiltonian H, one has

$$H = \frac{\vec{p}^{\,2}}{2m} + V(r) = \frac{1}{2m} \frac{1}{r} p_r^2 r + \frac{\vec{L}^2}{2m\,r^2} + V(r) . \qquad (11.6)$$

As usual, the problem is that of determining eigenvalues and eigenvectors of the Hamiltonian H.

As H, $\vec{L}^2$ and L_z commute with one another, we can search for the simultaneous eigenvectors $|\,E, l, m\,\rangle$ of these three observables.

The vectors $|\,E, l, m\,\rangle$ being eigenvectors of $\vec{L}^2$ belonging to the eigenvalue $l\,(l+1)\,\hbar^2$, thanks to (11.6) the eigenvalue equation

$$H\,|\,E, l, m\,\rangle = E\,|\,E, l, m\,\rangle$$

takes the form:

$$\left[\frac{1}{2m}\frac{1}{r}p_r^2\, r + \frac{\hbar^2\, l\,(l+1)}{2m\, r^2} + V(r)\right]|\,E,\,l,\,m\,\rangle = E\,|\,E,\,l,\,m\,\rangle\,. \qquad (11.7)$$

Let us now rewrite (11.7) in the Schrödinger representation recalling that

$$|\,E,\,l,\,m\,\rangle \xrightarrow{\text{SR}} \psi_{E\,l\,m}(r,\theta,\phi) = R_{E\,l}(r)\,Y_{l\,m}(\theta,\phi) \qquad (11.8)$$

(besides E and l, the **radial wavefunction** $R(r)$ could in principle depend also on the quantum number m; we will shortly see that this does not happen) one has:

$$-\frac{\hbar^2}{2m}\frac{1}{r}\frac{\partial^2}{\partial r^2}\left(r\,R_{E\,l}(r)\right)Y_{l\,m}(\theta,\phi) + \frac{\hbar^2\, l\,(l+1)}{2m\, r^2}\,R_{E\,l}(r)\,Y_{l\,m}(\theta,\phi)$$
$$+V(r)\,R_{E\,l}(r)\,Y_{l\,m}(\theta,\phi) \;=\; E\,R_{E\,l}(r)\,Y_{l\,m}(\theta,\phi)$$

that, since $Y_{l\,m}(\theta,\phi)$ can be factorized, boils down to

$$-\frac{\hbar^2}{2m}\frac{1}{r}\frac{\partial^2}{\partial r^2}\left(r\,R_{E\,l}(r)\right) + \frac{\hbar^2\, l\,(l+1)}{2m\, r^2}\,R_{E\,l}(r) + V(r)R_{E\,l}(r) = E\,R_{E\,l}(r) \quad (11.9)$$

that is a differential equation in which only the variable r appears.

One sees that in (11.9) only E and l, but not m, appear (do not confuse the quantum number m with the mass!): so it is true that the radial function only depends on the values of E and l. This fact expresses the existence of degeneracy on m: indeed, for any solution $R_{E\,l}(r)$ of (11.9) there exist $2l+1$ independent states with the same energy, whose wavefunctions are $R_{E\,l}(r)\,Y_{l\,m}(\theta,\phi)$ with $-l \le m \le +l$.

A further simplification is obtained if the **reduced radial wavefunction** $u_{E\,l}(r)$ is introduced, according to the definition:

$$u_{E\,l}(r) \equiv r\,R_{E\,l}(r)\,. \qquad (11.10)$$

Indeed, after multiplying (11.9) by r, one obtains:

$$-\frac{\hbar^2}{2m}\,u''_{E\,l}(r) + \left[\frac{\hbar^2\, l\,(l+1)}{2m\, r^2} + V(r)\right]u_{E\,l}(r) = E\,u_{E\,l}(r)\,. \qquad (11.11)$$

The eigenfunctions $\psi_{E\,l\,m}(r,\theta,\phi)$ of the Hamiltonian must belong to the domain of the operator $\vec{p}^{\,2}$, that in the Schrödinger representation $\big($see (6.29)$\big)$ is proportional to the Laplace operator $\Delta = \sum_i \partial^2/\partial x_i^2$: it can be proved that (in space-dimension $n \le 3$) the functions belonging to the domain of the Laplace operator are bounded; it then follows that the functions $R_{E\,l}(r)$ must be finite at $r=0$ whence, by (11.10), the reduced radial wavefunctions $u_{E\,l}(r)$ must vanish at $r=0$:

$$u_{E\,l}(0) = 0\,. \qquad (11.12)$$

Equation (11.11), supplemented with the boundary condition (11.12), is formally identical with the Schrödinger equation in one dimension for a particle constrained to move on the semi-axis $r \geq 0$, subject to the potential $V(r)$ plus the centrifugal potential $\hbar^2\, l\,(l+1)/2m\,r^2$.

If we are interested in the bound states of the system, we must impose that the wavefunctions $R_{E\,l}(r)\, Y_{l\,m}(\theta,\phi)$ be normalizable:

$$\int_0^\infty \left| R_{E\,l}(r)\right|^2 r^2\, \mathrm{d}r < \infty$$

(see (10.41)), whence

$$\int_0^\infty \left| u_{E\,l}(r)\right|^2 \mathrm{d}r < \infty \qquad (11.13)$$

i.e. $u_{E\,l}(r)$ must be L_2 with respect to the measure $\mathrm{d}r$, exactly as the wavefunctions of a one-dimensional problem.

In conclusion, the problem of a particle in a central force field has been brought back to a one-dimensional problem (on a half line) not only because the Schrödinger equation reduces to an ordinary differential equation, but also because the conditions to be required for the solutions are identical to those of a one-dimensional problem.

As an example, we will consider a particle subject to the "spherical well" potential:

$$V(r) = \begin{cases} -V_0 & 0 \leq r \leq a \\ 0 & r > a\,. \end{cases} \qquad (11.14)$$

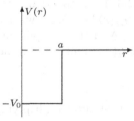

Fig. 11.1

The problem of finding eigenvalues and eigenfunctions with $l = 0$ ("s-states") – the centrifugal potential being absent in this case – is identical with the problem of the one-dimensional well discussed in Sect. 8.2, limited to the odd states *only*: indeed (11.11) (with $l = 0$) and the condition (11.12) demand that the reduced radial wavefunction $u_{E\,l}(r)$ for $0 \leq r \leq a$ be $A \sin k\,r$ (instead of a linear combination of $\sin k\,r$ and $\cos k\,r$), exactly as for the odd states of the one-dimensional well. The ground state, if it exists, certainly is a s-state because the centrifugal potential, being a positive definite operator, has the effect of raising the energies: we thus see that, contrary to the one-dimensional case in which a bound state always exists, in the present case – in which the ground state is the first *odd* state of the one-dimensional problem – the system possesses a bound state if and only if $2m\,V_0\,a^2/\hbar^2 > \pi^2/4$.

11.2 Two-Particle Systems: Separation of Variables

Let us now consider a system consisting of two particles that interact with each other by means of a potential that only depends on the relative distance (that is certainly true if the two particle are structureless).

Let m_1 and m_2 be the two masses and $(\vec{q}_1\,,\vec{p}_1)$, $(\vec{q}_2\,,\vec{p}_2)$ the respective canonical variables. If the system is isolated , its Hamiltonian is

$$H = \frac{\vec{p}_1^2}{2m_1} + \frac{\vec{p}_2^2}{2m_2} + V(|\vec{q}_1 - \vec{q}_2|) \ . \tag{11.15}$$

The standard procedure is that of making the change of variables that take from $(\vec{q}_1, \vec{p}_1)$ and $(\vec{q}_2, \vec{p}_2)$ to the new canonical variables $(\vec{q}, \vec{p})$, $(\vec{Q}, \vec{P})$ that respectively refer to the *relative* motion of the two particles and to the motion of the *center-of-mass*:

$$\begin{cases} \vec{q} = \vec{q}_1 - \vec{q}_2 \\ \vec{p} = \dfrac{m_2\,\vec{p}_1 - m_1\,\vec{p}_2}{m_1 + m_2} \end{cases} \qquad \begin{cases} \vec{Q} = \dfrac{m_1\,\vec{q}_1 + m_2,\vec{q}_2}{m_1 + m_2} \\ \vec{P} = \vec{p}_1 + \vec{p}_2 \ . \end{cases} \tag{11.16}$$

The meaning of $\vec{q}$ and $\vec{Q}$ is evident; $\vec{p}$ and $\vec{P}$ are defined in such a way as to be conjugated respectively to $\vec{q}$ and $\vec{Q}$, i.e.

$$[\,q_i\,,\,p_j\,] = i\,\hbar\,\delta_{ij}\,, \qquad\qquad [\,Q_i\,,\,P_j\,] = i\,\hbar\,\delta_{ij} \tag{11.17}$$

and all the other commutators are vanishing.

In terms of the new variables the Hamiltonian (11.15) has the form

$$H = \frac{\vec{P}^2}{2M} + \frac{\vec{p}^2}{2\mu} + V(|\vec{q}\,|) \tag{11.18}$$

where $M = m_1 + m_2$ is the total mass and $\mu = m_1\,m_2/(m_1 + m_2)$ is the reduced mass of the system.

The Hamiltonian (11.18) is one displaying *separate variables*:

$$H = H_1 + H_2$$

i.e. it is the sum of an operator (H_1) that depends only on the center-of-mass variables (indeed only on $\vec{P}$) and of an operator (H_2) that depends only on the relative variables; so H_1 and H_2 commute with each other and, as we will shortly see, there is even more to the point.

The eigenfunctions of H are obtained by solving the Schrödinger equation: since, due to (11.17)

$$p_i \xrightarrow{\text{SR}} -i\,\hbar\,\frac{\partial}{\partial x_i}\,, \qquad\qquad P_i \xrightarrow{\text{SR}} -i\,\hbar\,\frac{\partial}{\partial X_i}$$

and

$$\vec{p}^2 \xrightarrow{\text{SR}} -\hbar^2\,\Delta_x \equiv -\hbar^2\left(\frac{\partial^2}{\partial x_1^2} + \frac{\partial^2}{\partial x_2^2} + \frac{\partial^2}{\partial x_3^2}\right), \qquad \vec{P}^2 \xrightarrow{\text{SR}} -\hbar^2\,\Delta_X$$

(Δ_X is the Laplace operator with respect to the X_i variables), one has

$$\left(-\frac{\hbar^2}{2M}\,\Delta_X - \frac{\hbar^2}{2\mu}\,\Delta_x + V(|\vec{x}\,|)\right) \Psi(\vec{X}, \vec{x}) = E\,\Psi(\vec{X}, \vec{x}) \ . \tag{11.19}$$

In order to find all the eigenfunctions and eigenvalues of H it suffices to find a complete set of eigenfunctions of H: let us search for those *particular* solutions of (11.19) that have the form

$$\Psi(\vec{X}, \vec{x}) = \Phi(\vec{X})\,\psi(\vec{x}) \tag{11.20}$$

and we will see that they form a complete set (method of **separation of variables**).

Substituting (11.20) into (11.19):

$$\left[-\frac{\hbar^2}{2M}\Delta_X\Phi(\vec{X})\right]\psi(\vec{x})$$
$$+ \left[\left(-\frac{\hbar^2}{2\mu}\Delta_x + V(|\vec{x}\,|)\right)\psi(\vec{x})\right]\Phi(\vec{X}) = E\,\Phi(\vec{X})\,\psi(\vec{x})$$

and then dividing both members by $\Phi(\vec{X})\,\psi(\vec{x})$ one obtains:

$$-\frac{\hbar^2}{2M}\frac{\Delta_X\Phi(\vec{X})}{\Phi(\vec{X})} - \frac{\hbar^2}{2\mu}\frac{\Delta_x\psi(\vec{x})}{\psi(\vec{x})} + V(|\vec{x}\,|) = E\,. \tag{11.21}$$

The left hand side is the sum of two functions, one only depending on $\vec{X}$, the other only depending on $\vec{x}$, whereas the right hand side is a constant: the two functions in the left hand side must therefore be both constants, which constants we will call E_1 and E_2:

$$-\frac{\hbar^2}{2M}\frac{\Delta_X\Phi(\vec{X})}{\Phi(\vec{X})} = E_1\,, \qquad \frac{\hbar^2}{2\mu}\frac{\Delta_x\psi(\vec{x})}{\psi(\vec{x})} + V(|\vec{x}\,|) = E_2 \tag{11.22}$$

where $E_1 + E_2 = E$. Therefore:

$$\begin{cases} -\dfrac{\hbar^2}{2M}\Delta_X\,\Phi(\vec{X}) = E_1\,\Phi(\vec{X}) \\ -\dfrac{\hbar^2}{2\mu}\Delta_x\,\psi(\vec{x}) + V(|\vec{x}\,|)\,\psi(\vec{x}) = E_2\,\psi(\vec{x})\,. \end{cases} \tag{11.23}$$

In other words one has to solve the two eigenvalue equations

$$H_1\,|\,E_1\,\rangle = E_1\,|\,E_1\,\rangle\,, \qquad H_2\,|\,E_2\,\rangle = E_2\,|\,E_2\,\rangle$$

that refer to different degrees of freedom, i.e. to different systems.

Once a complete set $\Phi_n(\vec{X})$ of eigenfunctions of H_1 (complete in the space of the functions of $\vec{X}$) and a complete set $\psi_m(\vec{x})$ of eigenfunctions of H_2 (complete in the space of the functions of $\vec{x}$) have been found, their products $\Phi_n(\vec{X})\,\psi_m(\vec{x})$ for all the pairs m, n form a complete set (in the space of functions of both $\vec{X}$ and $\vec{x}$) of eigenfunctions of H, belonging to the eigenvalues $E_{nm} = E_{1\,n} + E_{2\,m}$ (m, n can be either discrete or even continuous indices): all this thanks to the fact that H (but it could be whatever operator) is one exhibiting 'separate variables'.

Since H_1 is the Hamiltonian of the free particle, a complete set of (improper) eigenfunctions of H_1 is given by $\big($see (6.45)$\big)$

$$\Phi_{\vec{K}}(\vec{X}) = \frac{1}{(2\pi\,\hbar)^{3/2}}\, e^{i\,\vec{K}\cdot\vec{X}}\,, \qquad\qquad \vec{K} \equiv \vec{P}/\hbar \in \mathbb{R}^3$$

so the problem is only that of determining eigenvalues and eigenvectors of H_2 that is formally identical with the Hamiltonian of a particle of mass μ in a central force field, for which the treatment of the previous section applies.

Let us now consider two examples.

Hydrogen-like atom

This is a system consisting of two particles (the nucleus endowed with electric charge $Z\,e$ and one electron) interacting via the Coulomb potential. The Hamiltonian describing only the relative motion is

$$H = \frac{\vec{p}^{\,2}}{2\,\mu_e} - \frac{Z\,e^2}{r}\,. \tag{11.24}$$

In the next section we will be concerned with the problem of finding eigenvalues and eigenfunctions relative to only the bound states of (11.24).

Diatomic molecule

In the schematization in which the two (pointlike) atoms stay at a *fixed* distance d (in the relative motion the radial degree of freedom is frozen: see Sect. 2.7) the kinetic energy $\vec{p}^{\,2}/2\mu$ is given only by the centrifugal term:

$$\frac{\vec{L}^{\,2}}{2\mu\,d^2} = \frac{\vec{L}^{\,2}}{2I}$$

where I is the moment of inertia of the molecule with respect to an axis passing through the center of mass and orthogonal to the segment joining the two atoms. The potential $V(r)$, r being fixed equal to d, is a constant that can be put equal to zero. In conclusion the Hamiltonian of the relative motion is

$$H = \frac{\vec{L}^{\,2}}{2I}$$

whose eigenvalues are

$$E_l = \frac{\hbar^2}{2I}\, l\,(l+1) \tag{11.25}$$

each of them $2l+1$ times degenerate. Comparison with (2.35) shows that Bohr theory provides the correct result only for large values of the quantum number $\big(n$ in (2.35), l in (11.25)$\big)$. Furthermore we now have precise information about the degeneracy of the energy levels of the molecule, i.e. on the g_i of (2.20).

11.3 Energy Levels of Hydrogen-like Atoms

We now want to determine the energy levels of the hydrogen-like atoms, i.e. the eigenvalues of the Hamiltonian (11.24). According to the discussion of Sect. 11.1, we must solve the equation:

$$-\frac{\hbar^2}{2\mu_e} u''(r) + \frac{\hbar^2 l\,(l+1)}{2\mu_e\, r^2} u(r) - \frac{Z\,e^2}{r} u(r) = E\,u(r) \tag{11.26}$$

with the boundary conditions (11.12) and (11.13) $\big($we have omitted the indices E and l in the reduced radial wavefunction $u(r)\big)$.

We immediately see that we can have discrete eigenvalues only for $E < 0$: indeed (11.26) is the Schrödinger equation in one dimension for a particle subject to the potential

$$U_l(r) = \begin{cases} \dfrac{\hbar^2 l\,(l+1)}{2\mu_e\, r^2} - \dfrac{Z\,e^2}{r} & r \geq 0 \\[2mm] \infty & r < 0\,. \end{cases}$$

Since $U_l(r) \to 0$ for $r \to \infty$ and, in addition, the minimum of $U_l(r)$ is negative (for any value of l), the discussion of Chap. 7 entails that discrete eigenvalues are possible only for $E < 0$.

The first step in the solution of (11.26) consists in determining the asymptotic behaviour of $u(r)$ for $r \to \infty$: this can be done by neglecting in (11.26) both the centrifugal and the Coulomb potential (that tend to zero for $r \to \infty$) with respect to $|E|$; since we are interested in the case $E < 0$, one has

$$u(r) \overset{r \to \infty}{\longrightarrow} c_1\, e^{-\kappa r} + c_2\, e^{\kappa r}$$

where

$$\kappa = \frac{\sqrt{2\mu_e\,|E|}}{\hbar}\,. \tag{11.27}$$

The condition (11.13) demands $c_2 = 0$ whence

$$u(r) \overset{r \to \infty}{\longrightarrow} e^{-\kappa r}\,. \tag{11.28}$$

Then we put

$$u(r) = f(r)\,e^{-\kappa r} \tag{11.29}$$

that, after insertion into (11.26) and a little algebra, gives:

$$f''(r) - 2\kappa\, f'(r) + \frac{2\mu_e\, Z\, e^2}{\hbar^2\, r} f(r) - \frac{l\,(l+1)}{r^2} f(r) = 0\,. \tag{11.30}$$

Let us put

$$f(r) = r^s \sum_{i=0}^{\infty} a_i\, r^i\,, \qquad a_0 \neq 0 \tag{11.31}$$

where the exposition of the factor r^s is needed to ensure $a_0 \neq 0$. Substituting (11.31) into (11.30) yields:

$$\sum_{i=0} \Big[(s+i)(s+i-1)\,a_i\,r^{s+i-2} - 2\kappa\,(s+i)\,a_i\,r^{s+i-1}$$

$$+ \frac{2\mu_{\mathrm{e}}\,Z\,e^2}{\hbar^2}\,a_i\,r^{s+i-1} - l\,(l+1)\,a_i\,r^{s+i-2}\Big] = 0 \,. \tag{11.32}$$

Let us rewrite the left hand side of (11.32) as a power series in r: to this end let us isolate the term with the lowest power $(s-2)$ and then change in the first and the fourth term the index i with $i+1$:

$$a_0\,[s\,(s-1) - l\,(l+1)]\,r^{s-2} + \sum_{i=0} \Big[(s+i+1)(s+i)\,a_{i+1} - 2\kappa\,(s+i)\,a_i$$

$$+ \frac{2\mu_{\mathrm{e}}\,Z\,e^2}{\hbar^2}\,a_i - l\,(l+1)\,a_{i+1}\Big]\,r^{s+i-1} = 0 \,. \tag{11.33}$$

Since (11.33) must hold for any value of $r \geq 0$, all the coefficients of the series must vanish; in particular, as $a_0 \neq 0$,

$$s\,(s-1) = l\,(l+1)$$

whence either $s = l+1$ or $s = -l$: the solution $s = -l$ is not acceptable because, being $l \geq 0$, it would violate the condition (11.12) $u(0) = 0$. Therefore:

$$f(r) = \sum_{i=0} a_i\,r^{i+l+1} \,. \tag{11.34}$$

Note that, when $l > 0$, the behaviour of $f(r)$ at the origin is determined by the centrifugal potential, not by the Coulomb potential, so it is the same for all potentials, provided they do not diverge at the origin faster than the centrifugal potential.

Equation (11.33) with $s = l+1$ becomes:

$$\sum_{i=0} \Big[(i+l+2)(i+l+1)\,a_{i+1} - 2\kappa\,(i+l+1)\,a_i$$

$$+ \frac{2\mu_{\mathrm{e}}\,Z\,e^2}{\hbar^2}\,a_i - l\,(l+1)\,a_{i+1}\Big]\,r^{i+l} = 0$$

whence

$$a_{i+1} = \frac{2\big(\kappa\,(i+l+1) - \mu_{\mathrm{e}}\,Z\,e^2/\hbar^2\big)}{(i+l+2)(i+l+1) - l\,(l+1)}\,a_i \,. \tag{11.35}$$

Equation (11.35) determines the a_i – i.e. $f(r)$ – by recursion once a_0 is given: actually a_0 is arbitrary because the Schrödinger equation is homogeneous.

We must now make sure that $f(r)$ does not diverge faster than $e^{\kappa r}$, otherwise (11.28) is no longer true, therefore (11.13) is no longer satisfied. The asymptotic behaviour of $f(r)$ is obtained by studying (11.31) for large values of i, in which case (11.35) gives:

$$a_{i+1} \approx a_i\,\frac{2\kappa}{i} \qquad \Rightarrow \qquad a_i \approx a_0\,\frac{(2\kappa)^i}{i!} \,, \qquad i \gg 1$$

that means that, for $r \to \infty$, $f(r) \approx e^{2\kappa r}$ and therefore, due to (11.29), $u(r) \to e^{\kappa r}$ in contradiction with (11.28). So, in general, the solutions of (11.26) that vanish at $r = 0$ diverge for $r \to \infty$ and therefore are not square summable, *unless* the summation in (11.31) is a sum involving a finite number of terms, instead of a series. As a consequence, in order that (11.28) be satisfied, it is necessary that $f(r)$ be a polynomial in r: this happens if and only if there exists a $\bar{\imath}$ such that $a_{\bar{\imath}+1} = 0$ and, by (11.35), this is the same as

$$\kappa\,(\bar{\imath}+l+1) = \frac{\mu_e\, Z\, e^2}{\hbar^2}\,. \tag{11.36}$$

As $\bar{\imath}$ and l are integer numbers, we put

$$\bar{\imath}+l+1 = n\,, \qquad\qquad n \geq 1 \tag{11.37}$$

so that (11.36), in view of (11.27), is equivalent to

$$E_n = -\frac{\mu_e\, Z^2\, e^4}{2\hbar^2}\frac{1}{n^2}\,, \qquad\qquad n \geq 1\,. \tag{11.38}$$

In other words: only for those energies given by (11.38) $f(r)$ is a polynomial and the conditions (11.12) and (11.13) are fulfilled.

The E_n are the eigenvalues we had in mind ro find and coincide with those given by Bohr theory. We have already discussed the experimental verifications of (11.38) that will not be discussed any further.

Let us now examine the degree of degeneracy of the E_n : by (11.37) a given value of n, i.e. of E_n , can be obtained by means of several choices of l:

$$\left.\begin{array}{ll} l = 0\,, & \bar{\imath} = n-1 \\ l = 1\,, & \bar{\imath} = n-2 \\ \cdots & \cdots\cdots \\ l = n-1\,, & \bar{\imath} = 0 \end{array}\right\} \quad \bar{\imath}+l+1 = n\,. \tag{11.39}$$

namely, there exist n different ways to obtain a given value of n: by (11.35) these n ways correspond to n different solutions of (11.30), i.e. of (11.26). These solutions, that we will denote by $u_{n\,l}(r)$, differ in the value of l that, as shown by (11.39), for a given value of n assumes all the (integer) values from 0 to $n-1$.

Therefore the energy levels of the hydrogen-like atoms are degenerate with respect to l: states with different values of the angular momentum are possible for a given energy. This degeneracy is typical of the systems endowed with an attractive potential proportional to $1/r$ and corresponds to the fact that, according to classical mechanics, the energy is a function exclusively of the major axis of the (elliptic) orbit of the particle, while the minor axis is proportional to the absolute value of the angular momentum: we will come again to this point at the end of the present chapter.

In addition, for any value of l there is the degeneracy on m (that classically corresponds to the fact that orbits only differing by the orientation in space have the same energy and angular momentum).

Therefore the overall degeneracy of the level E_n is

$$g_n = \sum_{l=0}^{n-1}(2l+1) = n^2 \,.$$

The simultaneous eigenvectors of H, $\vec{L}^2$ and L_z for hydrogen-like atoms are normally denoted by $|\,n,\,l,\,m\,\rangle$ and the corresponding eigenfunctions in the Schrödinger representation by $\psi_{nlm}(r,\theta,\phi)$. One has

$$\psi_{nlm}(r,\theta,\phi) = R_{nl}(r)\,Y_{lm}(\theta,\phi) = r^{-1}\,u_{nl}(r)\,Y_{lm}(\theta,\phi)$$

and, thanks to (11.29), (11.34), (11.36) and (11.37)

$$\psi_{nlm}(r,\theta,\phi) = r^l\,(a_0 + \cdots + a_{n-l-1}\,r^{n-l-1})\,\mathrm{e}^{-Z\,r/n\,a_\mathrm{B}}\,Y_{lm}(\theta,\phi) \quad (11.40)$$

where a_B is the Bohr radius defined by either (2.10) or (2.14): $a_\mathrm{B} = \hbar^2/(m_\mathrm{e}\,e^2)$ (actually, in (11.40) instead of a_B we should write $a_\mathrm{B}(m_\mathrm{e}/\mu_\mathrm{e})$: from now on we will neglect the difference between m_e and the reduced mass μ_e).

Equation (11.40) emphasizes that:

1. the larger the value of l, the faster ψ_{nlm} vanishes for $r \to 0$; so the higher the angular momentum, the less the probability of finding the electron close to the nucleus;
2. the behaviour of ψ_{nlm} for $r \to \infty$ is determined by the exponential $\mathrm{e}^{-Z\,r/n\,a_\mathrm{B}}$: as n increases, it decreases more and more slowly;
3. owing to the presence of a polynomial in r of degree $n-l-1$ (**Laguerre polynomial**), one expects that $u_{nl}(r)$ – and therefore $R_{nl}(r)$ – has $n-l-1$ nodes inside the region of definition, i.e. for $0 < r < \infty$. This result can be established by means of the oscillation theorem for one-dimensional systems enunciated in Sect. 7.4. Indeed, for given l, the $u_{nl}(r)$ are solutions of the one-dimensional Schrödinger equation (11.26), and for each value of l one has a different Schrödinger equation because the centrifugal potential changes. We can then apply the oscillation theorem to each of the above Schrödinger equations: ordering by increasing energies the $u_{nl}(r)$ with the same l, one realizes that u_{nl} is the $(n-l)$-th solution with the given value of l and that, by the oscillation theorem, it has $n-l-1$ nodes (the first is that with $n = l+1$, the second is that with $n = l+2$, $\cdots$, so $n-l$ is just the order number of the solution).

We report the radial functions $R_{nl}(r)$ for $n = 1$ and $n = 2$ and in Fig. 11.2 the corresponding graphs:

$$R_{10}(r) = \left(\frac{Z}{a_\mathrm{B}}\right)^{3/2} 2\,\mathrm{e}^{-Z\,r/a_\mathrm{B}}\,; \quad R_{20}(r) = \left(\frac{Z}{2a_\mathrm{B}}\right)^{3/2}\left(2 - \frac{Z\,r}{a_\mathrm{B}}\right)\mathrm{e}^{-Z\,r/2a_\mathrm{B}}$$

$$R_{21}(r) = \left(\frac{Z}{2a_\mathrm{B}}\right)^{3/2}\frac{Z\,r}{\sqrt{3}\,a_\mathrm{B}}\,\mathrm{e}^{-Z\,r/2a_\mathrm{B}}\,. \tag{11.41}$$

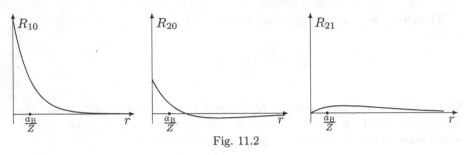

Fig. 11.2

Although it could seem the contrary, the three functions R_{10}, R_{20} and R_{21} of Fig. 11.2 are equally normalized: the reason is in the fact that the volume element in the normalization integral contains the factor r^2 that weighs more the region $r > a_B/Z$, where R_{20} and R_{21} decrease slower than R_{10}. We report in Fig. 11.3 the radial probability densities:

$$\rho_{nl}(r) = r^2 |R_{nl}(r)|^2 \int |Y_{lm}(\theta,\phi)|^2 \, d\Omega = r^2 |R_{nl}(r)|^2 \qquad (11.42)$$

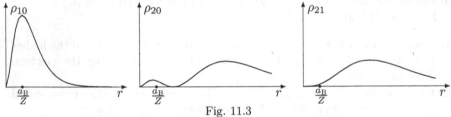

Fig. 11.3

that correspond to the three radial functions of Fig. 11.2.

In the first two chapters of the book we have discussed the problem of the atomic sizes and we have seen, in the particular case of the hydrogen atom, that Bohr theory attributes well definite sizes to the atom coinciding with those of the orbits of the electron in the allowed energy levels. In the present scheme the statement that the electron moves along a given orbit makes no longer sense: the state of the electron is described by a wavefunction ψ and the electron can be found in any point where $|\psi|^2$ is nonvanishing. It is however clear that talking about atomic sizes is still meaningful, at least in the sense of 'average sizes'. Referring to the hydrogen atom, several definitions of atomic sizes can be given: however all are equivalent with one another inasmuch as they all give, for the hydrogen atom in the ground state, a number of the order of magnitude of the Bohr radius a_B.

One can for example define the atomic radius as:

1. the radius of the sphere such that the probability of finding the electron inside is – say – 90 %;
2. the mean value of r: $\langle n, l, m \mid r \mid n, l, m \rangle = \frac{1}{2}[3n^2 - l(l+1)] a_B$;
3. the reciprocal of the mean value of r^{-1}: $\left(\langle n, l, m \mid r^{-1} \mid n, l, m \rangle\right)^{-1} = n^2 a_B$.

The last definition may appear a little odd: however it is the only one that exactly reproduces the result of Bohr theory, whereas the first two give a result that, besides n, also depends on l.

11.4 The Three-Dimensional Isotropic Harmonic Oscillator

The Hamiltonian of the three-dimensional isotropic harmonic oscillator is

$$H = \frac{\vec{p}^{\,2}}{2m} + \frac{1}{2}m\,\omega^2\,\vec{q}^{\,2} \,. \tag{11.43}$$

Since the potential is a central potential, the eigenstates of the Hamiltonian can be classified by means of the eigenvalues of H, $\vec{L}^2$ and L_z: $|\,E,\,l,\,m\,\rangle$, and the eigenvalues E can be determined, with the corresponding eigenfunctions, by solving (11.11) with the boundary condition (11.12). If we limit ourselves to the only states with $l = 0$, as in the case of the spherical well discussed in Sect. 11.1, (11.11) and (11.12) are the equations for the odd states of the one-dimensional oscillator: so we find that the s-states ($l = 0$) have energies $\frac{3}{2}\hbar\omega,\,\frac{7}{2}\hbar\omega,\,\cdots$, corresponding to the values $n = 1\,,\,3\,,\,\cdots$ of the one-dimensional oscillator.

Obviously in this way we find neither all the eigenfunctions of H (we find indeed only those with $l = 0$), nor all the eigenvalues, as the spacing between those we find is $2\hbar\omega$. So one should solve (11.11) without setting $l = 0$.

Equation (11.11) can be solved rather easily by a method similar to that used for the hydrogen atom. It is however simpler and more instructive to proceed in a different way. The Hamiltonian (11.43) can be written in the form

$$\begin{aligned} H &= \left(\frac{p_1^2}{2m} + \frac{1}{2}m\,\omega^2\,q_1^2\right) + \left(\frac{p_2^2}{2m} + \frac{1}{2}m\,\omega^2\,q_2^2\right) + \left(\frac{p_3^2}{2m} + \frac{1}{2}m\,\omega^2\,q_3^2\right) \\ &\equiv H_1 + H_2 + H_3 \end{aligned} \tag{11.44}$$

because in this way it appears that it is a separate variables one: in this way the problem is led back to the determination of the eigenvalues of three one-dimensional oscillators – identical with one another in the case at hand. We already know the solution of this problem. So the eigenvalues of H are given by (see (5.16))

$$\begin{aligned} E_{n_1\,n_2\,n_3} &= \left(n_1 + \tfrac{1}{2}\right)\hbar\omega + \left(n_2 + \tfrac{1}{2}\right)\hbar\omega + \left(n_3 + \tfrac{1}{2}\right)\hbar\omega = \left(n + \tfrac{3}{2}\right)\hbar\omega \equiv E_n \\ n &\equiv n_1 + n_2 + n_3\,, \qquad n_1\,,\,n_2\,,\,n_3 = 0\,,\,1\,,\,2\,,\,\cdots \end{aligned} \tag{11.45}$$

and the corresponding eigenvectors are characterized by the quantum numbers n_1, n_2, n_3:

$$|\,E_{n_1\,n_2\,n_3}\,\rangle = |\,n_1,\,n_2,\,n_3\,\rangle \,. \tag{11.46}$$

In the Schrödinger representation, putting $\widetilde{H}_n(x) = H_n(\sqrt{m\,\omega/\hbar}\,x)$ (see (8.4)), one has:

$$| n_1, n_2, n_3 \rangle \to \psi_{n_1 n_2 n_3}(x, y, z) = \tilde{H}_{n_1}(x)\, \tilde{H}_{n_2}(y)\, \tilde{H}_{n_3}(z)\, \mathrm{e}^{-(m\omega/2\hbar)\, r^2}.$$
$$(11.47)$$

It goes without saying that the separation of the variables allows for the solution of the problem even when the oscillator is not isotropic.

In the isotropic case the degree of degeneracy g_n of the level E_n is given by the number of the triples n_1, n_2, n_3 whose sum is n:

$$g_n = \sum_{n_1=0}^{n} (n+1-n_1) = \frac{1}{2}(n+1)\,(n+2)\,. \qquad (11.48)$$

If instead the eigenstates of the Hamiltonian are classified as $|\, E,\, l,\, m \,\rangle$, the degeneracy g_n must correspond to the degeneracy on m and – possibly – on l of the energy levels: let us therefore examine the way in which one passes from the classification $|\, n_1,\, n_2,\, n_3 \,\rangle$ to the classification $|\, E,\, l,\, m \,\rangle$.

Since the lowest energy level is nondegenerate ($n_1 = n_2 = n_3 = 0$), it necessarily is a state with $l = 0$. Indeed from (11.47) one sees that $\psi_{000}(x, y, z)$ is a spherically symmetric function: up to the normalization coefficient

$$\psi_{000}(x, y, z) = \mathrm{e}^{-(m\omega/2\hbar)\, r^2}\,. \qquad (11.49)$$

The level $n = 1$ is three times degenerate, so either it is a level with $l = 1$ (and $m = \pm 1, 0$) or $l = 0$ three times: the latter possibility is ruled out because, for example, the parity of the level is $(-1)^{n_1+n_2+n_3} = (-1)^n$.

The level $n = 2$ is six times degenerate and even:

$$| n_1, n_2, n_3 \rangle = |\,2,0,0\,\rangle,\ |\,0,2,0\,\rangle,\ |\,0,0,2\,\rangle,\ |\,1,1,0\,\rangle,\ |\,1,0,1\,\rangle,\ |\,0,1,1\,\rangle$$
$$(11.50)$$

so either $l = 0$ six times or $l = 2,\, l = 0$. The first possibility is excluded because it would entail that all the states with $n = 2$ are s-states while none of the six wavefunctions (11.47), corresponding to the states (11.50), is spherically symmetric. Therefore the level $n = 2$ is degenerate with respect to l. A question immediately arises: since none of the six wavefunction is spherically symmetric, where is the state with $l = 0$? The three states $|\,1,1,0\,\rangle,\ |\,1,0,1\,\rangle,\ |\,0,1,1\,\rangle$ all have $l = 2$ (even if they are not eigenstates of L_z): for example

$$\psi_{110}(x, y, z) \propto x\, y\, \mathrm{e}^{-(m\omega/2\hbar)\, r^2} \propto r^2 \sin^2\theta \, \sin 2\phi \; \mathrm{e}^{-(m\omega/2\hbar)\, r^2}$$

whose integration with respect to the solid angle variables gives 0, i.e. it is orthogonal to Y_{00} (that is a constant).

The other three states $\big(|\,2,0,0\,\rangle, \cdots \big)$ are superpositions of states with $l = 2$ and of the state with $l = 0$: indeed, since $\tilde{H}_2(x) \propto x^2 + a$,

$$\psi_{200} + \psi_{020} + \psi_{002} \propto (x^2 + y^2 + z^2 + 3a)\, \mathrm{e}^{-(m\omega/2\hbar)\, r^2}$$

is spherically symmetric, so it is the wavefunction of a state with $l = 0$; therefore

$$|E_2, l = 0\rangle = |2,0,0\rangle + |0,2,0\rangle + |0,0,2\rangle .$$

The level with $n = 3$ is 10 times degenerate and in a similar way one sees that there are 7 states with $l = 3$ and 3 states with $l = 1$.

In general, the n-th level contains states with values of the angular momentum $l = n, n - 2, \cdots$ down to either $l = 0$ or $l = 1$, depending on whether n respectively is even or odd.

We have seen two cases in which the energy levels exhibit degeneracy on l: the case of the Coulomb potential and that of the isotropic harmonic oscillator. These are also the only central force potentials that give rise, in classical mechanics, to closed orbits (the bounded ones, in the case of the Coulomb potential). Both the closed orbits of the classical case and the degeneracy (often improperly referred to as *accidental*) of the quantum case have their origins in some constants of motion that are peculiar to these two problems (and indeed we know, thanks to the degeneracy theorem of Sect. 7.2, that there is a strict relationship between the degeneracy of the Hamiltonian and the constants of motion): in the case of the Coulomb potential the components of the Lenz vector

$$\vec{N} = \frac{1}{2m} \, (\vec{p} \wedge \vec{L} - \vec{L} \wedge \vec{p}) - \frac{Z e^2}{r} \, \vec{r}$$

are constants of motion (they commute with H but not with $\vec{L}^2$), whereas in the case of the oscillator, defining – as in the case of the one-dimensional oscillator – the 'raising' and 'lowering' operators $\eta_i^\dagger$, η_i $(i = 1, 2, 3)$, $[\eta_i, \eta_j^\dagger] = \delta_{ij}$, all the nine operators $\eta_i^\dagger \eta_j$ commute with the Hamiltonian

$$H = \sum_{i=1}^{3} \hbar \omega \left(\eta_i^\dagger \eta_i + \frac{1}{2} \right) .$$

The three antisymmetric combinations $\eta_i^\dagger \eta_j - \eta_j^\dagger \eta_i$ are proportional to the components of the angular momentum $\vec{L}$, $\sum_i \eta_i^\dagger \eta_i$ is proportional to $H - \frac{3}{2} \hbar \omega$ and the remaining five (symmetric) combinations are those responsible for the degeneracy on l: indeed they do not commute with $\vec{L}^2$.

Chapter 12

Perturbations to Energy Levels

12.1 Perturbation Theory: Heuristic Approach

There are only few cases in which it is possible to exactly determine eigenvalues and eigenvectors of the Hamiltonian and, unfortunately, just in the most interesting cases (atoms with more than one electron, atoms in presence of an external either electric or magnetic field etc.) the Hamiltonian is so complicated that there is no hope to find the exact solution of the eigenvalue problem. So, in order that the general theory be not fruitless, it is necessary to develop as simple as possible methods capable to provide – albeit in an approximate manner – eigenvalues and eigenvectors of complicated Hamiltonians. We will discuss one of such methods – the most used – which is known under the name "perturbation theory for discrete energy levels".

Let us assume that the Hamiltonian H of the system can be written in the form

$$H = H_0 + H' \tag{12.1}$$

where H_0 is exactly solvable (i.e. its eigenvalues and eigenvectors are known) and H' is 'small'. What does the statement ' H' is small' mean? Indeed, H' is, in most cases, an unbounded operator and saying it is small has no clear meaning: to clarify the meaning of such a statement is one of the purposes of the following discussion.

The problem we are confronted with is that of establishing in an approximate way how the **perturbation** H' modifies the eigenvalues and eigenvectors of H_0, i.e. how the eigenvalues and eigenvectors of H can approximately be obtained from the knowledge of those of H_0. The modification caused by H' will, in general, be twofold: shift of the levels and removal, either partial or total, of their degeneracies.

Before carrying on with the general discussion, let us consider the example of a hydrogen atom in a static and uniform electric field $\vec{E}$. The Hamiltonian for the relative motion is ($-e\,\vec{q}$ is the dipole moment of the electron–proton system)

$$H = \frac{\vec{p}^{\,2}}{2\mu} - \frac{e^2}{r} + e\,\vec{E}\cdot\vec{q} \tag{12.2}$$

© Springer International Publishing Switzerland 2016
L.E. Picasso, *Lectures in Quantum Mechanics*, UNITEXT for Physics,
DOI 10.1007/978-3-319-22632-3_12

i.e. just of the form (12.1) if the positions

$$H_0 = \frac{\vec{p}^{\,2}}{2\mu} - \frac{e^2}{r}\,, \qquad H' = e\,\vec{E}\cdot\vec{q} \qquad (12.3)$$

are made. Let us firstly see what 'H' is small' may mean in the present case. If we limit ourselves to consider the modifications caused by H' to the *lowest* energy levels of H_0, we can say – in a sense to be clarified in the sequel – that H' is of the order of magnitude of $e\,E\,a_B$, with a_B the Bohr radius, so that if $E \simeq 10^4\,\mathrm{V/cm}$, then H' is of the order of $10^{-4}\,\mathrm{eV}$, which is positively small with respect to the *distances* among the *lowest* energy levels of H_0. It is worth remarking that, energies being always defined up to an additive constant, the only things with which a comparison makes sense are precisely the distances among energy levels.

We therefore expect that the presence of the electric field E:

1. modifies by a small amount ($\simeq 10^{-4}\,\mathrm{eV}$) the energy levels of the hydrogen atom;
2. removes, at least partially, the degeneracy of the energy levels: indeed, while H_0 commutes with all the components of $\vec{L}$ (which gives raise to the degeneracy on m), H only commutes with the component of $\vec{L}$ parallel to the direction of the electric field, so that the conditions that allow for the application of the degeneracy theorem no longer hold.

Let us go back to the general problem. Let us consider a representation in which H_0 is diagonal (see Sects. 6.1 and 6.2) and let $|n\rangle$ ($n = 1, \cdots$) be the basis of eigenvectors of H_0 that characterizes such a representation. The complete Hamiltonian H will then be represented by the (infinite-dimensional) matrix:

$$H_{nm} = E_n^0\,\delta_{nm} + H'_{nm} \qquad (12.4)$$

where the E_n^0 are the eigenvalues of H_0, called **unperturbed eigenvalues**.

In order to find the exact eigenvalues of H we should diagonalize the matrix H_{nm}: let us try to understand which approximations can be made to simplify the present problem – and, as a matter of fact, to convert it into a possible one.

The formal treatment will be presented in Sects. 12.3 and 12.4. We now limit ourselves to consider the simple (and fictitious) case of a system possessing only two independent states, i.e. a system that can be represented in a two-dimensional Hilbert space. Then H will be represented by the 2×2 matrix

$$H \to \begin{pmatrix} E_1^0 + H'_{11} & H'_{12} \\ H'_{21} & E_2^0 + H'_{22} \end{pmatrix}. \qquad (12.5)$$

In this case H can be exactly diagonalized (it suffices to solve the secular equation that is a second degree equation) and the eigenvalues are

$$E_{\pm} = \frac{1}{2} \Big[(E_1^0 + H'_{11} + E_2^0 + H'_{22}) + $$

$$\pm \sqrt{\left((E_1^0 + H'_{11}) - (E_2^0 + H'_{22}) \right)^2 + 4|H'_{21}|^2} \, \Big] \qquad (12.6)$$

(the fact that, due to the Hermiticity of H, $(H'_{21})^* = H'_{12}$ has been exploited). Equation (12.6) can be written in the following form

$$E_{\pm} = \frac{1}{2} \Big[(E_1^0 + H'_{11} + E_2^0 + H'_{22}) + $$

$$\pm \left((E_1^0 + H'_{11}) - (E_2^0 + H'_{22}) \right) \sqrt{1 + \left[\frac{2|H'_{21}|}{(E_1^0 + H'_{11}) - (E_2^0 + H'_{22})} \right]^2} \, \Big]. \quad (12.7)$$

If

$$|H'_{21}| \ll |(E_1^0 + H'_{11}) - (E_2^0 + H'_{22})| \qquad (12.8)$$

the term $[\cdots]^2$ in the argument of the square root can be neglected with respect to 1 and we get:

$$\begin{cases} E_+ \simeq E_1^0 + H'_{11} \\ E_- \simeq E_2^0 + H'_{22} \, . \end{cases} \qquad (12.9)$$

Since $\sqrt{1 + x^2} = 1 + \frac{1}{2}x^2 + \cdots$, the terms we have neglected in (12.9) are of the second order in the matrix elements of the perturbation H' (i.e. they are proportional to $|H'_{21}|^2$): therefore, *when condition* (12.8) *is fulfilled* and if we content ourselves to find the corrections of the first order in the matrix elements of H' to the unperturbed eigenvalues E_1^0 and E_2^0, we can 'set the off-diagonal matrix elements H'_{12} and H'_{21} equal to 0 in (12.5)' and one will say that *the first-order perturbative approximation* has been made. If instead H'_{21} is either of the same order of magnitude as $|(E_1^0 + H'_{11}) - (E_2^0 + H'_{22})|$ or even bigger (i.e. if (12.8) is not satisfied), the terms neglected in (12.9) are as much or more important than those one has kept, so no simplification is possible: in other words H'_{12} and H'_{21} 'cannot be set equal to 0'.

It usually happens that the matrix elements H'_{11} and H'_{22} are of the same order of magnitude as $|H'_{12}|$: in this case (12.8) takes the more expressive form

$$|H'_{12}| \ll |E_1^0 - E_2^0| \qquad (12.10)$$

i.e. the off-diagonal matrix elements H'_{12} and H'_{21} are negligible if, in absolute value, they are negligible with respect to difference between the unperturbed energies they refer to. If H'_{11}, H'_{22} and $|H'_{12}|$ are of the same order of magnitude, the hypothesis (12.10) concerning the validity of the first-order perturbative approximation, thanks to (12.9) can be rephrased by saying that the perturbative approximation is legitimate if the first order effect of the perturbation on the unperturbed energies E_1^0 and E_2^0 is small with respect to the difference $E_1^0 - E_2^0$. This provides us with an a posteriori criterion to check the validity of the perturbative calculation.

Let us now go back to the problem of diagonalizing (12.4): what we have learned about the two level system holds in general. In other words all the matrix elements such that

$$|H'_{nm}| \ll |E_n^0 - E_m^0| \qquad (12.11)$$

'can be set equal to 0'. After doing so, only those matrix elements of H' referring to eigenstates of H_0 with the same energy survive in (12.4) (i.e. the diagonal elements and those between the states of a *degenerate* energy level of H_0) and those that refer to eigenstates of H_0 belonging to 'close' energies ("quasi-degenerate" levels of H_0): in this way H_{nm} becomes a "block-diagonal" matrix, i.e. possibly nonvanishing within square blocks straddling the principal diagonal, and it is possible to diagonalize separately the blocks of which H is made out: if these are finite-dimensional, the diagonalization is performed by solving the secular equation that is an algebraic equation of finite degree.

Let us summarize: the off-diagonal matrix elements H_{nm} in (12.4) between states of different energies contribute to the exact eigenvalues by terms of the order

$$\left| \frac{H'_{nm}}{E_n^0 - E_m^0} \right|^2$$

(i.e. the second order) or higher. So if these terms are small, they can be neglected, i.e. we can neglect all the terms that satisfy (12.11) and in this way we correctly obtain the eigenvalues E_n perturbed up to the first order. There are other general aspects of this perturbative procedure that deserve a discussion, but instead of discussing them in abstract terms we prefer to present them in the concrete example we are going to discuss in the next section.

12.2 The Stark Effect in the Hydrogen Atom

Let us reconsider the example introduced in the previous section of an hydrogen atom in a uniform electric field $\vec{E}$.

The Hamiltonian is given by (12.2). If we now choose the z-axis parallel to the direction of $\vec{E}$, it takes the form

$$H = H_0 + e\,E\,z \qquad (12.12)$$

with H_0 defined by (12.3).

We choose the representation in which H_0 is diagonal as that given by the basis $|n, l, m\rangle$ of the simultaneous eigenvectors of H_0, $\vec{L}^2$ and L_z. It will become evident in a few lines that, having chosen the z-axis parallel to the electric field, it is convenient to have L_z diagonal instead of another component of $\vec{L}$ – which would be perfectly legitimate. We must first decide which matrix elements of the perturbation $H' = e\,E\,z$ can be neglected and which ones cannot. For sure we cannot neglect those between states with the same n (i.e. the same energy). Regarding the matrix elements of H' between states with different n, we can proceed in two different ways.

1. Evaluation of the order of magnitude of these matrix elements and comparison with the differences $E_n^0 - E_m^0$.
2. First set to 0 all the matrix elements of H' referring to states with different n, then accept a posteriori only those results for which the energy shift due to the perturbation is small with respect to the distances from the other levels.

Let us proceed according to the first choice. However, since the following is not a rigorous discussion, an a posteriori control as described in point 2 will be appropriate.

We are interested in the matrix elements of H' between states with different energies, let us say E_n^0 and $E_{n'}^0$: they will depend on n and n' and also on the values of the angular momentum l and l'. It is not worth to calculate exactly such matrix elements: indeed, given we only have to make sure that $|H'_{nm}| \ll |E_n^0 - E_{n'}^0|$, we may take into consideration only the nearest level, i.e. $n' = n + 1$, and in this case one can expect that the matrix elements of z are of the order of magnitude of the radius of the n-th Bohr orbit, namely:

$$|\langle n \cdots | z | n+1 \cdots \rangle| \simeq n^2 \, a_B \, ,$$

whence

$$|H'_{n\,n+1}| \simeq e \, E \, a_B \, n^2 \, . \tag{12.13}$$

As a matter of fact, a more convincing way to obtain (12.13) is the following, in which the first inequality can be shown in the same way as (4.45) and the second is obvious:

$$|\langle n | z | n' \rangle|^2 = \langle n | z | n' \rangle \langle n' | z | n \rangle \leq \langle n | z^2 | n \rangle < \langle n | r^2 | n \rangle$$

holding for *any* n, n'. At this point, since we expect that $\big(\langle n | r^2 | n \rangle\big)^{1/2}$ is of the order of the n-th Bohr orbit, (12.13) is found again in the stronger form:

$$|H'_{nn'}| < e \, E \, \sqrt{\langle n | r^2 | n \rangle} \simeq e \, E \, a_B \, n^2$$

(just to satisfy the curious reader: $\langle n | r^2 | n \rangle = \frac{1}{2} n^2 \big[5n^2 + 1 - 3l\,(l+1)\big] \, a_B^2$).

Equation (12.13) shows that the larger n, the larger the matrix elements $H'_{n\,n+1}$. Since, in addition, the distance between adjacent levels decreases, it is clear that the perturbative approximation will be legitimate only for the energy levels with n not too large. Note however that, for electric fields even rather strong as e.g. $E \simeq 10^4 \, \text{V/cm}$, the right hand side of (12.13) is about $n^2 \times 10^{-4} \, \text{eV}$, so only for the levels with $n \simeq 10$ the matrix elements of the perturbation are comparable with the distances between the unperturbed levels.

As a consequence, in the perturbative approximation the matrix that represents the perturbation (12.12) in the basis $| n, l, m \rangle$ is block-diagonal only for n not too large, whereas from a given n on (its value depending on the value of E and on the desired degree of accuracy) H is a unique block of infinite dimension (see Fig. 12.1).

The finite blocks are matrices of dimension 1×1, 4×4, 9×9, $\cdots$, because $1, 4, 9$, etc. are the degeneracies of the energy levels of the hydrogen atom.

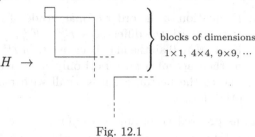

$H \to$ $\quad$ blocks of dimensions 1×1, 4×4, 9×9, $\cdots$

Fig. 12.1

As a consequence, the perturbative calculation cannot be made for all the levels of hydrogen: we will limit ourselves to the study of the effect of the electric field on the first two levels.

As far as the lowest energy level is concerned, being nondegenerate, the first-order shift due to the perturbation simply is the mean value of H' in the state $|1,0,0\rangle$. We thus have to calculate

$$\langle 1,0,0 \mid z \mid 1,0,0 \rangle . \tag{12.14}$$

There are several ways to see that this matrix element is zero. They are based on arguments of general character known as **selection rules**.

Angular Momentum Selection Rule: *the matrix elements of x, y and z between s (i.e. $l = 0$) states are vanishing.*

Indeed, if $|A\rangle$ and $|B\rangle$ are two s states, i.e. with $l = 0$, let us consider

$$\langle A \mid x_i \mid B \rangle , \qquad i = 1, 2, 3 .$$

The wavefunction of the state $|B\rangle$ only depends on r: let it be $f(r)$. That of the state $x_i|B\rangle$ therefore is $x_i f(r)$, that is the wavefunction of a p state ($l = 1$) (see (10.42) and (10.43)). Therefore the state $|A\rangle$ and the state $x_i|B\rangle$ are orthogonal to each other, inasmuch as belonging to different eigenvalues of the operator $\vec{L}^2$ – which proves the statement.

In the case at hand $|A\rangle = |B\rangle = |1,0,0\rangle$ and the mean value (12.14) vanishes.

Space Inversion Selection Rule: *the matrix elements of x, y and z between states with the same parity are vanishing.*

Indeed, if $|A\rangle$ and $|B\rangle$ are two states with the same eigenvalue $w = \pm 1$ of the space-inversion operator I, recalling the properties of the space-inversion operator given in Sect. 7.2:

$$I = I^\dagger = I^{-1} , \qquad I\, x_i\, I^{-1} = -x_i \quad \Leftrightarrow \quad I\, x_i\, I^\dagger = -x_i$$

one has

$$\langle A \mid x_i \mid B \rangle = \langle A \mid I^\dagger\, I\, x_i\, I^\dagger\, I \mid B \rangle = w^2 \langle A \mid I\, x_i\, I^\dagger \mid B \rangle$$
$$= \langle A \mid I\, x_i\, I^\dagger \mid B \rangle = -\langle A \mid x_i \mid B \rangle$$

whence $\langle A \mid x_i \mid B \rangle = 0$.

In the case at hand $|A\rangle = |B\rangle = |1,0,0\rangle$ and the state $|1,0,0\rangle$ is an eigenstate of I (belonging to the eigenvalue $w = +1$), so, once more, the mean value (12.14) vanishes.

In conclusion the perturbation does not produce any effect, to the first order, on the lowest energy level.

Let us now consider the first excited level: $n = 2$.
Its degeneracy is 4 (one state with $l = 0$, three independent states with $l = 1$). We thus have to determine, and then diagonalize, the 4×4 matrix that represents the Hamiltonian (12.12) in the basis $|n, l, m\rangle$ with $n = 2$. So we have to calculate the matrix elements $\langle 2, l, m \mid H \mid 2, l', m' \rangle$. The matrix elements of H_0 are known: H_0 is diagonal and the four matrix elements all equal E_2^0; there remain the matrix elements of H' i.e. of z.

Let us start with pinpointing those elements that certainly are vanishing: $\langle 2,0,0 \mid z \mid 2,0,0\rangle = 0$ for the same reasons why $\langle 1,0,0 \mid z \mid 1,0,0\rangle$ is vanishing;
$\langle 2,1,m \mid z \mid 2,1,m'\rangle = 0$ for any m, $m' = +1, 0, -1$ owing to the space-inversion selection rule (the states with $l = 1$ all have parity $w = -1$);
$\langle 2,0,0 \mid z \mid 2,1,\pm 1 \rangle = 0$: this follows from another selection rule that we will now demonstrate.

L_z Selection Rule: *the matrix elements of z between eigenstates belonging to different eigenvalues of L_z are vanishing.*

Let us, indeed, consider $\langle m \mid z \mid m'\rangle$, where $|m\rangle$ and $|m'\rangle$ are eigenvectors of L_z corresponding to the eigenvalues m and m'. The dependence on ϕ of the wavefunction of the state $|m'\rangle$ is $e^{im'\phi}$; that of the state $z|m'\rangle$ still is $e^{im'\phi}$ since $z = r\cos\theta$. As a consequence, $z|m'\rangle$ still is an eigenvector of L_z belonging to the eigenvalue m', so for $m \neq m'$ one has $\langle m \mid z \mid m'\rangle = 0$.

The same argument can be given a more abstract form and the same conclusion can be drawn by using the lemma of p. 87 (Sect. 4.10) after observing that $[z, L_z] = 0$.

The use of the last selection rule is made possible by having chosen as a basis in which H_0 is diagonal precisely that in which L_z is diagonal. Of course also $\langle 2,1,\pm 1 \mid z \mid 2,0,0\rangle = 0$ both owing to the L_z selection rule and because such matrix elements are complex conjugate of the previous ones.

There remains to consider the matrix element $\langle 2,0,0 \mid z \mid 2,1,0\rangle$ and its complex conjugate $\langle 2,1,0 \mid z \mid 2,0,0\rangle$. None of the three selection rules applies: not both the states $|2,0,0\rangle$ and $|2,1,0\rangle$ are s states, they do not have the same parity and they do have the same m. Nothing prevents the existence of a further selection rule enabling one to show that also this matrix elements is vanishing.

However it is now important to note that the three selection rules we have discussed and utilized refer to the quantum numbers (l, w, m) that characterize the states $|n, l, m\rangle$ we are considering, i.e. to the eigenvalues of observables that commute with H_0: once the set of the quantum numbers we are able to attribute to such states is exhausted, the possibility of discovering new selection rules gives out. This shows (we will come back in the sequel to this point) the importance of being able to classify the states by means of – possibly overabundant – quantum numbers (in the just treated case we already knew that the parity w is determined by the knowledge of l).

After the above considerations we are left with the direct calculation as the only tool to determine the matrix element. One has

$$\langle 2,0,0 \mid z \mid 2,1,0 \rangle =$$
$$= \int_0^{2\pi} d\phi \int_{-1}^{+1} d\cos\theta \int_0^\infty Y_{00}^*(\theta,\phi)\, R_{20}^*(r)\, r\, \cos\theta\, Y_{10}(\theta,\phi)\, R_{21}(r)\, r^2\, dr$$

(the asterisks are unnecessary, as both Y_{00} and R_{20} are real). The calculation of the above integrals is not very instructive, in any event it can be made by aid of (10.42) for the spherical harmonics and of (11.41) for the radial functions. The result is

$$\langle 2,1,0 \mid z \mid 2,0,0 \rangle = 3\, a_B .$$

To summarize, the 4×4 matrix representing H in the basis $\mid n, l, m \rangle$ with $n = 2$ is

$$\qquad\qquad |2,0,0\rangle \quad |2,1,0\rangle \quad |2,1,+1\rangle \quad |2,1,-1\rangle$$

$$H \xrightarrow{(n=2)} \begin{pmatrix} E_2^0 & 3eEa_B & 0 & 0 \\ 3eEa_B & E_2^0 & 0 & 0 \\ 0 & 0 & E_2^0 & 0 \\ 0 & 0 & 0 & E_2^0 \end{pmatrix} . \qquad (12.15)$$

The diagonalization of this matrix is straightforward: one only has to diagonalize the upper left 2×2 block, since the remaining 2×2 block already is diagonal. The eigenvalues of (12.15) are

$$E_2^0 - 3e\,E\,a_B , \qquad E_2^0 + 3e\,E\,a_B , \qquad E_2^0 , \qquad E_2^0 .$$

It it clear, as predicted in the previous section, that the perturbation has only partially removed the degeneracy of the level $n = 2$: from a level 4 times degenerate in the absence of the external field one passes to a nondegenerate level with energy $E_2^0 + 3e\,E\,a_B$, a level with degeneracy 2 and energy E_2^0 and, finally, to a nondegenerate level with energy $E_2^0 - 3e\,E\,a_B$ (Fig. 12.2).

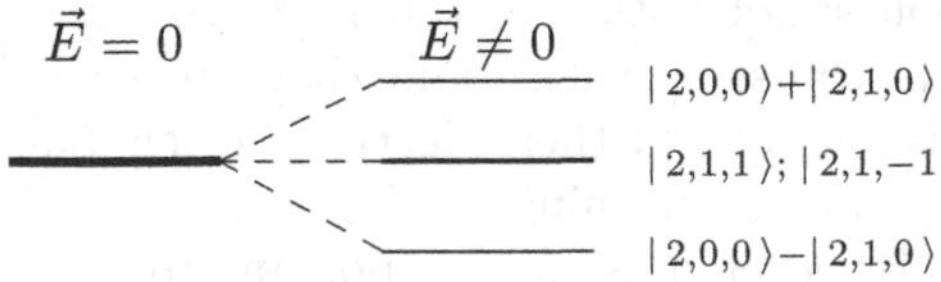

The level shift due to the perturbation is $\Delta E = 3\,e\,E\,a_B$ that – as we have already seen – is of the order of 10^{-4} eV for a field E of the order of 10^4 V/cm. The distance between the $n = 2$ unperturbed level and that clos-

Fig. 12.2

est to it, i.e. $n = 3$, is about $2\,$eV: in these condition the validity of the perturbative calculation is fully guaranteed.

Let us now find the approximate eigenvectors of H corresponding to the found eigenvalues. It is easy to check that the normalized eigenvectors of matrix (12.15) are

$$\frac{1}{\sqrt{2}}\begin{pmatrix} +1 \\ -1 \\ 0 \\ 0 \end{pmatrix} \; ; \quad \frac{1}{\sqrt{2}}\begin{pmatrix} +1 \\ +1 \\ 0 \\ 0 \end{pmatrix} \; ; \quad \frac{1}{\sqrt{|\alpha|^2 + |\beta|^2}}\begin{pmatrix} 0 \\ 0 \\ \alpha \\ \beta \end{pmatrix} , \quad \alpha, \beta \in \mathbb{C} . \quad (12.16)$$

The first two respectively correspond to the eigenvalues $E_2^0 \mp 3eE\,a_{\rm B}$ and, since we are in the representation $|\,n, l, m\,\rangle$ with $n = 2$, they represent the normalized vectors

$$\frac{1}{\sqrt{2}}\big(|\,2, 0, 0\,\rangle - |\,2, 1, 0\,\rangle\big) \; ; \qquad \frac{1}{\sqrt{2}}\big(|\,2, 0, 0\,\rangle + |\,2, 1, 0\,\rangle\big) . \qquad (12.17)$$

Finally, for any α and β, the last one corresponds to the degenerate eigenvalue E_2^0 and represents all the vectors having the form

$$\alpha\,|\,2, 1, +1\,\rangle + \beta\,|\,2, 1, -1\,\rangle , \qquad |\alpha|^2 + |\beta|^2 = 1 . \qquad (12.18)$$

Observe that, while the approximate eigenvalues of H depend linearly on the matrix elements of the perturbation H', the approximate eigenvectors are independent of the perturbation and still are eigenvectors of H_0: one then obtains the eigenvalues approximated to first order in correspondence with eigenvectors approximated to the order zero.

It is interesting to note that the vectors (12.17), i.e. those relative to the 'shifted' eigenvalues, are – from among all the vectors with $n = 2$ – those respectively having the maximum and the minimum mean value of the component, parallel to the electric field, of the electric dipole operator $\vec{D}$ of the electron-proton system:

$$\vec{D} = -e\,\vec{q} \qquad (\,\vec{q} = \vec{q}_{\rm e} - \vec{q}_{\rm p}\,)$$

(we leave the verification of this statement to the reader). These mean values exactly equal $\pm 3\,e\,a_{\rm B}$, i.e. the result we have found. The mean value of D_z is instead vanishing in the states described by (12.18): this follows, for example, by application of the space-inversion selection rule. Therefore the effect of the electric field on them is vanishing as well.

The analogy with the corresponding classical problem appears evident: there exists a first-order effect of the electric field on the level $n = 2$ since, from among the states of the energy level $n = 2$ there exist states that have – in the sense of the mean value – a nonvanishing dipole moment; in other words, since the hydrogen atom in the $n = 2$ level possesses an *intrinsic* dipole moment (but not in the ground state).

It is also important to note that the existence of such states endowed with an intrinsic dipole moment is due to the fact that in hydrogen there is the degeneracy on l: thanks to the space-inversion selection rule, in order to obtain a nonvanishing mean value of the dipole operator $\vec{D}$, it is necessary to superpose states with opposite parities , i.e. with different l – exactly as in the case of the sates (12.17): the eigenfunctions in the Schrödinger representation of these states are such that the corresponding $|\psi|^2$ exhibit an asymmetry with respect to the plane $z = 0$: this provides an intuitive explanation for the existence of a dipole moment for such states.

Let us discuss, as the final point, how the effect of the electric field on the level $n = 2$ of hydrogen is experimentally observed. In the absence of the field, the transitions between the levels $n = 1$ and $n = 2$ give rise to the first line of the Lyman series; as seen above, the effect of the electric field is that of producing three levels with $n = 2$ and of leaving the level $n = 1$ unaffected. As a consequence one should observe three lines in the transitions between the levels $n = 1$ and $n = 2$: the central one coincides with the first line of the Lyman series; the other two are symmetrically placed with respect to this one with frequency shifts given respectively by $\Delta\nu = \pm 3\,e\,E\,a_B/h$.

We shall see in the next chapter that the above three lines exhibit different features, as far as polarization is concerned, and for this reason can be simultaneously observed only in particular experimental conditions.

The effect just discussed is known as the **Stark Effect**, after the name of one of its discoverers.

12.3 First-Order Perturbation Theory: Formal Treatment

This section does not contain any new result besides those enunciated in the two previous sections; such results will be often repeated for the sake of completeness, we here only present a formal derivation of the fact that the terms of H' we have neglected do not contribute to the first-order shift of the energy levels.

Let us start again from (12.1):

$$H = H_0 + H'$$

and let us put

$$H' = H_0' + H'' \tag{12.19}$$

where the only nonvanishing matrix elements of H_0' are those between states with the same energy: as a consequence H_0 and H_0' respectively have, in any representation in which H_0 is diagonal, the structure reported in Figs. 6.1a and 6.1b.

The decomposition (12.19) is unique, i.e. independent of the choice of the basis, and one has

$$[\,H_0\,,\,H_0'\,] = 0\,.$$

Indeed, if $\mathcal{P}_n$ is the projection operator onto the eigenspace of H_0 corresponding to the eigenvalue E_n^0, one has $\left(\sum_n \mathcal{P}_n = \mathbb{1}\right)$:

$$H' \equiv \sum_{nm} \mathcal{P}_n\,H'\,\mathcal{P}_m = \sum_n \mathcal{P}_n\,H'\,\mathcal{P}_n + \sum_{n\neq m} \mathcal{P}_n\,H'\,\mathcal{P}_m \overset{\text{def}}{=} H_0' + H''\,.$$

We now exploit the fact that the separation (12.1) of H into a 'solvable' and a 'small' term is arbitrary and set

$$H = (H_0 + H_0') + H'' . \tag{12.20}$$

For sure, if H' is 'small', also H'' is such (indeed, it may happen that H' is not 'small', but H'' is such), so that – if $H_0 + H_0'$ is 'solvable' – the decomposition (12.20) is at least as legitimate as (12.1).

Let us observe that $(H_0 + H_0')_{nm}$ is a block matrix the dimensions of whose blocks equal the degrees of degeneracy of the several eigenvalues E_n^0 of H_0 : at this point, in order to diagonalize such a matrix, one can separately diagonalize the single blocks of which it is made out, and if these (or at least those relative to the energy levels we are interested in) are finite-dimensional, the diagonalization is achieved by solving the relative secular equations, that are algebraic equations of finite degree. It is in this sense that we say that also $H_0 + H_0'$ is solvable. Obviously, since within any block H_0 is a multiple of the identity, it suffices to diagonalize the blocks representative of H_0'. As a consequence, the diagonalization of the block corresponding to the generic eigenvalue E_n^0 of H_0 provides the eigenvalues

$$E_{n,\nu}^{(1)} = E_n^0 + \Delta_{n,\nu}^{(1)} , \qquad \nu = 1, \cdots, g_n = \text{degree of degeracy of } E_n^0 \tag{12.21}$$

where the $\Delta_{n,\nu}^{(1)}$ are the eigenvalues of the $g_n \times g_n$ block of H_0'.

It is easy to realize that the corrections $\Delta_{n,\nu}^{(1)}$ to the unperturbed eigenvalues E_n^0 are of the first order in H' : this means that, if we multiply H' by an arbitrary parameter λ, the $\Delta_{n,\nu}^{(1)}$ are proportional to λ. Indeed, if $H' \to \lambda H'$, also $H_0' \to \lambda H_0'$ and, as a consequence, also the eigenvalues $\Delta_{n,\nu}^{(1)}$ of H_0' are multiplied by λ. In addition we will show that the term H'' so far neglected does not contribute to the first-order corrections to the unperturbed eigenvalues E_n^0 : therefore we can conclude that the $E_{n,\nu}^{(1)}$ given by (12.21) are the eigenvalues of H corrected up to the first order in H'.

For this reason one says that (12.21) is the first-order approximation. The effect of the perturbation H' is evident already to the first order: shift of energy levels (not all the $\Delta_{n,\nu}^{(1)}$, in general, vanish) and (at least partial) removal of the possible degeneracies of H_0 (for fixed n, in general not all the $\Delta_{n,\nu}^{(1)}$ are equal to one another).

Corresponding to the approximate eigenvalues $E_{n,\nu}^{(1)}$ of H we will have (approximate) eigenvectors $| E_{n,\nu}^{(1)} \rangle$: they are the eigenvectors of $H_0 + H_0'$ and it should be evident (because of the block–diagonalization procedure) that they also are eigenvectors of H_0 (but not viceversa!); actually, from the fact that H_0 and H_0' commute with each other it follows that H_0 and $H_0 + H_0'$ have a complete set of simultaneous eigenvectors.

Let us now show what we have just stated, namely that H'' does not contribute to the first order of the eigenvalues of H. To this end we multiply (only) H'' by the real parameter λ and put

$$H_\lambda = (H_0 + H_0') + \lambda H'' \,.$$

The eigenvalue equation for H_λ is

$$\left[(H_0 + H_0') + \lambda H''\right] | E_{n,\nu}(\lambda) \rangle = E_{n,\nu}(\lambda) | E_{n,\nu}(\lambda) \rangle \qquad (12.22)$$

where, for $\lambda \to 0$: $E_{n,\nu}(\lambda) \to E_{n,\nu}^{(1)}$ and $| E_{n,\nu}(\lambda) \rangle \to | E_{n,\nu}^{(1)} \rangle$.

The statement that $E_{n,\nu}(\lambda)$ does not depend, to the first order, on λ (and therefore on H'') is the same as saying that

$$\left. \frac{\mathrm{d} E_{n,\nu}(\lambda)}{\mathrm{d}\lambda} \right|_{\lambda=0} = 0 \,. \qquad (12.23)$$

In order to show that this is true, we take the derivative of (12.22) with respect to λ and put $\lambda = 0$. Calling

$$| \mathrm{d}_\lambda E(0) \rangle \equiv \left. \frac{\mathrm{d}}{\mathrm{d}\lambda} | E_{n,\nu}(\lambda) \rangle \right|_{\lambda=0} , \qquad \frac{\mathrm{d} E(0)}{\mathrm{d}\lambda} \equiv \left. \frac{\mathrm{d} E_{n,\nu}(\lambda)}{\mathrm{d}\lambda} \right|_{\lambda=0}$$

one has

$$(H_0 + H_0') | \mathrm{d}_\lambda E(0) \rangle + H'' | E_{n,\nu}^{(1)} \rangle = \frac{\mathrm{d} E(0)}{\mathrm{d}\lambda} | E_{n,\nu}^{(1)} \rangle + E_{n,\nu}^{(1)} | \mathrm{d}_\lambda E(0) \rangle \qquad (12.24)$$

and by taking the scalar product of both sides with the vector $| E_{n,\nu}^{(1)} \rangle$:

$$\langle E_{n,\nu}^{(1)} | (H_0 + H_0' - E_{n,\nu}^{(1)}) | \mathrm{d}_\lambda E(0) \rangle + \langle E_{n,\nu}^{(1)} | H'' | E_{n,\nu}^{(1)} \rangle = \frac{\mathrm{d} E(0)}{\mathrm{d}\lambda} \,.$$

The first term is vanishing (the 'bra' $\langle E_{n,\nu}^{(1)} |$ is an eigenvector of $H_0 + H_0'$), the second also is vanishing due to the very definition of H''. Therefore

$$\frac{\mathrm{d} E(0)}{\mathrm{d}\lambda} \equiv \left. \frac{\mathrm{d} | E_{n,\nu}(\lambda) \rangle}{\mathrm{d}\lambda} \right|_{\lambda=0} = 0 \,.$$

This completes the treatment of perturbation theory to the first order which, to sum up, consists in setting $H'' = 0$, i.e. in the neglecting of all the matrix elements of H' between eigenstates of H_0 corresponding to different unperturbed energies.

12.4 Second-Order Perturbation Theory and Quasi-Degenerate Levels

We have now to face the problem of establishing when the perturbative approximation is meaningful: to this end we will calculate the first nonvanishing contribution of H'' to the eigenvalues $E(\lambda)$, i.e. – in the language of the pervious section – the contribution of order λ^2 of $\lambda H''$.

Taking two derivatives of (12.22) with respect to λ, setting $\lambda = 0$ and recalling (12.23) one has

$$(H_0 + H_0') \, | \, \mathrm{d}_\lambda^2 E(0) \, \rangle + 2H'' \, | \, \mathrm{d}_\lambda E(0) \, \rangle = \frac{\mathrm{d}^2 E(0)}{\mathrm{d}\lambda^2} \, | \, E_{n,\nu}^{(1)} \, \rangle + E_{n,\nu}^{(1)} \, | \, \mathrm{d}_\lambda^2 E(0) \, \rangle \, .$$

Let us now take the scalar product of both sides with the vector $| \, E_{n,\nu}^{(1)} \, \rangle$: the contributions due to the first and the last term are equal to each other, therefore

$$\frac{1}{2} \frac{\mathrm{d}^2 E(0)}{\mathrm{d}\lambda^2} \equiv \frac{1}{2} \frac{\mathrm{d}^2 E_{n,\nu}(\lambda)}{\mathrm{d}\lambda^2} \bigg|_{\lambda=0} = \langle \, E_{n,\nu}^{(1)} \, | \, H'' \, | \, \mathrm{d}_\lambda E(0) \, \rangle \, .$$

By use of the completeness relation $\sum_{k,\kappa} | \, E_{k,\kappa}^{(1)} \, \rangle \langle \, E_{k,\kappa}^{(1)} \, | = \mathbb{1}$ one has

$$\langle \, E_{n,\nu}^{(1)} \, | \, H'' \, | \, \mathrm{d}_\lambda E(0) \, \rangle = \sum_{k,\kappa}' \langle \, E_{n,\nu}^{(1)} \, | \, H'' \, | \, E_{k,\kappa}^{(1)} \, \rangle \langle \, E_{k,\kappa}^{(1)} \, | \, \mathrm{d}_\lambda E(0) \, \rangle$$

where $\sum_{k,\kappa}'$ means that the terms with $k = n$ are excluded from the sum inasmuch as, by the definition of H'', $\langle \, E_{n,\nu}^{(1)} \, | \, H'' \, | \, E_{n,\mu}^{(1)} \, \rangle = 0$. In order to obtain the expression of $\langle \, E_{k,\kappa}^{(1)} \, | \, \mathrm{d}_\lambda E(0) \, \rangle$, one takes the scalar product of (12.24) with $| \, E_{k,\kappa}^{(1)} \, \rangle$ (here $k \neq n$):

$$\left(E_{n,\nu}^{(1)} - E_{k,\kappa}^{(1)} \right) \langle \, E_{k,\kappa}^{(1)} \, | \, \mathrm{d}_\lambda E(0) \, \rangle = \langle \, E_{k,\kappa}^{(1)} \, | \, H'' \, | \, E_{n,\nu}^{(1)} \, \rangle$$

whence

$$\frac{1}{2} \frac{\mathrm{d}^2 E_{n,\nu}(\lambda)}{\mathrm{d}\lambda^2} \bigg|_{\lambda=0} = \sum_{k,\kappa}' \frac{| \langle \, E_{n,\nu}^{(1)} \, | \, H'' \, | \, E_{k,\kappa}^{(1)} \, \rangle |^2}{E_{n,\nu}^{(1)} - E_{k,\kappa}^{(1)}} \, .$$

The first nonvanishing contribution of $\lambda H''$ is $\frac{1}{2} \lambda^2 \times \left(\mathrm{d}^2 E_{n,\nu}(\lambda)/\mathrm{d}\lambda^2 \right)\big|_{\lambda=0}$, therefore that of H'' is

$$\Delta_{n,\nu}^{(2)} = -\sum_{k,\kappa}' \frac{| \langle \, E_{n,\nu}^{(1)} \, | \, H'' \, | \, E_{k,\kappa}^{(1)} \, \rangle |^2}{E_{k,\kappa}^{(1)} - E_{n,\nu}^{(1)}} = -\sum_{k,\kappa}' \frac{| \langle \, E_{n,\nu}^{(1)} \, | \, H' \, | \, E_{k,\kappa}^{(1)} \, \rangle |^2}{E_{k,\kappa}^{(1)} - E_{n,\nu}^{(1)}} \quad (12.25)$$

(we have replaced H'' with H' as in $\sum'$ only the matrix elements with $k \neq n$ intervene). The minus sign in (12.25) is there because we have changed the sign of the denominator: in a while we shall explain why.

Usually the perturbation H' contains a parameter (not necessarily a dimensionless one: see e.g. the factor $e E$ in (12.3)) that determines its order of magnitude and that is the analogue of the parameter λ we have introduced in a formal way. In this case (12.25) contains (not meaningful) terms that – with respect to the parameter of the perturbation – are of order higher than the second: this is so since in the denominators there also are terms of the first order. If in (12.25) the terms of higher order are neglected, the expression for the energy shifts of the second order in the parameter of the perturbation is

$$\Delta_{n,\nu}^{(2)} = -\sum_{k,\kappa}' \frac{| \langle \, E_{n,\nu}^{(1)} \, | \, H' \, | \, E_{k,\kappa}^{(1)} \, \rangle |^2}{E_k^0 - E_n^0} \, . \quad (12.26)$$

The minus sign emphasizes the fact that the second order contribution to the energy of the ground state $|E_0^0\rangle$ always is negative ($E_k^0 > E_0^0$). In general (12.26) shows that, for the second order effect on a given level E_n^0, the contribution of the upper levels ($E_k^0 > E_n^0$) is negative whereas that of the lower levels ($E_k^0 < E_n^0$) is positive: one says that the levels 'repel' each other.

The condition guaranteeing that, for a given energy level, the first-order approximation is meaningful is:

1. if $\Delta^{(1)} \neq 0$: $\Delta^{(2)} \ll \Delta^{(1)}$;
2. if $\Delta^{(1)} = 0$: $\Delta^{(2)} \ll \delta E_0$ where δE_0 is the distance between the unperturbed level and the one closest to it.

In any event, roughly, the condition is that the matrix elements of H' be much smaller than δE_0.

It may happen that the levels of H_0 exhibit a **multiplet** structure, i.e. groups of levels E_1, E_2, $\cdots$ close to one another, but far from other multiplets (for example in atoms one has **fine-structure** multiplets). In such cases the condition $|H'_{nm}| \ll \delta E_0$ may not be fulfilled, owing to the presence of small δE_0's, and not even the second order may be sufficient because higher orders (the third, fourth, ...) are powers of the ratios $|H'_{nm}/\delta E_0|$ that are not small. In such cases one proceeds by exploiting once more the arbitrariness in the separation of H into a solvable term and a small term. Let us put:

$$H_0 = H_0^0 + \delta H_0 \tag{12.27}$$

where H_0^0 is degenerate on every multiplet and δH_0 has, within the multiplet, the eigenvalues δE_1, δE_2, $\cdots$ (e.g.: $E_1 = E_0^0 + \delta E_1$, $E_2 = E_0^0 + \delta E_2$, $\cdots$). The decomposition (12.27) is not unique, but – as we shall see in a while – this is not a problem; in any event note that $[H_0^0, \delta H_0] = 0$. Now we put

$$H \equiv H_0 + H' = H_0^0 + (\delta H_0 + H')$$

and treat $(\delta H_0 + H')$ as a perturbation to the first order: from the physical point of view this procedure is just the inclusion into the perturbation of those terms of H_0 that are responsible for the multiplet structure.

In this way we have gone back to the case of Sect. 12.3, in which the unperturbed Hamiltonian, now H_0^0, is degenerate; then we proceed according to the decomposition (12.19):

$$(\delta H_0 + H') = (\delta H_0 + H_0') + H''$$

and the analogue of (12.20) is

$$H = (H_0^0 + \delta H_0 + H_0') + H'' \equiv (H_0 + H_0') + H''$$

and now it is clear that the problem of the non-uniqueness of the decomposition (12.27) is not a real one.

From now on one can proceed as in Sect. 12.3, i.e. one diagonalizes the "unperturbed Hamiltonian" $H_0 + H_0'$ that is block diagonal: the dimensions of the blocks are given by the degree of degeneracy of the eigenvalues of H_0^0, i.e. they are those of the multiplets; the only difference with respect to the case of Sect. 12.3 is that now H_0 is not a multiple of the identity within each block, namely within each multiplet; so it is not sufficient to diagonalize H_0', but it is necessary to diagonalize the blocks of $H_0 + H_0'$ (or of $\delta H_0 + H_0'$).

Now the first-order result is acceptable because the neglected terms are of the order of $(|H_{nm}/\delta E_0^0|)^2$, where now the denominators δE_0^0 are large because they are the distances *between*, not *within*, the multiplets.

To sum up, one takes the matrix H_{nm}, sets to 0 all the off-diagonal terms between states belonging to different multiplets, then diagonalizes what has survived.

Let us make an example: if, in the Hamiltonian of the hydrogen atom, we consider the relativistic expression for the kinetic energy, the result is that the degeneracy on l is removed: e.g. the states with $n = 2$ and $l = 0, 1$ are no longer degenerate in energy and one has $E_{2,1}^0 - E_{2,0}^0 \simeq 4 \times 10^{-5}$ eV (this is not the only relativistic correction to the energy levels of the hydrogen atom: as we will see in Chap. 16, the existence of spin entails other effects that are relativistic in nature).

If we now consider, as in Sect. 12.2, the effect of an electric field on the levels $n = 2$, we are in the just described situation of quasi-degenerate levels, and the solution of the problem is provided by the diagonalization of the matrix

$$\begin{pmatrix} E_{2,0}^0 & 3eEa_{\rm B} & 0 & 0 \\ 3eEa_{\rm B} & E_{2,1}^0 & 0 & 0 \\ 0 & 0 & E_{2,1}^0 & 0 \\ 0 & 0 & 0 & E_{2,1}^0 \end{pmatrix}. \qquad (12.28)$$

that differs from (12.15) in that the diagonal terms are not all equal to one another. Apart from the eigenvalue $E_{2,1}^0$ twice degenerate, the other eigenvalues of (12.28) are:

$$\frac{1}{2}\left[(E_{2,0}^0 + E_{2,1}^0) \pm \sqrt{(\delta E_0)^2 + 4 \times (3eEa_{\rm B})^2} \right], \qquad \delta E_0 \equiv (E_{2,1}^0 - E_{2,0}^0)$$

and, as it was already clear a priori, with $|3eEa_{\rm B}|$ of the same order of δE_0, an expansion in powers of $3eEa_{\rm B}/\delta E_0$ would be completely meaningless.

Chapter 13

Electromagnetic Transitions

13.1 Introduction

As the reader will remember, the incapability o classical physics to explain
the line structure of atomic spectra was one of the main motivations that
led physicist to the search for a new theory: one of the first attempts in this
direction was due to Bohr who, just to overcome the difficulties connected with
the problem of spectra, postulated the existence of discrete energy levels and
the well known relationship (2.5) connecting the emitted/absorbed frequencies
and the energy levels.

At this point we think it is convenient to point out that Bohr theory could
not be considered the 'new theory' but, rather, just the first fundamental
step in that direction: indeed some facts (hydrogen spectrum, specific heats
of solids, etc.) were very well explained, other facts only had a qualitative
explanation (in general, Bohr results are correct only for large values of the
quantum numbers), eventually some other facts had no explanation at all
(typically: intensity of spectral lines, selection rules, etc.).

It is however from a conceptual point of view that Bohr theory is lacking: the foundations of classical physics are incompatible with Bohr hypotheses and must, as a consequence, be modified – but Bohr does not say how.
Furthermore, in Bohr theory many questions have no answer: here are two
examples.

(i) Do there exist states other than those with energies E_0, E_1, $\cdots$?
(ii) In the time interval of about 10^{-8} s during which the radiating electron
goes from the energy level E_n to the level E_{n-1}, does it pass through all
the intermediate energies?

And so on. On the other hand, a proof of the fact that, in Bohr's time,
the situation was not considered satisfactory is given by the fact that de
Broglie felt the necessity of proposing another hypothesis in a spirit completely
different from that of Bohr hypotheses. And the fact that not even after de
Broglie 'wave-like' hypothesis the theory was in a satisfactory status is shown

© Springer International Publishing Switzerland 2016   219
L.E. Picasso, *Lectures in Quantum Mechanics*, UNITEXT for Physics,
DOI 10.1007/978-3-319-22632-3_13

by all the problems that the experimental verifications of de Broglie hypothesis raised (and which were the subject of the discussions presented in Chap. 3).

Quantum mechanics not only provides the overcoming of such conceptual difficulties, but also – and mainly – it is a theory capable both to frame in a single scheme the positive aspects of old quantum theories, and to provide correct results in many cases in which the latter were unable to say anything.

In particular we have already extensively seen that both the existence of energy levels and the wave-like properties of particles are consequences of the new theory (essentially of the superposition principle and of the quantization rules): the same thing can be said about the second Bohr hypothesis: it is to this subject that we will dedicate the present chapter.

13.2 Perturbation Theory for Time Evolution

Let us assume we have an atom in the stationary state $| E_i^0 \rangle$. At time $t = t_0$ we send electromagnetic radiation on the atom: the electromagnetic field interacts with the atom so that the state $| E_i^0 \rangle$ ceases being a stationary state (the Hamiltonian has changed) and will therefore evolve in time. Let us denote, as usual, by $| E_i^0, t \rangle$ the state of the atom at time t. The experimental conditions we will assume for the discussion of the interaction between atom and radiation are – as we will specify in the next section – those of incoherent radiation of low intensity: under such conditions the atom rarely interacts with radiation and, if it does so, this interaction lasts for short time intervals ($\tau \simeq 10^{-7} \div 10^{-8}$ s) and for most of the time the state of the atom evolves freely. It is therefore meaningful – from the physical point of view – to ask how much is the probability to find, at time t, the atom in given stationary state $| E_f^0 \rangle$: if the vectors $| E_i^0 \rangle$ and $| E_f^0 \rangle$ are normalized, such probability is given by

$$p_{fi}(t) = \left| \langle E_f^0 \mid E_i^0, t \rangle \right|^2 . \tag{13.1}$$

$p_{fi}(t)$ is named **transition probability** from the initial state (at time $t = t_0$) $| E_i^0 \rangle$ to the state $| E_f^0 \rangle$, called the final state.

The problem then is that of calculating $| E_i^0, t \rangle$, i.e. the time evolution of $| E_i^0 \rangle$. This is obtained by solving the Schrödinger equation (9.7): to this end it is necessary to know the Hamiltonian H of the atom in presence of the external electromagnetic field.

Let $\vec{A}(\vec{x}, t)$ be the vector potential of the electromagnetic field defined by the equations:

$$\vec{E} = -\frac{1}{c} \frac{\partial \vec{A}}{\partial t} , \qquad\qquad \vec{B} = \nabla \wedge \vec{A}$$

(we have set the scalar potential $\varphi = 0$: this is legitimate for a radiation field, which is the case we are interested in). It is known from analytical mechanics that, if H_0 is the Hamiltonian of the atom, the field being absent:

$$H_0 = \sum_{\alpha=1}^{Z} \frac{\vec{p}_\alpha^2}{2m} + V(\vec{q}_1, \cdots, \vec{q}_Z) \tag{13.2}$$

(we have supposed the nucleus of infinite mass, fixed at the origin), the Hamiltonian of the atom in presence of the field is obtained by replacing the electron momenta $\vec{p}_\alpha$ by the expressions (the electron charge is $-e$)

$$\vec{p}_\alpha + \frac{e}{c}\vec{A}(\vec{q}_\alpha,t)\,, \qquad \alpha = 1,\cdots,Z$$

(the interaction of the electromagnetic field with the nucleus – even if not supposed of infinite mass – can be neglected since of order m/M with respect to that with the electrons). Therefore

$$H = \sum_{\alpha=1}^{Z} \frac{1}{2m}\left[\vec{p}_\alpha + \frac{e}{c}\vec{A}(\vec{q}_\alpha,t)\right]^2 + V(\vec{q}_1,\cdots,\vec{q}_Z)\,. \qquad (13.3)$$

Note that (but it is no news: the same can be said for the Coulomb potential due to the nucleus of the hydrogen atom), although $A(\vec{x},t)$ is an external field, in the expression (13.3) for the Hamiltonian it acts as an operator on the atomic states inasmuch as $\vec{x}$ is replaced by the operators $\vec{q}_\alpha$ of the electrons. By expanding the square in (13.3) one has

$$H = H_0 + \frac{e}{2mc}\sum_{\alpha=1}^{Z}\left[\vec{p}_\alpha\cdot\vec{A}(\vec{q}_\alpha,t) + A(\vec{q}_\alpha,t)\cdot\vec{p}_\alpha\right] + \frac{e^2}{2mc^2}\sum_{\alpha=1}^{Z}\vec{A}^2(\vec{q}_\alpha,t)\,.$$

The expression $\vec{p}\cdot\vec{A}+\vec{A}\cdot\vec{p}$ can be simplified by choosing the vector potential in such a way that

$$\nabla\cdot\vec{A} = 0\,.$$

Indeed, in this case by (4.54) one has

$$\vec{p}\cdot\vec{A} - \vec{A}\cdot\vec{p} = \sum_{i=1}^{3}[p_i\,,A_i] = -\mathrm{i}\,\hbar\,\nabla\cdot\vec{A} = 0$$

whence

$$\vec{p}\cdot\vec{A} + \vec{A}\cdot\vec{p} = 2\,\vec{p}\cdot\vec{A}$$

and H takes the form

$$H = H_0 + \frac{e}{mc}\sum_{\alpha=1}^{Z}\vec{p}_\alpha\cdot\vec{A}(\vec{q}_\alpha,t) + \frac{e^2}{2mc^2}\sum_{\alpha=1}^{Z}\vec{A}^2(\vec{q}_\alpha,t)\,. \qquad (13.4)$$

It is this H that determines the time evolution of the state $|\,E_i^0,\,t\,\rangle$: however the methods of solution of the Schödinger equation discussed in Sect. 9.1 cannot be applied in the present case, since H explicitly depends on time through the vector potential $\vec{A}(\vec{x},t)$. We will therefore take advantage of a method known as *time-dependent perturbation theory*.

Let

$$H = H_0 + H'(t) \qquad (13.5)$$

with H_0 independent of time, as in (13.4). Let $U(t, t_0)$ be the time evolution operator of the system whose Hamiltonian is (13.5) and $U_0(t, t_0)$ that of the "unperturbed" system whose Hamiltonian is H_0. Thanks to (9.10):

$$\frac{\mathrm{d}}{\mathrm{d}t} U(t, t_0) = -\frac{\mathrm{i}}{\hbar} H(t)\, U(t, t_0) \,, \qquad \frac{\mathrm{d}}{\mathrm{d}t} U_0(t, t_0)^\dagger = \frac{\mathrm{i}}{\hbar} U_0(t, t_0)^\dagger\, H_0 \,. \tag{13.6}$$

Since
$$U_0(t, t_0) \,|\, E_{\mathrm{f}}^0 \,\rangle = \mathrm{e}^{-\mathrm{i}\, E_{\mathrm{f}}^0 (t - t_0)/\hbar}|\, E_{\mathrm{f}}^0 \,\rangle$$

by (13.1) and owing to $\left|\exp\left(\mathrm{i}\, E_{\mathrm{f}}^0 (t - t_0)/\hbar\right)\right| = 1$, one has

$$p_{\mathrm{f}\,\mathrm{i}}(t) = \left|\langle\, E_{\mathrm{f}}^0 \,|\, U(t, t_0) \,|\, E_{\mathrm{i}}^0 \,\rangle\right|^2 = \left|\langle\, E_{\mathrm{f}}^0 \,|\, U_0(t, t_0)^\dagger\, U(t, t_0) \,|\, E_{\mathrm{i}}^0 \,\rangle\right|^2 \,.$$

By putting
$$\widetilde{U}(t, t_0) \equiv U_0(t, t_0)^\dagger\, U(t, t_0) \tag{13.7}$$

one has
$$p_{\mathrm{f}\,\mathrm{i}}(t) = \left|\langle\, E_{\mathrm{f}}^0 \,|\, \widetilde{U}(t, t_0) \,|\, E_{\mathrm{i}}^0 \,\rangle\right|^2 \,. \tag{13.8}$$

The operator $\widetilde{U}(t, t_0)$ is called time evolution operator in the **interaction picture**. The interaction picture is something midway between the Schrödinger and the Heisenberg pictures for time evolution: states are evolved by the operator $\widetilde{U}(t, t_0)$ whereas the observables are evolved by the operator $U_0(t, t_0)$:

$$\overline{\xi}_t \equiv \langle\, A, 0 \,|\, U(t)^\dagger\, \xi\, U(t) \,|\, A, 0 \,\rangle = \langle\, A, 0 \,|\, \widetilde{U}(t)^\dagger\, U_0(t)^\dagger\, \xi\, U_0(t)\, \widetilde{U}(t) \,|\, A, 0 \,\rangle \,.$$

By (13.7) it is evident that, if $H' = 0$, then $\widetilde{U} = \mathbb{1}$, i.e. $U = U_0$. Therefore, if the perturbation H' is 'small', $\widetilde{U}$ only slightly differs from the identity, namely a perturbative expansion is therefore possible for $\widetilde{U}$. By (13.6) one has

$$\frac{\mathrm{d}}{\mathrm{d}t} \widetilde{U}(t, t_0) = \left(\frac{\mathrm{d}}{\mathrm{d}t} U_0(t, t_0)^\dagger\right) U(t, t_0) + U_0(t, t_0)^\dagger \left(\frac{\mathrm{d}}{\mathrm{d}t} U(t, t_0)\right)$$

$$= \frac{\mathrm{i}}{\hbar} U_0(t, t_0)^\dagger\, (H_0 - H)\, U(t, t_0)$$

$$= -\frac{\mathrm{i}}{\hbar} U_0(t, t_0)^\dagger\, H'(t)\, U_0(t, t_0)\, \widetilde{U}(t, t_0)$$

so that, putting
$$\widetilde{H}'(t) = U_0(t, t_0)^\dagger\, H'(t)\, U_0(t, t_0) \tag{13.9}$$

one has
$$\frac{\mathrm{d}}{\mathrm{d}t} \widetilde{U}(t, t_0) = -\frac{\mathrm{i}}{\hbar} \widetilde{H}'(t)\, \widetilde{U}(t, t_0) \,, \qquad \widetilde{U}(t_0, t_0) = \mathbb{1} \,. \tag{13.10}$$

The perturbative expansion for $\widetilde{U}$ begins by setting $\widetilde{U} = \mathbb{1}$ in the right hand side of (13.10):

$$\frac{\mathrm{d}}{\mathrm{d}t}\,\widetilde{U}^{(1)}(t,t_0) = -\frac{\mathrm{i}}{\hbar}\,\widetilde{H}'(t) \quad \Rightarrow \quad U^{(1)}(t,t_0) = \mathbb{1} - \frac{\mathrm{i}}{\hbar}\int_{t_0}^{t}\widetilde{H}'(t')\,\mathrm{d}t'. \quad (13.11)$$

The second order is obtained by setting $\widetilde{U} = \widetilde{U}^{(1)}(t,t_0)$ in the right hand side of (13.10), and so on for higher orders.

In the sequel we will only consider the first-order approximation (13.11): in this case we are allowed to (indeed, we must) neglect the last term in the Hamiltonian (13.4), the term proportional to e^2, inasmuch as it is of the same order as the second order contribution of the term proportional to e. As we will explain in Sect. 13.4, from the physical point of view this means that we will only consider those transition in which one single photon at a time can be either emitted or absorbed.

If $\langle\, E_{\mathrm{f}}^0 \mid E_{\mathrm{i}}^0\,\rangle = 0$, by inserting (13.11) into (13.8) and in view of (13.9), one has

$$p_{\mathrm{f\,i}}(t) = \frac{1}{\hbar^2}\left|\int_{t_0}^{t}\langle\, E_{\mathrm{f}}^0 \mid \widetilde{H}'(t') \mid E_{\mathrm{i}}^0\,\rangle\,\mathrm{d}t'\right|^2$$

$$= \frac{1}{\hbar^2}\left|\int_{t_0}^{t}\langle\, E_{\mathrm{f}}^0 \mid H'(t') \mid E_{\mathrm{i}}^0\,\rangle\,\mathrm{e}^{\mathrm{i}(E_{\mathrm{f}}^0 - E_{\mathrm{i}}^0)(t'-t_0)/\hbar}\,\mathrm{d}t'\right|^2. \quad (13.12)$$

13.3 Semiclassical Theory of Radiation

We will use (13.12) in the problem of emission and absorption of electromagnetic radiation by an atom or, in any event, by a system with discrete energy levels. The theory we will present is called *semiclassical* because, while the atom is considered as a quantum system, the electromagnetic field is considered from the classical point of view: the vector potential $\vec{A}\,(\vec{x},t)$ and its time derivative are the canonical variables of the electromagnetic field and for them commutation relations analogous to (4.51) should hold; instead, in the semiclassical approximation, they are considered as quantities that commute with each other. We will come back on the meaning and the limitations of the semiclassical approximation in the next section.

With the specification

$$H'(t) = \frac{e}{m\,c}\sum_{\alpha=1}^{Z}\vec{p}_\alpha \cdot \vec{A}\,(\vec{q}_\alpha,t)$$

equation (13.12) becomes:

$$p_{\mathrm{f\,i}}(t) = \left(\frac{e}{m\,c\,\hbar}\right)^2\left|\sum_{\alpha=1}^{Z}\int_{t_0}^{t}\langle\, E_{\mathrm{f}}^0 \mid \vec{p}_\alpha\cdot\vec{A}\,(\vec{q}_\alpha,t') \mid E_{\mathrm{i}}^0\,\rangle\,\mathrm{e}^{\mathrm{i}(E_{\mathrm{f}}^0 - E_{\mathrm{i}}^0)(t'-t_0)/\hbar}\,\mathrm{d}t'\right|^2.$$
$$(13.13)$$

Let us assume that the radiation incident on the atom is produced by a lamp: this means that $\vec{A}\,(\vec{x},t)$ is not a unique coherent wave, but rather a statistical mixture of photons that, from the classical point view, we describe as a set of 'wave packets', not necessarily all with the same frequency ν, with relative phases that are distributed in a totally random way.

Each packet has a finite *time duration* of the order of the **lifetime** τ of the transition responsible for the radiation emitted by the source (as we shall see, typically $\tau \simeq 10^{-9} \div 10^{-7}\,\mathrm{s}$); as a consequence – see (6.50) – the spectral width $\Delta\nu$ of the packets is of the order of $1/\tau$ and the spatial extension of the order of $c \times \tau \simeq 30\,\mathrm{cm} \div 30\,\mathrm{m}$.

Let us consider a packet of frequency $\nu = \omega/2\pi$; in the time interval $(t_0, t_0 + \tau)$ in which, in the region occupied by the atom, the field is nonvanishing, it has the form $\big($see (3.8)$\big)$

$$\vec{A}(\vec{x}, t) = \frac{A}{2} \left(\vec{e}\, \mathrm{e}^{\mathrm{i}\left(\vec{k}\cdot\vec{x} - \omega\,(t-t_0)\right)} + \vec{e}^{\,*}\, \mathrm{e}^{-\mathrm{i}\left(\vec{k}\cdot\vec{x} - \omega\,(t-t_0)\right)} \right), \qquad \omega > 0 \quad (13.14)$$

where $\vec{e}$ is the complex unit vector describing the polarization of the wave $\big($the same that in (3.12) was denoted by $e_{\vartheta\,\varphi}$ $\big)$. Let us put

$$\begin{aligned} F_\alpha &\equiv \frac{e\,A}{2m\,c\,\hbar} \langle E_{\mathrm{f}}^0 \mid (\vec{p}_\alpha \cdot \vec{e})\, \mathrm{e}^{\,\mathrm{i}\,\vec{k}\cdot\vec{q}_\alpha} \mid E_{\mathrm{i}}^0 \rangle \\[2mm] G_\alpha &\equiv \frac{e\,A}{2m\,c\,\hbar} \langle E_{\mathrm{f}}^0 \mid (\vec{p}_\alpha \cdot \vec{e}^{\,*})\, \mathrm{e}^{-\mathrm{i}\,\vec{k}\cdot\vec{q}_\alpha} \mid E_{\mathrm{i}}^0 \rangle \end{aligned} \qquad (13.15)$$

and

$$\omega_{\mathrm{f}\,\mathrm{i}} \equiv (E_{\mathrm{f}}^0 - E_{\mathrm{i}}^0)/\hbar \,.$$

Then (13.13) writes

$$\begin{aligned} P_{\mathrm{f}\,\mathrm{i}}(t) &= \left| \sum_{\alpha=1}^{Z} \left(F_\alpha \int_{t_0}^{t} \mathrm{e}^{\mathrm{i}\,(\omega_{\mathrm{f}\,\mathrm{i}}-\omega)(t'-t_0)}\,\mathrm{d}t' + G_\alpha \int_{t_0}^{t} \mathrm{e}^{\mathrm{i}\,(\omega_{\mathrm{f}\,\mathrm{i}}+\omega)(t'-t_0)}\,\mathrm{d}t' \right) \right|^2 \\[2mm] &= \left| \sum_{\alpha=1}^{Z} \left(F_\alpha\, \frac{\mathrm{e}^{\mathrm{i}\,(\omega_{\mathrm{f}\,\mathrm{i}}-\omega)(t-t_0)} - 1}{\mathrm{i}\,(\omega_{\mathrm{f}\,\mathrm{i}} - \omega)} + G_\alpha\, \frac{\mathrm{e}^{\mathrm{i}\,(\omega_{\mathrm{f}\,\mathrm{i}}+\omega)(t-t_0)} - 1}{\mathrm{i}\,(\omega_{\mathrm{f}\,\mathrm{i}} + \omega)} \right) \right|^2. \quad (13.16) \end{aligned}$$

Let us take $t = t_0 + \tau$ and analyze the structure of (13.16): it contains terms of the type

$$\left| \frac{\mathrm{e}^{\mathrm{i}\,(\omega_{\mathrm{f}\,\mathrm{i}}-\omega)\tau} - 1}{\mathrm{i}\,(\omega_{\mathrm{f}\,\mathrm{i}} - \omega)} \right|^2 = 4\,\frac{\sin^2[\frac{1}{2}(\omega_{\mathrm{f}\,\mathrm{i}} - \omega)\,\tau]}{(\omega_{\mathrm{f}\,\mathrm{i}} - \omega)^2} = \tau^2\,\frac{\sin^2[\frac{1}{2}(\omega_{\mathrm{f}\,\mathrm{i}} - \omega)\,\tau]}{[\frac{1}{2}(\omega_{\mathrm{f}\,\mathrm{i}} - \omega)\,\tau]^2} \quad (13.17)$$

$$\left| \frac{\mathrm{e}^{\mathrm{i}\,(\omega_{\mathrm{f}\,\mathrm{i}}+\omega)\tau} - 1}{\mathrm{i}\,(\omega_{\mathrm{f}\,\mathrm{i}} + \omega)} \right|^2 = 4\,\frac{\sin^2[\frac{1}{2}(\omega_{\mathrm{f}\,\mathrm{i}} + \omega)\,\tau]}{(\omega_{\mathrm{f}\,\mathrm{i}} + \omega)^2} = \tau^2\,\frac{\sin^2[\frac{1}{2}(\omega_{\mathrm{f}\,\mathrm{i}} + \omega)\,\tau]}{[\frac{1}{2}(\omega_{\mathrm{f}\,\mathrm{i}} + \omega)\,\tau]^2} \quad (13.18)$$

and those that come out of the double products ("interference terms").

The function $\sin^2 x/x^2$ is well known: it is practically nonvanishing for $|x| \lesssim 1$; this means that (13.17) is nonvanishing when ω is a neighborhood of $\omega_{\mathrm{f}\,\mathrm{i}}$ of width $\simeq 1/\tau$ (therefore, since $\omega > 0$, in this case $\omega_{\mathrm{f}\,\mathrm{i}} > 0$), whereas (13.18) is nonvanishing when ω is a neighborhood of $-\omega_{\mathrm{f}\,\mathrm{i}}$ of width

$\simeq 1/\tau$ (in this case $\omega_{f\,i} < 0$). Since in the region of the optical transitions $1/\tau \simeq 10^8 \div 10^9\,\mathrm{s}^{-1}$ and $|\omega_{f\,i}| \simeq 10^{14} \div 10^{15}\,\mathrm{s}^{-1}$, it is evident that the two terms given respectively by (13.17) and (13.18) can never simultaneously contribute to the transition probability: for a given $\omega_{f\,i} \gtrless 0$ the region in which one of the two terms is nonvanishing is separated by $\simeq 2|\omega_{f\,i}| \simeq 10^6 \times 1/\tau$ from the region in which the other one contributes.

As a consequence, the interference terms, that are of the order of the square root of the product of (13.17) with (13.18)), *never* contribute to optical transitions.

We have just seen that, in order that either (13.17) or (13.18) be nonvanishing, it is necessary that $\omega \simeq |\omega_{f\,i}|$: in the first case $E_f^0 > E_i^0$ and one has *absorption* of radiation, in the second case $E_f^0 < E_i^0$ and one has *induced* (or *stimulated*) *emission*.

This result demonstrates the second Bohr hypothesis:

In order that the transition $|E_i^0\rangle \to |E_f^0\rangle$ *be possible, it is necessary that the frequency* ν *of the radiation be* (better: *in the radiation be present the frequency* ν) *such that*

$$\nu = \frac{|E_f^0 - E_i^0|}{h}\,. \tag{13.19}$$

Notice that the Bohr's rule is here presented only as a necessary condition for the transition to be possible: we will see that the occurrence of (13.19) does not necessarily imply that the transition takes place.

In the light of what we have just seen, we can proceed by discussing separately the case of absorption and that of stimulated emission.

Let us start with absorption. By (13.16) and (13.17)

$$\pi_{f\,i}(\omega;\tau) = \Big| \sum_{\alpha=1}^{Z} F_\alpha \Big|^2 \, \tau^2 \, \frac{\sin^2[\frac{1}{2}(\omega_{f\,i} - \omega)\,\tau]}{[\frac{1}{2}(\omega_{f\,i} - \omega)\,\tau]^2} \tag{13.20}$$

is the probability that the single wave packet with frequency $\nu = \omega/2\pi$ induces the transition. As the several packets – both those with the same frequency and those with a different frequency – are statistically independent, the transition probability in the time interval (t_0, t) is the sum of the probabilities due to the single packets; it is therefore evident that, if the intensity of the incident radiation is constant, the probability is proportional to $t - t_0$, i.e. to the number of packets that have arrived. In this case it is possible to define the *transition probability rate* $w_{f\,i}^{\mathrm{ind}}$ (*induced* probability transition per unit time): if $N_\tau(\omega)\,\mathrm{d}\omega$ is the number of packets with frequency in the interval $\omega, \omega + \mathrm{d}\omega$ that arrive in the time interval τ (for the sake of simplicity we have assumed that all the packets have the same time duration τ), then

$$p_{f\,i}(\tau) = \int_0^\infty \pi_{f\,i}(\omega;\tau)\,N_\tau(\omega)\,\mathrm{d}\omega \quad \Rightarrow \quad w_{f\,i}^{\mathrm{ind}} = \frac{1}{\tau} \int_0^\infty \pi_{f\,i}(\omega;\tau)\,N_\tau(\omega)\,\mathrm{d}\omega\,.$$

In order to calculate $N_\tau(\omega)\,\mathrm{d}\omega$, let us consider the radiation in the frequency interval $\omega, \omega + \mathrm{d}\omega$ and let $I(\omega)\,\mathrm{d}\omega$ be the intensity. The intensity is the time average over a period of the square modulus of the Poynting vector; for the single packet given by (13.14)

$$I_1(\omega) = \frac{c}{4\pi}\,\overline{\vec{E}^2} = \frac{1}{4\pi c}\,\overline{\left(\frac{\partial \vec{A}}{\partial t}\right)^2} = \frac{\omega^2 A^2}{8\pi c}$$

whence

$$N_\tau(\omega)\,d\omega = \frac{I(\omega)\,d\omega}{I_1} = \frac{8\pi c}{\omega^2 A^2}\,I(\omega)\,d\omega \tag{13.21}$$

and the transition probability rate is given by

$$w_{\mathrm{fi}}^{\mathrm{ind}} = \frac{8\pi c}{A^2}\left|\sum_{\alpha=1}^{Z} F_\alpha\right|^2 \int_0^\infty \frac{I(\omega)}{\omega^2}\,\frac{\sin^2\left[\frac{1}{2}(\omega_{\mathrm{fi}}-\omega)\,\tau\right]}{\left[\frac{1}{2}(\omega_{\mathrm{fi}}-\omega)\,\tau\right]^2}\,\tau\,d\omega\;.$$

Thanks to the property of the function $\sin^2 x/x^2$, only the region $\omega \simeq \omega_{\mathrm{fi}}$ contributes to the integral: therefore $I(\omega)/\omega^2$ can be taken out of the integral and the lower limit can be taken to $-\infty$ so that, by putting $x = \frac{1}{2}\,(\omega-\omega_{\mathrm{fi}})\,\tau$ one has

$$w_{\mathrm{fi}}^{\mathrm{ind}} = \frac{8\pi c}{A^2}\left|\sum_{\alpha=1}^{Z} F_\alpha\right|^2 \frac{I(\omega_{\mathrm{fi}})}{\omega_{\mathrm{fi}}^2}\int_{-\infty}^{+\infty} 2\,\frac{\sin^2 x}{x^2}\,dx\;.$$

The integral of $\sin^2 x/x^2$ gives π, so, in view of the first of (13.15) one has:

$$w_{\mathrm{fi}}^{\mathrm{ind}} = \frac{4\pi^2 e^2\,I(\omega_{\mathrm{fi}})}{m^2 c\,\hbar^2\,\omega_{\mathrm{fi}}^2}\left|\langle\, E_{\mathrm{f}}^0\mid \sum_{\alpha=1}^{Z}(\vec{p}_\alpha \vec{e})\,\mathrm{e}^{+\mathrm{i}\,\vec{k}\cdot\vec{q}_\alpha}\mid E_{\mathrm{i}}^0\,\rangle\right|^2, \qquad E_{\mathrm{f}}^0 > E_{\mathrm{i}}^0\;. \tag{13.22}$$

and in the case of stimulated emission little must be changed:

$$w_{\mathrm{fi}}^{\mathrm{ind}} = \frac{4\pi^2 e^2\,I(\omega_{\mathrm{fi}})}{m^2 c\,\hbar^2\,\omega_{\mathrm{fi}}^2}\left|\langle\, E_{\mathrm{f}}^0\mid \sum_{\alpha=1}^{Z}(\vec{p}_\alpha \cdot \vec{e}^{\,*})\,\mathrm{e}^{-\mathrm{i}\,\vec{k}\cdot\vec{q}_\alpha}\mid E_{\mathrm{i}}^0\,\rangle\right|^2, \qquad E_{\mathrm{f}}^0 < E_{\mathrm{i}}^0\;.$$

$$\tag{13.23}$$

Thanks to the transversality of electromagnetic waves, $\vec{p}_\alpha \cdot \vec{e}$ commutes with $\vec{k}\cdot\vec{q}_\alpha$ ($\vec{e}$ is orthogonal to $\vec{k}$), therefore

$$\left(\vec{p}_\alpha \cdot \vec{e}\;\mathrm{e}^{\mathrm{i}\,\vec{k}\cdot\vec{q}_\alpha}\right)^\dagger = \vec{p}_\alpha \cdot \vec{e}^{\,*}\;\mathrm{e}^{-\mathrm{i}\,\vec{k}\cdot\vec{q}_\alpha}$$

and it is possible to express both (13.22) and (13.23) in a unique formula

$$w_{ba}^{\mathrm{ind}} = \frac{4\pi^2\,e^2\,I(\omega_{ba})}{m^2 c\,\hbar^2\,\omega_{ba}^2}\left|\langle\, E_b^0\mid \sum_{\alpha=1}^{Z}(\vec{p}_\alpha \cdot \vec{e})\,\mathrm{e}^{\mathrm{i}\,\vec{k}\cdot\vec{q}_\alpha}\mid E_a^0\,\rangle\right|^2, \qquad E_a^0 < E_b^0 \tag{13.24}$$

that holds both for absorption $\mid E_a^0\,\rangle \to \mid E_b^0\,\rangle$ and emission $\mid E_b^0\,\rangle \to \mid E_a^0\,\rangle$.

It is evident that, restricting to the first order, the absorption and emission probabilities between two energy levels are identical with each other. It is also clear that (13.19) alone is not sufficient to guarantee that the transition between the two states $|E_a^0\rangle$ and $|E_b^0\rangle$ be possible: it is necessary that $w_{ba}^{\text{ind}} \neq 0$.

The treatment above is correct for incoherent radiation with spectral intensity $I(\omega)$ practically constant in the frequency interval $|\omega_{fi}| \pm 1/\tau$; furthermore, due to the use of perturbation theory, the result for the probability transition $p_{fi}(t)$ is reliable only until $p_{fi}(t) \ll 1$, i.e. for $t \ll 1/w_{fi}^{\text{ind}}$: we postpone to Sect. 13.5 the numerical analysis of the conditions of validity of the theory presented above.

13.4 Spontaneous Emission

The result expressed by (13.23) has a main flaw: the probability for emission is proportional to the intensity of the incident radiation, therefore, in particular, it is vanishing in the absence of radiation. Instead it is known from everyday experience (think of an ordinary lamp) that, if an atom is excited from the state $|E_i^0\rangle$ to the state $|E_f^0\rangle$ in a nonradiative way, i.e. without use of an external electromagnetic field (e.g. by collisions with different atoms or electrons), the atom decays to lower energy states by emitting radiation. One therefore has a *spontaneous emission* rate, i.e. a nonvanishing emission probability even in the absence of an external electromagnetic field, in contradiction with (13.23).

This fact indeed is a limitation of the semiclassical treatment of the interaction between charges and electromagnetic radiation: since the postulates of quantum mechanics apply to any kind of system – not only to particles – also the electromagnetic field requires a quantum mechanical treatment, based on quantization rules analogous to (4.51). As a matter of fact, only in this way the original hypothesis of quantization of the electromagnetic field – i.e. the existence of photons – arises as a consequence of a general theoretical setting and the first-order approximation for the electromagnetic transitions corresponds to considering transitions in which only one photon at a time may be exchanged between atom and radiation.

Considering the electromagnetic field as a classical system is the same as considering it as an *external field*: therefore the semiclassical treatment of the interaction between charges and field may correctly explain the influence of the field on the atom (absorption and stimulated emission), but not the influence of the atom on the field (spontaneous emission).

Only in the framework of a complete quantum approach – quantum electrodynamics – it is possible to treat also spontaneous emission of radiation correctly. We will here derive the spontaneous emission rate by means of a very beautiful argument that is statistical in nature and is due – as usual! – to Einstein who, beyond showing the necessity of spontaneous emission, also succeeds in calculating its rate.

Let us consider a *black body*, i.e. a cavity in which radiation and matter (atoms) are in thermodynamic equilibrium at temperature T. Let $|E_a^0\rangle$ and $|E_b^0\rangle$ be two atomic states ($E_a^0 < E_b^0$); owing to the thermodynamic equilibrium, the number N_a of atoms in the state $|E_a^0\rangle$ and the number N_b

of atoms in the state $|\,E_b^0\,\rangle$ stay constant over time. Since the properties of the black body are independent of the properties of the matter it is made of, we can – for the sake of simplicity – assume that our atoms only have the two energy levels E_a^0 and E_b^0: in this case the fact that N_a and N_b stay constant entails that the number of transitions $|\,E_a^0\,\rangle \to |\,E_b^0\,\rangle$ must equal the number of transitions $|\,E_b^0\,\rangle \to |\,E_a^0\,\rangle$: if $w(a \to b)$ and $w(b \to a)$ are the transition rates, it must happen that

$$N_a \times w(a \to b) = N_b \times w(b \to a) \,. \qquad (13.25)$$

If, in agreement with (13.22) and (13.23), $w(a \to b) = w(b \to a)$, then also $N_a = N_b$, in contrast with the Maxwell–Boltzmann distribution according to which $N_b/N_a = \exp[-(E_b^0 - E_a^0)/k_{\mathrm B}T]$. Then, with w_{ba}^{ind} given by (13.24), let us put

$$w(a \to b) \equiv w_{ba}^{\mathrm{ind}} \,, \qquad\qquad w(b \to a) = w_{ba}^{\mathrm{ind}} + w_{ba}^{\mathrm{sp}}$$

and, by aid of (13.25), one has

$$w_{ba}^{\mathrm{sp}} = w_{ba}^{\mathrm{ind}} \left(\frac{N_a}{N_b} - 1 \right) = w_{ba}^{\mathrm{ind}} \left(\mathrm{e}^{(E_b^0 - E_a^0)/k_{\mathrm B}T} - 1 \right). \qquad (13.26)$$

If we consider only the radiation either absorbed or emitted by the atoms around a direction within the solid angle $\mathrm{d}\Omega$ and in a given state of polarization described by the complex unit vector $\vec{e}$, then (13.25), and therefore (13.26), keeps on holding true inasmuch as it guarantees that the radiation in the cavity stays isotropic and unpolarized.

Let us take the expression (13.24) for w_{ba}^{ind}: the spectral intensity $I(\omega)$ we must use in (13.24) is

$$I(\omega) = \frac{1}{2}\, u(\omega)\, c\, \frac{\mathrm{d}\Omega}{4\pi} \qquad (13.27)$$

where $u(\omega)$ is the spectral density of energy of the black body at temperature T and is given by the **Planck distribution** $(u(\omega)\,\mathrm{d}\omega = \tilde{u}(\nu)\,\mathrm{d}\nu)$:

$$u(\omega) = \frac{1}{2\pi}\, \tilde{u}(\nu) = \frac{\hbar\,\omega^3}{\pi^2\,c^3\,(\mathrm{e}^{\hbar\,\omega/k_{\mathrm B}T} - 1)} \,. \qquad (13.28)$$

The factor $\frac{1}{2}$ in (13.27) is due to the fact that the density of energy of the radiation in an assigned state of polarization is $\frac{1}{2}\,u(\omega)$.

Equations (13.25)$\div$(13.28) give (we now write $w_{ab}^{\mathrm{sp}}\,\mathrm{d}\Omega$ in place of w_{ab}^{sp})

$$w_{ab}^{\mathrm{sp}}\,\mathrm{d}\Omega = \frac{e^2\,\omega_{ba}}{2\pi\,\hbar\,m^2\,c^3}\, \left| \langle\,E_b^0\,|\, \sum_{\alpha=1}^{Z}(\vec{p}_\alpha \cdot \vec{e})\, \mathrm{e}^{\mathrm{i}\,\vec{k}\cdot\vec{q}_\alpha}\, |\,E_a^0\,\rangle \right|^2 \mathrm{d}\Omega \,. \qquad (13.29)$$

So the emission rate is different from the absorption rate: w_{ab}^{sp} does not depend on the intensity of the incident radiation and, as a consequence, is the probability per unit time and per unit solid angle of *spontaneous* emission in the direction of the vector $\vec{k}$ with polarization $\vec{e}$.

Note that (13.24) and (13.29) are very similar to each other: only the prefactor of the $|\cdots|^2$ is different. However, while in (13.24) $\vec{e}$ and $\vec{k}$ refer to the radiation *sent* on the atom, in (13.29) they refer to the observational conditions: the direction of the vector $\vec{k}$, whose modulus is $2\pi\nu/c$, is the direction in which one wants to observe the emitted radiation; $\vec{e}$ describes the polarization state accepted by the detector. The meaning of (13.29) is, in conclusion, that of probability per unit of time to observe spontaneously emitted radiation in a given direction and in a given polarization state.

In the next section we will give a numerical estimate of the total probability w_{ab}^{tot} of spontaneous emission between two levels (i.e. (13.29) integrated over the solid angle and summed on the polarization states).

If on a given atom there arrives some electromagnetic radiation, both stimulated and spontaneous emissions intervene: however, given that the radiation due to stimulated emission has the same features as the incident radiation, it will be possible to observe the spontaneously emitted radiation by means of observations in directions different from that of the incident radiation.

We believe that many readers already have raised the following question: how does one reconcile the statement that, for an *isolated* atom, the excited states $|\,E_n^0\,\rangle$ are stationary states with the existence of spontaneous emission that, over a time of the order of $1/(w_{ab}^{\text{tot}})$, enables the atom to perform a transition to a lower energy state?

The point is that the states $|\,E_n^0\,\rangle$ are eigenstates of the atomic Hamiltonian H_{at}, whereas if one is interested in spontaneous emission, the system necessarily is 'atom + electromagnetic field', independently of whether there are photons around or not: the Hamiltonian of the system – regardless of the state the system is in – is:

$$H = H_{\text{at}} + H_{\text{em}} + H_{\text{int}} \tag{13.30}$$

where H_{em} is the Hamiltonian of the field (i.e. the energy properly expressed by means of canonical variables) and H_{int} is the interaction between atom and field. A basis of eigenvectors of H_{em} is provided by states with 0 photons, 1 photon, 2 photons, $\cdots$, N photons, $\cdots$. The state with $N = 0$, namely that in which there are no photons, is called the **vacuum state** (and, since the number of photons is not just 'a number', but rather is a dynamical variable, i.e. an observable, the vacuum state is a state like all the others). If H_{int} were not there, the system would be one with 'separate variable' – much as the two particle system discussed in Sect. 11.2 – and it would be legitimate to look only at the atom and forget about the electromagnetic field (or viceversa). In particular the state $|\,\text{e.m. field; atom}\,\rangle = |\,0 \text{ photons}; E_n^0\,\rangle$ would be a stationary state. The matter is different if H_{int} is taken into account: H_{int} has nonvanishing matrix elements between the states $|\,N \text{ photons}; E_n^0\,\rangle$ and $|\,N \pm 1 \text{ photons}; E_m^0\,\rangle$, and in particular between $|\,0 \text{ photons}; E_n^0\,\rangle$ and $|\,1 \text{ photon}; E_m^0\,\rangle$ that exactly are those involved in the spontaneous emission.

Therefore the eigenstates of the Hamiltonian (13.30) no longer display 'separate variables', and no doubt are more complicated:

$$\alpha \,|\, 0 \text{ photons; } E_n^0 \,\rangle + \beta \,|\, 1 \text{ photon; } E_m^0 \,\rangle + \cdots$$

and, in addition (with the exception of the state $|\, 0 \text{ photons; } E_0^0 \,\rangle$), all are improper states: therefore, strictly speaking, for the system 'atom + field', the only stationary state is the lowest energy one $|\, 0 \text{ photons; } E_0^0 \,\rangle$.

13.5 Electric-Dipole Transitions

As we have seen in the previous sections, electromagnetic transitions are determined by the matrix elements

$$\frac{e}{m} \,\langle\, E_b^0 \,|\, \sum_{\alpha=1}^{Z} (\vec{p}_\alpha \cdot \vec{e}) \, e^{i \, \vec{k} \cdot \vec{q}_\alpha} \,|\, E_a^0 \,\rangle \tag{13.31}$$

where $|\vec{k}| = 2\pi/\lambda$. Since the transitions we are interested in usually correspond to energy differences of the order of the electronvolt, the wavelength λ of the electromagnetic radiations is of the order of some thousand ångström ($\lambda[\text{Å}] = 12400/\Delta E[\text{eV}]$), therefore at least three orders of magnitude larger than the atomic size. This means that the vector potential is practically uniform in the region occupied by the atom: under such conditions it is legitimate to expand the field given by (13.14)

$$\vec{A}\,(\vec{x}, t) = \frac{A}{2}\, \vec{e}\,(1 + i\,\vec{k} \cdot \vec{x} + \cdots)\, e^{-i\omega(t-t_0)} + \text{c.c.} \tag{13.32}$$

and neglect, in comparison to 1, the terms $\vec{k} \cdot \vec{x} + \cdots$.
Since

$$\vec{E}(\vec{x}, t) = -\frac{1}{c}\frac{\partial \vec{A}}{\partial t} = \frac{i\omega A}{2c}\, \vec{e}\,(1 + i\,\vec{k} \cdot \vec{x} + \cdots)\, e^{-i\omega(t-t_0)} + \text{c.c.} \tag{13.33}$$

the approximation $e^{i\,\vec{k}\cdot\vec{x}} \simeq 1$ in (13.32) is the same as considering an electric field uniform in the region occupied by the atom, and a vanishing magnetic field $\vec{B} = \nabla \wedge \vec{A}$: for this reason such an approximation is called **electric-dipole approximation**. In this approximation the matrix element (13.31) becomes

$$\frac{e}{m} \,\langle\, E_b^0 \,|\, \sum_{\alpha=1}^{Z} (\vec{p}_\alpha \cdot \vec{e}) \,|\, E_a^0 \,\rangle \tag{13.34}$$

and, as we shall see, when the matrix element (13.34) is nonvanishing, the neglected terms $\vec{k} \cdot \vec{x} + \cdots$ are expected to add a contribution starting from the term $(\vec{k} \cdot \vec{x})^2$, therefore smaller by a factor of order $(a_B/\lambda)^2$, a_B being the Bohr radius.

By taking advantage of the identity

$$\vec{p}_\alpha = i\,\frac{m}{\hbar}\,[\,H_0\,,\vec{q}_\alpha\,]$$

H_0 being given by (13.2) (i.e. $\vec{p} = m\,\dot{\vec{q}}$ for the electrons of the unperturbed atom), and since $|\,E_a^0\,\rangle$ and $|\,E_b^0\,\rangle$ are eigenvectors of H_0, (13.34) becomes

$$\frac{i}{\hbar}\,(E_b^0 - E_a^0)\langle\,E_b^0\,|\sum_{\alpha=1}^{Z} e\,\vec{q}_\alpha\cdot\vec{e}\,|\,E_a^0\,\rangle = -i\,\omega_{ba}\,\langle\,E_b^0\,|\,\vec{D}\cdot\vec{e}\,|\,E_a^0\,\rangle \qquad (13.35)$$

where

$$\vec{D} = -e\sum_{\alpha=1}^{Z}\vec{q}_\alpha \qquad (13.36)$$

is the **electric-dipole moment** operator of the atom (the $\vec{q}_\alpha$ are taken with respect to the position of the nucleus).

So, in the electric-dipole approximation, (13.24) takes the form

$$w_{ba}^{\text{ind}} = \frac{4\pi^2}{\hbar^2 c}\,I(\omega_{ba})\big|\langle\,E_b^0\,|\,\vec{D}\cdot\vec{e}\,|\,E_a^0\,\rangle\big|^2, \qquad E_a^0 < E_b^0 \qquad (13.37)$$

and (13.29) $\ (E_b \equiv E_i,\ E_a \equiv E_f)$

$$w_{ab}^{\text{sp}} = \frac{\omega_{ba}^3}{2\pi\,\hbar\,c^3}\,\Big|\langle\,E_b^0\,|\,\vec{D}\cdot\vec{e}^*\,|\,E_a^0\,\rangle\Big|^2, \qquad |\,E_b^0\,\rangle \to |\,E_a^0\,\rangle. \qquad (13.38)$$

When for two states $|\,E_b^0\,\rangle$, $|\,E_a^0\,\rangle$ it happens that $\langle\,E_b^0\,|\,\vec{D}\,|\,E_a^0\,\rangle \neq 0$, one says that the transition between the two states is an **electric-dipole transition**.

The term $\vec{k}\cdot\vec{x}$ in (13.32) gives rise to both the magnetic-dipole and the electric-quadrupole interactions, whose matrix elements are smaller than those of the electric-dipole by a factor of the order $k\,a_B \simeq a_B/\lambda$ and the corresponding transition probabilities are, as a rule, smaller by a factor $(k\,a_B)^2$: in the next section we will see that when a transition is an electric-dipole transition it cannot also be either a magnetic-dipole or an electric-quadrupole transition, therefore there never occurs interference between the matrix elements of the first with those relative to the latter ones.

At this point, thanks to (13.37), we can estimate the stimulated transition rate: let us consider a transition corresponding to $\Delta E \equiv E_b - E_a = 2\,\text{eV}$ $\Rightarrow \lambda \simeq 6000\,\text{Å}$ (yellow light); for $I(\omega_{ba})$ we will take the spectral intensity of a black body, given by (13.27) and (13.28), at the temperature $T = 3000\,\text{K}$, that is the temperature typical of a lamp used in a laboratory (to a good approximation a lamp emits as a black body); the most critical term is the matrix element of the dipole operator that obviously depends on the particular transition and that, furthermore, appears squared in (13.37): as a typical order of magnitude we will take $|\langle\,E_b^0\,|\,\vec{D}\cdot\vec{e}\,|\,E_a^0\,\rangle| \simeq e\,a_B$. With such data:

$$w_{ba}^{\text{ind}} = \frac{4\pi^2\,e^2\,a_B^2}{\hbar^2\,c}\,I(\omega_{ba}) = \frac{4\pi^2\,e^2\,a_B^2}{\hbar^2\,c}\,\frac{8\pi\,\hbar c}{\lambda^3\,(e^{\hbar\,\omega_{ba}/k_B T} - 1)}$$

$$= \alpha\,\frac{32\pi^3}{e^8 - 1}\Big(\frac{a_B}{\lambda}\Big)^2\,\frac{c}{\lambda} \qquad (13.39)$$

where

$$\alpha = \frac{e^2}{\hbar c} \simeq \frac{1}{137} \qquad \left(\alpha^{-1} = 137.035\,999\,76(50)\right) \qquad (13.40)$$

is the **fine-structure constant** and is dimensionless. By inserting numerical values one has

$$w_{ba}^{\text{ind}} \simeq 10^4\,\text{s}^{-1} \qquad (13.41)$$

and this means that perturbation theory provides us with acceptable results for times smaller than $10^{-5} \div 10^{-4}$ s that, on the atomic scale ($t \simeq h/\Delta E \simeq 10^{-15}$ s) are long times.

The estimate of the spontaneous emission rate goes along in a similar way: after integrating (13.38) with respect to the solid angle and summing over the polarizations one obtains an expression in which the only important difference with respect to (13.39) is the absence of the denominator $e^8 - 1 \simeq 3000$. The lifetime τ of an excited state is the reciprocal of the sum of the transition rates to all the levels with a lower energy and, at least for the low lying levels, is of the order of $10^{-9} \div 10^{-8}$ s. In Sect. 13.9 we will calculate the lifetime of the states with $n = 2$, $l = 1$ of the hydrogen atom.

13.6 Selection Rules I

We will be concerned only with electric-dipole transitions and will give *necessary* conditions in order that (13.37) (or (13.38)) be nonvanishing. Such condition are named **selection rules for electric-dipole transitions**: as we have already seen in Sect. 12.2, in general selection rules for an *operator* are conditions on the states $|A\rangle$, $|B\rangle$, i.e. on the quantum number that characterize them, in order that the matrix element of the operator between $|A\rangle$ and $|B\rangle$ *may* be different from zero.

In the particular case we are interested in, the operator is any component of the electric-dipole operator $\vec{D}$ and the states between which we consider the matrix elements are the stationary states $|E_a^0\rangle$, $|E_b^0\rangle$, $\cdots$ of the isolated atom.

We must then establish which are the quantum numbers that characterize the stationary states of the atom: what we are going to say now only has a provisional character, because the introduction of the electron spin shall force us to re-examine the classification we give now. For the same reason also the discussion on the selection rules that concern the electric-dipole transition will be made in two steps.

The quantum numbers that characterize the stationary states of the atom are the eigenvalues of a complete set of compatible observables that includes the Hamiltonian: for an *isolated* atom the invariance of the Hamiltonian under both rotations and spatial inversion guarantees that, no matter how complicated the expression of the Hamiltonian may be, we may certainly include, along with he Hamiltonian H_0, the square of the total angular momentum $\vec{L}^2$, one component of $\vec{L}$ – say L_z –, and the space-inversion operator I.

Therefore we can characterize the stationary sates of the isolated atom in the following way:

$$| E, L, M, w \cdots \rangle \tag{13.42}$$

in which $\cdots$ stands for possible other quantum numbers to be determined case by case (we are using the capital letters L and M since we are considering many electrons atoms). The selection rules we are going to discuss refer to the quantum numbers L, M, w.

Space-Inversion Selection Rule: the matrix elements of D_x, D_y and D_z between states with the same parity are vanishing.

Indeed the components of $\vec{D}$ are odd operators under spatial inversion:

$$I \, D_i \, I^{-1} = -D_i$$

so the demonstration of the statement is the same as that in Sect. 12.2 for the operators x_i.

An equivalent way to express the above selection rule is the following:

In order that the matrix elements of D_x, D_y and D_z between eigenstates of the space-inversion operator I may be nonvanishing, it is necessary that the states have opposite parities:

$$\langle \cdots w'' \cdots | D_i | \cdots w' \cdots \rangle \neq 0 \qquad \Rightarrow \qquad w'' \cdot w' = -1 . \tag{13.43}$$

Magnetic-dipole and the electric-quadrupole transitions, instead, may occur only between states with the same parity: indeed, owing to the extra factor $\vec{k} \cdot \vec{x}$ (with respect to the electric-dipole term), the terms responsible for such transitions are even under spatial inversion. Therefore – as we have anticipated – there is no interference between the electric-dipole transitions and either the magnetic-dipole or the electric-quadrupole ones.

The next selection rules concern the angular momentum: since they are an exclusive consequence of the commutation rules (10.9) among the components of $\vec{L}$ and those of any *vector*, they hold not only for $\vec{D}$, but for *any* vector operator $\vec{V}$.

Selection Rule on L_z: let us introduce the operators

$$D_+ = D_x + i\, D_y , \qquad D_- = D_x - i\, D_y , \qquad D_z .$$

Let us consider the matrix elements of D_+, D_- and D_z between eigenstates of L_z. The following necessary conditions, in order that the above matrix elements be nonvanishing, apply:

$$\begin{aligned}
\langle \cdots M'' \cdots | D_+ | \cdots M' \cdots \rangle \neq 0 &\quad \Rightarrow \quad \Delta M = +1 \\
\langle \cdots M'' \cdots | D_- | \cdots M' \cdots \rangle \neq 0 &\quad \Rightarrow \quad \Delta M = -1 \\
\langle \cdots M'' \cdots | D_z | \cdots M' \cdots \rangle \neq 0 &\quad \Rightarrow \quad \Delta M = 0
\end{aligned} \tag{13.44}$$

where $\Delta M \equiv M'' - M'$.

In other words D_+, D_- and D_z have nonvanishing matrix elements between eigenstates of L_z only if, respectively:

$$M'' = M' + 1\,, \qquad M'' = M' - 1\,, \qquad M'' = M'\,.$$

Equations (13.44) are an immediate consequence of the commutation rules among the components of $\vec{L}$ an those of $\vec{D}$, given by (10.9). Indeed (10.9) entails that

$$[L_z\,,\,D_\pm] = \pm\hbar\,D_\pm\,, \qquad\qquad [L_z\,,\,D_z] = 0 \qquad\qquad (13.45)$$

analogous to (10.15). Let us consider the matrix elements of (13.45) between $|\,M''\,\rangle$ and $|\,M'\,\rangle$:

$$\langle\,M''\,|\,[L_z\,,\,D_\pm]\,|\,M'\,\rangle = \pm\hbar\,\langle\,M''\,|\,D_\pm\,|\,M'\,\rangle\,, \quad \langle\,M''\,|\,[L_z\,,\,D_z]\,|\,M'\,\rangle = 0$$

whence

$$(M'' - M' \mp 1)\langle\,M''\,|\,D_\pm\,|\,M'\,\rangle = 0\,, \quad (M'' - M')\langle\,M''\,|\,D_z\,|\,M'\,\rangle = 0$$

and in conclusion:

$$\begin{aligned}
\langle\,M''\,|\,D_\pm\,|\,M'\,\rangle \neq 0 &\quad\Rightarrow\quad M'' - M' = \pm 1 \\
\langle\,M''\,|\,D_z\,|\,M'\,\rangle \neq 0 &\quad\Rightarrow\quad M'' - M' = 0
\end{aligned}$$

i.e. (13.44). An equivalent way of demonstrating (13.44) consists in observing that (13.45) express the fact that D_+ and D_- respectively behave as raising and lowering operators of one unit for the eigenvalue of L_z, whereas D_z, owing to the lemma of p. 87 (Sect. 4.10) does not change the eigenvalue of L_z.

For the Cartesian components D_x and D_y, that are obtained as linear combinations of D_+ and D_-, owing to (13.44) one has:

$$\begin{aligned}
\langle\cdots M''\cdots\,|\,D_x\,|\cdots M'\cdots\rangle \neq 0 &\quad\Rightarrow\quad \Delta M = \pm 1 \\
\langle\cdots M''\cdots\,|\,D_y\,|\cdots M'\cdots\rangle \neq 0 &\quad\Rightarrow\quad \Delta M = \pm 1\,.
\end{aligned} \qquad (13.46)$$

Note that, however, (13.44) contain more detailed information than (13.46).

As we have already pointed out, (13.44) give only necessary conditions for the transition between the states $|\,E_i^0\,\rangle$ and $|\,E_f^0\,\rangle$ to take place; however when $\langle\,E_f^0\,|\,\vec{D}\,|\,E_i^0\,\rangle \neq 0$, i.e. when the transition is possible, then (13.44) can be read the opposite way: so, for instance, if the transition occurs between states with the same M (i.e. $\Delta M = 0$), then only D_z has nonvanishing matrix element between $|\,E_i^0\,\rangle$ and $|\,E_f^0\,\rangle$. Similar conclusions if the transition occurs between states with $\Delta M = +1$, or with $\Delta M = -1$.

Let us finally discuss the:

Selection Rule on $\vec{L}^2$: *the components D_i of $\vec{D}$ have nonvanishing matrix elements between eigenstates of $\vec{L}^2$ only if the latter correspond to eigenvalues differing at most by one unit, provided the eigenvalues themselves are not both vanishing:*

$$\langle \cdots L'' \cdots \mid D_i \mid \cdots L' \cdots \rangle \neq 0 \;\Rightarrow\; \Delta L = \pm 1, 0, \quad L' + L'' \neq 0 \;. \quad (13.47)$$

The demonstration of this selection rule follows from the identity (that we will not derive)

$$\left[\vec{L}^2, \left[\vec{L}^2, D_i \right] \right] = 2\hbar^2 \left(\vec{L}^2 D_i + D_i \vec{L}^2 \right) - 4\hbar^2 (\vec{L} \cdot \vec{D}) L_i \qquad (13.48)$$

and is similar to that of (13.44). Equation (13.48) is general in the sense that it follows only from the commutation rules (10.9); it will be used in the sequel.

Selection rules account for the experimentally verified fact that electric-dipole electromagnetic transitions are possible only between particular pairs of states: those on space-inversion I and $\vec{L}^2$, that are the same for all the components of $\vec{D}$ (and therefore hold for any polarization of the incident wave, i.e. regardless of what the polarization unit vector in (13.37) and (13.38) may be), establish that transitions are possible only between states with opposite parities and such that $\Delta L = \pm 1, 0$ – always maintaining that the transitions from states with $L = 0$ to states with $L = 0$ are forbidden. It turns out that the latter, called $0 \to 0$ transitions, are prohibited not only in the electric-dipole approximation but, in general, for any type of (one photon) electromagnetic transition. For this reason one says that the $0 \to 0$ transitions are ***rigorously forbidden transitions***: we now give the demonstration of this fact that is, in turn, a consequence of the transversality of the electromagnetic fields with respect to the direction of propagation of the wave. Let

$$(\vec{p} \cdot \vec{e}) \, e^{i \, \vec{k} \cdot \vec{q}} \qquad (13.49)$$

be the generic addend in (13.24) (or in (13.29)). Let us choose the z-axis parallel to the direction of $\vec{k}$. Then $\vec{e}$, that by the transversality condition $\nabla \cdot \vec{A} = 0$ is orthogonal to $\vec{k}$, has the z component equal to 0: in this way (13.49) takes the form

$$(p_x \, e_x + p_y \, e_y) \, e^{i k z} \;. \qquad (13.50)$$

Let us now apply the operator (13.50) to a state with $L = 0$: $e^{i k z}$ commutes with L_z, therefore $e^{i k z} \mid L = 0 \rangle$ still is a vector with $M = 0$. To the latter vector we apply $p_x \, e_x + p_y \, e_y$: for p_x and p_y, that are the components of a vector, the selection rule (13.46) we have found for D_x and D_y apply: i.e. they may have nonvanishing matrix elements only between $e^{i k z} \mid L = 0 \rangle$, that has $M = 0$, and states with $M = \pm 1$. As a consequence the matrix elements of the operator (13.49) between states with $L = 0$ are always vanishing. The above argument holds for any addend in the sum (13.24): we have thus shown that the probability transition between S states ($L = 0$) is vanishing.

Let us go back to the electric-dipole transitions. The selection rules (13.44) on L_z are different for the different components of $\vec{D}$ and, as a consequence, the variation of M in a given transitions depends on the polarization of the incident wave. So, for example:

– for incident radiation polarized linearly in the direction of the z-axis, $\vec{e} = (0, 0, 1)$, so that $\vec{D} \cdot \vec{e} = D_z$ and therefore $\Delta M = 0$ (for an isolated atom the direction of the z-axis is arbitrary and may be taken parallel to the direction of the linear polarization of the radiation, whatever it is);

– for incident radiation polarized circularly in the x-y plane and that, therefore, propagates either along the z-axis or opposite to it, if the electromagnetic field rotates counterclockwise $\left(\cos\vartheta = \sin\vartheta \text{ and } \varphi = \pi/2 \text{ in } (3.12)\right)$, $\vec{e} = (1, i, 0)$ so that $\vec{D} \cdot \vec{e} = D_x + i\, D_y$, $\vec{D} \cdot \vec{e}^* = D_x - i\, D_y$ and, as a consequence, $\Delta M = +1$ in absorption and $\Delta M = -1$ in emission; if instead the electromagnetic field rotates clockwise $(\varphi = -\pi/2)$, $\vec{e} = (1, -i, 0)$ so that $\vec{D} \cdot \vec{e} = D_x - i\, D_y$, $\vec{D} \cdot \vec{e}^* = D_x + i\, D_y$ and, as a consequence, $\Delta M = -1$ in absorption and $\Delta M = +1$ in emission.

The transitions with $\Delta M = 0$ are called **π transitions**, whereas those with $\Delta M = \pm 1$ are called **σ transitions**.

For polarizations different from the previous ones one can proceed in a similar way; for an *isolated* atom, however, it is sometimes possible to bring the discussion back to one of the just discussed cases by a suitable choice of the z-axis (or, equivalently, by making use of selection rules analogous to (13.44) for a suitable component of $\vec{L}$).

The subject of the previous discussion has been the determination of the variations that the quantum number M may have in transitions induced by radiation with an assigned polarization: in Sect. 13.8 we will discuss the related issue that shows up in spontaneous emission, i.e. that of determining the polarization of the radiation emitted in a given direction as a consequence of a transition with an assigned variation of M.

Note that, in the case of an *isolated* atom, the selection rules (13.44) do not entail limitations for the existence of transitions between two energy levels, unless both have $L = 0$: indeed any level is degenerate on M (all the components of $\vec{L}$ commute with H_0) so that it is always possible to satisfy (13.44) i.e. there always exist states belonging to the two levels such that the difference in the value of M is either 0 or ± 1. We will see the effect of (13.44) in the case of atoms in external (either electric or magnetic) fields that remove the degeneracy on M.

Finally, consider the case of hydrogen: since it is a system with only one electron, parity is related with the value of l by the relationship $w = (-1)^l$, so that (13.43) and (13.47) are summarized in the single selection rule

$$\Delta l = \pm 1 \qquad\qquad (13.51)$$

because the transitions with $\Delta l = 0$ violate (13.43). Moreover, owing to the degeneracy on l, transitions between any pair of levels may always occur without violating (13.51).

13.7 Atom in Magnetic Field: the Normal Zeeman Effect

In Sect. 12.2 we studied the effect of an electric field on the energy levels of the hydrogen atom: this practically is the only case in which an electric field produces an observable effect on atomic spectra (the Stark effect), inasmuch as the existence of a first-order effect is due to the existence of unperturbed states with the same energy but opposite parities, which happens for hydrogen (where there is the degeneracy on l), but not in other atoms, where the complexity of the Hamiltonian removes any degeneracy not imposed by symmetry reasons: as a matter of fact, only the degeneracy on M, due to rotational invariance, survives. This is the reason for the relatively scarce interest in the exploitation of electric fields in experimental investigations of the atomic structure.

By far more important and usual is instead the use of magnetic fields as tool of investigation on every kind of atoms. For this reason we begin in this section the study of the effects of a constant and uniform magnetic field on the levels of an arbitrary atom.

The Hamiltonian of an atom in the presence of a magnetic field is given by (13.4), where H_0 is the Hamiltonian of the isolated atom and $\vec{A}(\vec{x})$ is the vector potential associated with the field $\vec{B}$:

$$\vec{B} = \nabla \wedge \vec{A}$$

whose solution, even imposing $\nabla \cdot \vec{A} = 0$ is not unique; for the problem we are presently interested in the following choice is convenient:

$$\vec{A} = -\frac{1}{2}\,\vec{x} \wedge \vec{B}\ .$$

In this case one has

$$\vec{p}_\alpha \cdot \vec{A}(\vec{q}_\alpha) = -\frac{1}{2}\,\vec{p}_\alpha \cdot \vec{q}_\alpha \wedge \vec{B} = -\frac{1}{2}\,\vec{p}_\alpha \wedge \vec{q}_\alpha \cdot \vec{B} = \frac{1}{2}\vec{L}_\alpha \cdot \vec{B}$$

so that (13.4) becomes

$$H = H_0 + \frac{e}{2m\,c}\,\vec{L}\cdot\vec{B} + \frac{e^2}{8m\,c^2}\sum_{\alpha=1}^{Z}(\vec{q}_\alpha \wedge \vec{B})^2 \tag{13.52}$$

where

$$\vec{L} = \sum_{\alpha=1}^{Z}\vec{L}_\alpha$$

is the total orbital angular momentum of the atom.

Let us examine the order of magnitude of the effects produced by the second and the third term in (13.52). Of course it depends on the magnitude of B: in a laboratory it is rather easy to produce magnetic fields of the order of 10^4 gauss. We will always take this as reference value in the following calculations.

The second term in (13.52) is of the order of

$$\frac{e\,\hbar}{2m\,c}\,B\ . \tag{13.53}$$

The ratio $e\,\hbar/2m\,c$ has the dimensions of a magnetic moment and takes the name of **Bohr magneton** μ_{B} whose value is

$$\mu_{\mathrm{B}} = \frac{e\,\hbar}{2m\,c} \simeq 0.93 \times 10^{-20}\,\mathrm{erg/gauss} \simeq 5.8 \times 10^{-9}\,\mathrm{eV/gauss}\,. \qquad (13.54)$$

As a consequence, for $B \simeq 10^4$ gauss, the effect of the second term in (13.52) is of the order of 10^{-4} eV. The third term in (13.52) (which is responsible for the atomic diamagnetism) brings a contribution of the order of

$$Z\,\frac{e^2}{8m\,c^2}\,B^2\,a^2 \qquad (13.55)$$

where by a we mean the typical size of the atom, i.e. the Bohr radius $a_{\mathrm{B}} = \hbar^2/m\,e^2$. As a consequence, the ratio of (13.55) over (13.53) may conveniently be written, up to numerical factors, as the ratio of the magnetic energy (13.53) over the atomic unit of energy $e^2/a_{\mathrm{B}} = 27.2$ eV:

$$Z\,\frac{e^2}{8m\,c^2}\,B^2 a_{\mathrm{B}}^2/\mu_{\mathrm{B}}B = \frac{1}{2}Z \times \left(\frac{e\,\hbar\,B}{2m\,c}\right)\Big/\left(\frac{e^2}{a_{\mathrm{B}}}\right) \simeq Z\,B \times 10^{-10}$$

so that, for $B \simeq 10^4$ gauss, the effect of the third term in (13.52) is about $Z \times 10^{-6}$ times smaller than that due to the second. For this reason we feel authorized to neglect the diamagnetic term in (13.52). In this case

$$H = H_0 + \frac{e}{2m\,c}\,\vec{L}\cdot\vec{B} \qquad (13.56)$$

or also

$$H = H_0 - \vec{\mu}_L \cdot \vec{B} \qquad (13.57)$$

where

$$\vec{\mu}_L = -\frac{e}{2m\,c}\,\vec{L} \qquad (13.58)$$

is the **orbital magnetic moment** of the atom (the adjective is there to distinguish it from the spin magnetic moment we will introduce later on): indeed, classically an electron going a circular orbit of radius r with velocity v is equivalent to a coil carrying the current $i = -e\,v/(2\pi\,r)$ and the magnetic moment associated with such a coil is orthogonal to the plane of the orbit and has the magnitude

$$|\vec{\mu}_L| = \frac{|i\,\pi\,r^2|}{c} = \frac{e}{2m\,c}\,m\,r\,v = \frac{e}{2m\,c}\,|\vec{L}|\,.$$

The problem is now to determine the eigenvalues and the eigenvectors of (13.56). We point out that having neglected the terms of order B^2 in (13.52) makes it possible to exactly solve the problem, once the eigenvectors and the eigenvalues of H_0 are supposed known. Indeed, since H_0 commutes with all the components of $\vec{L}$, it also commutes with the term $e/(2m\,c)\,\vec{L}\cdot\vec{B}$ and, therefore, with H. The Hamiltonians H and H_0 have, as a consequence, a complete set of simultaneous eigenvectors, i.e. there exists a set of eigenvectors of H_0 that also are eigenvectors of H. If we take, for the sake of simplicity, the z-axis in the direction of $\vec{B}$, (13.56) takes the form

$$H = H_0 + \frac{eB}{2mc} L_z \; . \tag{13.59}$$

Let us consider the simultaneous eigenvectors of H_0, $\vec{L}^2$ and L_z:

$$| E^0, L, M \rangle \; . \tag{13.60}$$

They are eigenvectors also of H:

$$H | E^0, L, M \rangle = \left(E^0 + \frac{e\hbar B}{2mc} M \right) | E^0, L, M \rangle \; .$$

Therefore the eigenvalues of H are

$$E = E^0 + \mu_B B M \; , \qquad -L \le M \le +L \tag{13.61}$$

where L is the angular momentum of the energy level E^0. Therefore the magnetic field totally removes the degeneracy on M: any energy level E^0 of the isolated atom, to which corresponds a well defined value of L (because, with the exception of the hydrogen, the levels of any atom are not degenerate on L), splits up – owing to the magnetic field – into $2L + 1$ levels, called **Zeeman sublevels**: the distance between two adjacent sublevels always is

$$\Delta E = \mu_B B$$

independently of the unperturbed level E^0 one considers.

Note that, if the eigenvalues of H_0 are nondegenerate on L, i.e. – contrary to what happens for the hydrogen atom – to any unperturbed energy level there corresponds a unique value of L, then L is a redundant quantum number in (13.60) or – which is the same – the set H_0, $\vec{L}^2$, L_z is a more than complete set of compatible observables. However the importance of specifying in (13.60) also the quantum number L should be evident:

- the selection rules (13.47) exactly refer to this quantum number;
- when L is known, the degeneracy of the level is known, i.e. the number of Zeeman sublevels into which the level splits up in presence of a magnetic field.

As a consequence of the splitting of the energy levels, a splitting of the spectral lines is expected: indeed, let $E_i^0 \to E_f^0$ be a transition between two levels of an atom in absence of magnetic field. In presence of a magnetic field we will have the transitions:

$$E^0 + \mu_B B \, M' \quad \to \quad E^0 + \mu_B B \, M''$$

corresponding to the energy jumps

$$E_f^0 - E_i^0 + \mu_B B \, \Delta M \; , \qquad \Delta M = M'' - M' \; .$$

But the selection rules (13.34) demand $\Delta M = \pm 1, 0$ so any line of the spectrum splits up into three lines: if $\nu_0 = |E_f^0 - E_i^0|/h$ is the frequency of the line in absence of the field, the frequencies of the three lines in presence of the field are:

$$\nu = \begin{cases} \nu_0 & \Delta M = 0: & \pi \text{ lines} \\ \nu_0 \pm \mu_\mathrm{B} B/h & \Delta M = \pm 1: & \sigma \text{ lines}. \end{cases} \qquad (13.62)$$

In Fig. 13.1 the Zeeman transitions between the sublevels with $L = 2$ and those with $L = 1$ are reported: note that the nine transitions permitted by the selection rules give rise to only three different lines, because the distance between the adjacent sublevels with $L = 2$ is the same as that between the sublevels with $L = 1$.
The result we have found is in agreement with Larmor's theorem: indeed the quantity $\mu_\mathrm{B} B/h$

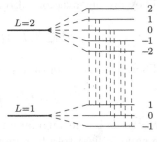

Fig. 13.1

$= e B/(4\pi\, m\, c)$ is just the Larmor frequency.
This shows that the splitting of spectral lines (**Zeeman effect**), that was already known before the advent of quantum mechanics, can be – and indeed was – explained even in classical terms. However the point is that experiments only partially confirm that any line splits up into to three lines, in agreement with (13.62), and in this case one says one has **normal Zeeman effect**. In many cases a splitting in more than three, not equally spaced, lines is observed: in the latter case it is said that one has **anomalous Zeeman effect**.

We will see that the anomalous Zeeman effect, as well as other discrepancies between theory and experiment that show up, for example, in the spectra of alkali atoms, are explained by attributing to the electron an intrinsic magnetic moment and, correspondingly, an intrinsic angular momentum, the spin.

If the *normal* Zeeman effect occurs, the possibility of observing all the three lines depends on the experimental conditions. Let us consider the case of absorption: if, for example, radiation is sent on the atom in the direction of $\vec{B}$ (i.e. in the direction of the z-axis), owing to the transversality of electromagnetic waves, the z component of the polarization vector $\vec{e}$ is vanishing: $\vec{e} = (e_x, e_y, 0)$, whence $\vec{D} \cdot \vec{e} = D_x\, e_x + D_y\, e_y$ and, as a consequence of (13.44) or (13.46), in general $\Delta M = +1$ and $\Delta M = -1$, i.e. only the central π line is missing; but if the radiation is circularly polarized, only one of the two σ lines is observed. If instead the radiation propagates along a direction orthogonal to $\vec{B}$ and is polarized parallel to $\vec{B}$ (the z direction), then $\vec{e} = (0, 0, 1)$, $\vec{D} \cdot \vec{e} = D_z$, therefore $\Delta M = 0$ and only the π line is observed.

In conclusion one way, but not the only one, to produce all the three transitions $\Delta M = 0, \pm 1$ is that of sending on the atom unpolarized radiation, provided the direction of propagation is not parallel to that of the field, e.g. along an orthogonal direction.

In the next section we will study the inverse problem, i.e. that of determining the polarization of the light emitted in the Zeeman transitions: the problem is relevant only for spontaneous emission because, as we have al-

ready said in Sect. 13.4, the radiation due to stimulated emission has the
same features – in particular the polarization – as the incident radiation.

13.8 Polarization and Angular Distribution of Emitted Radiation

Let us consider the *spontaneous* emission of radiation in the transition between
two well defined states: for example between two nondegenerate energy levels,
as in the Zeeman effect, and let us assume we observe the radiation emitted
along the direction pinpointed by the unit vector $\vec{n}$.

The intensity $I(\vec{n}, \vec{e})$ of the radiation detected by an observer in a given
state of polarization $\vec{e}$ orthogonal (the usual transversality of electromagnetic
waves) to $\vec{n}$ is proportional to the emission probability per unit time: in the
electric-dipole approximation, by (13.38)

$$I(\vec{n}, \vec{e}) \propto |\langle E_{\text{f}}^0 \mid \vec{D} \cdot \vec{e}^* \mid E_{\text{i}}^0 \rangle|^2 , \qquad\qquad \vec{e} \cdot \vec{n} = 0 . \qquad (13.63)$$

The radiation observed in a given direction is polarized: indeed, what we are
considering is a measurement process and, thanks to von Neumann postulate
(Sect. 4.4), after the measurement the state of the system is well determined:
in the case of interest, as discussed in Sect. 13.4, the system consists of both the
atom and the emitted photon. Since the measurement consists in detecting
the state of the atom and the state of motion of the photon (energy and
direction of propagation), it follows that also the polarization state $\vec{e}$ of the
photon is determined: it is (always thanks to von Neumann postulate) the
state such that the emission probability, i.e. the right hand side of (13.63), is
a maximum. The latter is obtained in the following way: putting

$$\vec{d} = \langle E_{\text{f}}^0 \mid \vec{D} \mid E_{\text{i}}^0 \rangle \qquad (13.64)$$

we must find the maximum of $|\vec{d} \cdot \vec{e}^*|$ with the constraints $\vec{e} \cdot \vec{n} = 0$ and
$\vec{e}^* \cdot \vec{e} = 1$: one can take advantage of the method of Lagrange multipliers by
maximizing

$$(\vec{e}^* \cdot \vec{d})(\vec{d}^* \cdot \vec{e}) - \lambda (\vec{e}^* \cdot \vec{n}) - \mu (\vec{e}^* \cdot \vec{e} - 1) .$$

By taking the derivative with respect to $\vec{e}^*$ one obtains

$$(\vec{d}^* \cdot \vec{e}) \vec{d} - \lambda \vec{n} - \mu \vec{e} = 0 \qquad \Rightarrow \qquad \vec{e} \propto (\vec{d}^* \cdot \vec{e}) \vec{d} - \lambda \vec{n} .$$

and imposing $\vec{e} \cdot \vec{n} = 0$ we get $\lambda = (\vec{d}^* \cdot \vec{e})(\vec{d} \cdot \vec{n})$, whence

$$\vec{e} \propto \vec{d} - (\vec{d} \cdot \vec{n}) \vec{n} \qquad \Rightarrow \qquad \vec{e} = \frac{\vec{d} - (\vec{d} \cdot \vec{n}) \vec{n}}{\sqrt{|\vec{d}|^2 - |\vec{d} \cdot \vec{n}|^2}} , \qquad \vec{e}^* \cdot \vec{e} = 1 \quad (13.65)$$

i.e. $\vec{e}$ is proportional to the projection of $\vec{d}$ onto the plane orthogonal to $\vec{n}$.
Let us examine some examples.

Suppose we have an atom in a magnetic field $\vec{B}$ parallel to the z-axis.
Consider a π transition ($\Delta M = 0$): then by (13.46) $\vec{d} = (0, 0, d)$ is – up to a

phase factor – a real vector, therefore also $\vec{e}$, given by (13.65), is real, hence the radiation is linearly polarized for any $\vec{n}$. If we choose $\vec{n}$ in the x-y plane, then $\vec{d} \cdot \vec{n} = 0$ and $\vec{e} = \vec{d}/d$, namely the direction of polarization is that of the z-axis. If $\vec{n} \parallel \vec{d}$, the vector $\vec{e}$ is undetermined, but (13.63) is vanishing: there is no emission of radiation in the direction of the $\vec{B}$ field.

Let us consider a σ transition, for example with $\Delta M = +1$: in such a case (13.44) give $d_z = 0$ and $d_x - \mathrm{i}\,d_y = 0$ i.e. $\vec{d} = (d/\sqrt{2})\,(1, \mathrm{i}, 0)$. If we choose $\vec{n} \parallel \vec{B}$: $\vec{n} = (0, 0, 1)$, then by (13.65) $\vec{e} = (1, \mathrm{i}, 0)/\sqrt{2}$ which means that the radiation is counterclockwise circularly polarized in the x-y plane. If $\vec{n}$ is orthogonal to $\vec{B}$, for example $\vec{n} = (1, 0, 0)$, then $\vec{e} = (0, \mathrm{i}, 0)$ and the radiation is linearly polarized parallel to the y direction. If $\vec{n}$ is neither orthogonal nor parallel to $\vec{B}$, then the radiation is elliptically polarized. In general the content of (13.65) can be summarized by a simple rule: the polarization of the emitted radiation in a generic direction $\vec{n}$ is the same one would have, according to classical radiation theory, for a dipole that

- oscillates parallel to the z-axis for transitions with $\Delta M = 0$;
- rotates in the x-y plane either counterclockwise or clockwise respectively for transitions with $\Delta M = +1$ and $\Delta M = -1$.

Indeed, according to the classical theory, the electric field associated with the radiation emitted by an accelerated charge is in any point proportional to the component of the acceleration orthogonal to the direction of observation $\vec{n}$: roughly speaking, the electric field 'follows' the projection of the motion of the charge onto the plane orthogonal to $\vec{n}$.

Let us now apply the above rule to determine the polarization of the radiation emitted in the transitions $n = 2 \to n = 1$ of hydrogen in an electric field $\vec{E}$ (Stark effect).

The levels with $n = 2$ are reported in Fig. 12.2: first of all, let us note that all the three transitions from the levels with $n = 2$ to the level with $n = 1$ are permitted by the selection rules (13.51) $\Delta l = \pm 1$: observe however that the states (12.17) (those with energies $E_2^0 \pm 3e\,E\,a_{\mathrm{B}}$) are not eigenstates of $\vec{L}^2$ and only the component with $l = 1$ contributes to the transition probability. As far as polarization is concerned, the two side lines correspond to $\Delta m = 0$, i.e. they are π lines: they can be observed from any direction except the z-axis (the direction of the electric field) and always are linearly polarized. The central line is originated by a degenerate level in which both $m = +1$ and $m = -1$: since normally the atoms are randomly distributed in all the states of the type $\alpha\,|\,m = +1\,\rangle + \beta\,|\,m = -1\,\rangle$, each of which classically corresponds to a dipole effecting a harmonic motion in the x-y plane, the line appears completely unpolarized if it is observed from the direction of the electric field; if instead it is observed from a direction orthogonal to the field, e.g. from the direction of the x-axis, it appears linearly polarized along the direction of the y-axis. Also in this case, much as in the conditions of the Zeeman effect, in order to observe all the lines it suffices to make the observation from any direction different from that of the electric field.

We now consider the problem of finding the angular distribution of the radiation emitted in a particular transition, i.e. the dependence of the transition rate on the direction of observation. To this end, substituting the unit vector $\vec{e}$ given by (13.65) into (13.38) we obtain:

$$w_{\mathrm{fi}}^{\mathrm{sp}}(\vec{n})\,\mathrm{d}\Omega = \frac{\omega_{\mathrm{fi}}^3}{2\pi\,\hbar\,c^3}\left(|\vec{d}\,|^2 - |\vec{d}\cdot\vec{n}\,|^2\right)\mathrm{d}\Omega \qquad (13.66)$$

that coincides $\big($as it must be, see (4.15)$\big)$ with the sum of the transition probabilities to any two polarization states $\vec{e}_1$, $\vec{e}_2$ orthogonal to each other and to the vector $\vec{n}$:

$$|\vec{d}\cdot\vec{e}_1|^2 + |\vec{d}\cdot\vec{e}_2|^2 = |\vec{d}\cdot\vec{e}_1|^2 + |\vec{d}\cdot\vec{e}_2|^2 + |\vec{d}\cdot\vec{n}\,|^2 - |\vec{d}\cdot\vec{n}\,|^2 = |\vec{d}\,|^2 - |\vec{d}\cdot\vec{n}\,|^2$$

(usually one says that the 'sum over final states' is made).

Equation (13.66) reproduces the classical result, according to the rule we have given above for the correspondence between the type of transition and the type of motion of the classical dipole.

So, for example, in the case of a π transition, taken the z-axis along the direction of the (real) vector $\vec{d}$: $\vec{d} = |\vec{d}\,|\,(0,0,1)$, one has $|\vec{d}\cdot\vec{n}\,|^2 = |\vec{d}\,|^2\cos^2\theta$, then from (13.66)

$$w_{\mathrm{fi}}^{\mathrm{sp}}(\vec{n})\,\mathrm{d}\Omega \equiv w_{\mathrm{fi}}^{\mathrm{sp}}(\theta,\phi)\,\mathrm{d}\Omega = \frac{\omega_{\mathrm{fi}}^3\,|\vec{d}\,|^2}{2\pi\,\hbar\,c^3}\sin^2\theta\,\mathrm{d}\Omega\,, \qquad \Delta M = 0 \qquad (13.67)$$

and the probability transition integrated over the solid angle is

$$w_{\mathrm{fi}}^{\mathrm{tot}} = \frac{\omega_{\mathrm{fi}}^3\,|\vec{d}\,|^2}{2\pi\,\hbar\,c^3}\,2\pi\int_{-1}^{+1}\sin^2\theta\,\mathrm{d}\cos\theta = \frac{4\,\omega_{\mathrm{fi}}^3\,|\vec{d}\,|^2}{3\,\hbar\,c^3}\,. \qquad (13.68)$$

In the case of a σ transition ($\Delta M = \pm 1$), $\vec{d} = (|\vec{d}\,|/\sqrt{2})\,(1,\pm\mathrm{i},0)$, $|\vec{d}\cdot\vec{n}\,|^2 = \frac{1}{2}|\vec{d}\,|^2(n_x^2 + n_y^2) = \frac{1}{2}|\vec{d}\,|^2(1 - \cos^2\theta)$, so, from (13.66),

$$w_{\mathrm{fi}}^{\mathrm{sp}}(\theta,\phi)\,\mathrm{d}\Omega = \frac{\omega_{\mathrm{fi}}^3\,|\vec{d}\,|^2}{2\pi\,\hbar\,c^3}\frac{1 + \cos^2\theta}{2}\,\mathrm{d}\Omega\,, \qquad \Delta M = \pm 1 \qquad (13.69)$$

and also in this case the integration over the angles leads to (13.68): in the next section we will see that the $|\vec{d}\,|^2$ appearing in (13.69) is the same that appears in (13.68).

13.9 The Lifetime of the $n = 2$ Energy Level of Hydrogen Atom

We can use (13.68) to calculate the lifetime of the level $n = 2$ of the hydrogen atom: the state $n = 2$, $l = 0$ cannot decay to the sate $n = 1$ (strongly forbidden $0 \to 0$ transition); the transition rate from the state $|2\,1\,0\rangle$ to the state $|1\,0\,0\rangle$ is given by (13.68) with $\vec{d} = (0,0,d)$, where

$$d = -e\,\langle 1,0,0 \mid z \mid 2,1,0\rangle$$
$$= -e\int Y_{00}(\theta,\phi)\,\cos\theta\,Y_{10}(\theta,\phi)\,\mathrm{d}\Omega\int_0^{\infty} R_{10}(r)\,r\,R_{21}(r)\,r^2\,\mathrm{d}r$$

where the spherical harmonics are given by (10.42) and the radial functions by (11.41). The integration over the angles gives $1/\sqrt{3}$, that with respect to r gives $(a_B/\sqrt{6}) \times (4/3)^4$, therefore $d = -0.74\,e\,a_B$, whence ($\lambda = 1216\,\text{Å}$)

$$w_{fi}^{tot} = \alpha \, \frac{32\pi^3}{3} \left(\frac{0.74\,a_B}{\lambda} \right)^2 \frac{c}{\lambda} = 6.2 \times 10^8\,\text{s}^{-1} \quad \Rightarrow$$

$$\tau \equiv (w_{fi}^{tot})^{-1} = 1.6 \times 10^{-9}\,\text{s} . \tag{13.70}$$

The same result, i.e. the same lifetime, is obtained for any state with $n = 2$, $l = 1$: the integration over the angles in (13.66) can be carried out by noting that

$$\int |\vec{d} \cdot \vec{n}\,|^2 \, d\Omega = d_i^* \left(\int n_i\,n_j \, d\Omega \right) d_j = d_i^* \, (\alpha\,\delta_{ij}) d_j$$

$$= d_i^* \left(\frac{4\pi}{3}\,\delta_{ij} \right) d_j = \frac{4\pi}{3} \, |\vec{d}\,|^2 \tag{13.71}$$

indeed the integration of the tensor $n_i\,n_j$ with respect to the angles gives an isotropic – i.e. invariant under rotations – tensor, therefore proportional to δ_{ij}; moreover, since the trace of the tensor $n_i\,n_j$ is 1, and that of δ_{ij} is 3, the proportionality constant α is $4\pi/3$.

Therefore the integration of $|\vec{d}\,|^2 - |\vec{d}\cdot\vec{n}\,|^2$ with respect to the angles always gives $(8\pi/3)\,|\vec{d}\,|^2$, therefore in any event (13.68) is obtained.

We still have to show that $|\vec{d}\,|^2$ is independent of the initial state $|\,2, 1, \cdots \rangle = \alpha\,|\,m = 0 \rangle + \beta\,|\,m = 1 \rangle + \gamma\,|\,m = -1 \rangle$.

A direct verification is possible, but we prefer to present an argument of general character that we will use in the sequel. To this end we firstly show the following

Theorem: *the mean value of a scalar operator S on states of given angular momentum L is independent of the state.*

By 'states of given angular momentum' we mean the states belonging to the $(2L + 1)$-dimensional manifold $\mathcal{V}_L$ generated by the vectors $|\,\Sigma', L, M \rangle$ where, according to the notation of Sect. 10.2, $\Sigma' (\equiv \sigma_1', \sigma_2' \cdots)$ is the generic eigenvalue of the scalar observable $\Sigma (\equiv \sigma_1, \sigma_2 \cdots)$ that, together with $\vec{L}^2$ and L_z, forms a complete set of compatible observables.

It suffices to show that the matrix $S_{M,M'} = \langle \Sigma', L, M \mid S \mid \Sigma', L, M' \rangle$ is a multiple of the identity.

Indeed, by the lemma of p. 87 (Sect. 4.10), since $[\,L_z\,, S\,] = 0$ and in $\mathcal{V}_L$ the eigenvalues of L_z are nondegenerate, $S_{M,M'}$ is diagonal; furthermore, by (10.18) and (10.25)

$$L_- |\,\Sigma', L, M \rangle = \hbar \, \sqrt{L(L+1) - M^2 + M} \, |\,\Sigma', L, M - 1 \rangle \tag{13.72}$$

(the vectors $|\,\Sigma'\,L\,M \rangle$ and $|\,\Sigma'\,L\,M - 1 \rangle$ in (13.72) are normalized); in addition

$$L_+\,S\,L_- = S\,L_+\,L_- = S\,(\vec{L}^2 - L_3^2 + \hbar\,L_3) \tag{13.73}$$

therefore by (13.72) (Σ' omitted)

$$\langle L, L \mid L_+ \, S \, L_- \mid L, L \rangle = \langle L, L - 1 \mid S \mid L, L - 1 \rangle \times \hbar^2 \left(L(L+1) - L^2 + L \right)$$

and by (13.73)

$$\langle L, L \mid L_+ S \, L_- \mid L, L \rangle = \langle L, L \mid S \mid L, L \rangle \times \hbar^2 \left(L(L+1) - L^2 + L \right) \ .$$

As a consequence $S_{L,L} = S_{L-1,L-1}$ and, by iteration of the procedure, one shows that all the matrix elements $S_{M,M}$ are equal to one another.

What we have just shown is nothing but the degeneracy theorem of the eigenvalues of scalar operators restricted to an eigenspace of $\vec{L}^2$; indeed, the restriction of the operator S to $\mathcal{V}_L$ is $S_L = \mathcal{P}_L \, S \, \mathcal{P}_L$ ($\mathcal{P}_L$ is the projector onto $\mathcal{V}_L$) and still is a scalar operator because $\mathcal{P}_L$ commutes with $\vec{L}$:

$$L_z \mathcal{P}_L = L_z \Big(\sum_{M=-L}^{+L} | \Sigma', L, M \rangle \langle \Sigma', L, M | \Big) = \sum_{M=-L}^{+L} M | \Sigma', L, M \rangle \langle \Sigma', L, M |$$

$$= \Big(\sum_{M=-L}^{+L} | \Sigma', L, M \rangle \langle \Sigma', L, M | \Big) L_z = \mathcal{P}_L L_z$$

and for the L_x and L_y components the same argument applies: it is sufficient to express $\mathcal{P}_L$ by means of the eigenvectors of L_x and L_y, instead of those of L_z.

Let us now use this theorem to demonstrate the independence of $|\vec{d}|^2$ of the initial state $| n = 2, l = 1, \cdots \rangle$:

$$|\vec{d}|^2 = \sum_{j=1}^{3} \langle 2, 1, \cdots \mid D_j \mid 1, 0, \cdots \rangle \times \langle 1 \, 0 \cdots \mid D_j \mid 2, 1, \cdots \rangle$$

$$= \langle 2, 1, \cdots \mid S \mid 2, 1, \cdots \rangle$$

where

$$S \equiv \sum_{j=1}^{3} D_j \, \mathcal{P}_0 \, D_j \qquad \Big(\mathcal{P}_0 = | 1, 0, 0 \rangle \langle 1, 0, 0 | \Big)$$

is a scalar operator: since $[L_i, \mathcal{P}_0] = 0$ (indeed, $L_i \, \mathcal{P}_0 = \mathcal{P}_0 \, L_i = 0$) and by use of the properties (4.48) of commutators:

$$[L_i, S] = \sum_{j=1}^{3} [L_i, D_j \, \mathcal{P}_0 \, D_j] = i \, \hbar \sum_{j \, k} \epsilon_{ijk} \Big(D_k \, \mathcal{P}_0 \, D_j + D_j \, \mathcal{P}_0 \, D_k \Big) = 0$$

(ϵ_{ijk} is antisymmetric in j and k whereas $D_k \, \mathcal{P}_0 \, D_j + D_j \, \mathcal{P}_0 \, D_k$ is symmetric). Alternatively, we could have noticed that S being the trace of the tensor (operator) $D_i \, \mathcal{P}_0 \, D_j$, is invariant under rotations.

The same argument can be used to show that, in general, the total transition probability from a state $|E_n^0, L, \cdots\rangle$ of a level E_n^0 to the states $|E_m^0, L', \cdots\rangle$ of a (degenerate) level E_m^0 is independent of the initial state. Such transition probability is the 'sum over the final states', i.e. the sum of the transition probabilities to an orthogonal basis of states of the level E_m^0, for example $|E_m^0, L', M'\rangle$ $(-L' \leq M' \leq L')$. The justification of such a statement again makes use of the von Neumann postulate, but in the present case it is more complicated inasmuch as it requires to explicitly consider the 'atom+photon' system: the measurement process consists in detecting only the final energy E_m^0 of the atom, but – the energy level being degenerate if $L' \neq 0$ – the state of the 'atom+photon' system after the measurement is not a factorized ('separate variable') state, namely the atom in a certain state and the photon in a certain state, but rather is a superposition of states $|\text{atom; photon}\rangle$:

$$\sum_{M'=-L'}^{L'} c_{M'} |E_m^0, L', M'; \gamma_{M'}\rangle \tag{13.74}$$

where $\gamma_{M'}$ is the state of the photon when the atom makes a transition from the initial state $|E_n^0, L, \cdots\rangle$ to the final state $|E_m^0, L', M'\rangle$. The transition probability we are after is the probability that the system makes a transition to the state (13.74) and the latter, always thanks to (4.15), is the sum of the probabilities to the orthogonal states $|E_m^0, L', M'; \gamma_{M'}\rangle$ and each of the latter is given by (13.68). Therefore:

$$w_{fi}^{tot} = \sum_{M'=-L'}^{L'} \frac{4\,\omega_{fi}^3}{3\,\hbar\,c^3} \langle E_n^0, L, \cdots | \vec{D} | E_m^0, L', M'\rangle\langle E_m^0, L', M' | \vec{D} | E_m^0, L, \cdots\rangle$$

$$= \frac{4\,\omega_{fi}^3}{3\,\hbar\,c^3} \langle E_n^0, L, \cdots | S | E_m^0, L, \cdots\rangle, \qquad S = \sum_j D_j\,\mathcal{P}_{L'}\,D_j$$

where $\mathcal{P}_{L'}$ is the projector onto the states of angular momentum L' (and energy E_m^0), and as a consequence commutes with the operators L_i, whence S is a scalar operator. In conclusion, we have shown that the probability transition from *any state* of the level E_n^0 to the *level* E_m^0 is independent of the initial state.

Chapter 14

Introduction to Atomic Physics

14.1 The Central-Field Approximation

The present and the following chapters aim at showing that the general theory expounded so far allows one to understand the general features of atomic energy levels. To this end, however, the attribution of an intrinsic angular momentum or *spin* to the electron will play an essential role: we have already mentioned that the theory, as it is at the present point of development, is not able to explain the existence of the anomalous Zeeman effect and we have also cited the discrepancies between theory and experiment that show up as a fine structure in the spectra of alkali atoms: these facts, and others that we will discuss, require the introduction of spin.

Let us begin the present chapter with a general introduction to the problem of the structure of atomic energy levels: on this subject, that we will discuss in an essentially qualitative way, we will come back several times in order to incorporate the effects due to the spin and to the Pauli principle. Part of the chapter will be dedicated to the study of the alkali atoms which, in many respects, are the simplest to treat after the hydrogen-like atoms.

The Hamiltonian of whatever atom with Z electron is, in the nonrelativistic approximation,

$$ H = \sum_{\alpha=1}^{Z} \left(\frac{\vec{p}_\alpha^{\,2}}{2m} - \frac{Z\,e^2}{r_\alpha} \right) + \sum_{\alpha>\beta} \frac{e^2}{r_{\alpha\beta}} \tag{14.1} $$

in which the variables with the index α refer to the α-th electron and $r_{\alpha\beta} = |\vec{q}_\alpha - \vec{q}_\beta|$.

The meaning of the several terms in (14.1) should be clear: the first term is the kinetic energy of the Z electrons, the second is the interaction of each electron with the nucleus whose electric charge is $Z\,e$; the third is the energy relative to the Coulombic repulsion of any pair of electrons (the sum is made for $\alpha > \beta$ so that the pairs of electrons are not counted twice).

The Hamiltonian (14.1) is too complicated: in no case, not even for $Z = 2$, it is possible to exactly solve the eigenvalue problem, so it is necessary to resort to approximation methods. Let us rewrite (14.1) as

© Springer International Publishing Switzerland 2016
L.E. Picasso, *Lectures in Quantum Mechanics*, UNITEXT for Physics,
DOI 10.1007/978-3-319-22632-3_14

$$H = \sum_{\alpha=1}^{Z} H_\alpha^0 + \sum_{\alpha>\beta} \frac{e^2}{r_{\alpha\beta}} \qquad (14.2)$$

in which H_α^0 is the Hamiltonian of a hydrogen-like atom with charge Z.

If the term relative to the electronic repulsion were absent, the problem would be exactly solvable: in each H_α^0 only the variables relative to the α-th electron appear, so not only they commute with one another, but the problem is a 'separate variables' one (see Sect. 11.2). In this situation a complete set of eigenvectors of $\sum_\alpha H_\alpha^0$ is provided by the simultaneous eigenvectors of the H_α^0's in which the state of each electron (as long as spin is ignored) is characterized by the three quantum numbers $(n\,l\,m)$:

$$| (n_1\,l_1\,m_1), (n_2\,l_2\,m_2), \cdots, (n_Z\,l_Z\,m_Z) \rangle . \qquad (14.3)$$

The eigenvalue of $\sum_\alpha H_\alpha^0$ corresponding to the eigenvector (14.3) is

$$E = E_{n_1} + E_{n_2} + \cdots + E_{n_Z} , \qquad E_n = -Z^2 \frac{e^2}{2n^2\,a_B} . \qquad (14.4)$$

In this approximation, i.e. the one in which the interaction among the electrons is neglected, each energy level is largely degenerate. First of all, since the energy of each electron only depends on n, there is the degeneracy on l: we will ignore this degeneracy, i.e. we will consider the energies of states with the same n and different l as being different from one another. The reason for doing so is that such a high degeneracy would complicate the discussion in a useless way, given that the most immediate effect of the electronic repulsion is just that of removing the degeneracy on l: since the degeneracy on l is peculiar of the Coulomb potential, it suffices – from a formal point of view – to add a central potential $U(r_\alpha)$ to each H_α^0, whose effect is that of removing the degeneracy on l, and then add $-\sum_\alpha U(r_\alpha)$ to the term $\sum_{\alpha>\beta} e^2/r_{\alpha\beta}$: the choice of $U(r_\alpha)$ is not relevant for our purposes here, whereas it is indeed relevant for anyone that intends to make accurate calculations on atomic structure.

What we have just described is known as **central-field approximation** or **independent-electron approximation** (we could also call it '0-th order approximation'): each electron has its own quantum numbers $(n\,l\,m)$, that constitute what is also called an **atomic orbital** (or simply **orbital**). Each electron is then characterized by occupying a given orbital.

On the basis of what we have said, one should conclude that, in the independent-electron approximation, the ground state of any atom is that in which any electron is in the orbital $n = 1$, $l = m = 0$. But in nature things go differently: it is well known, in effect, that in order to explain the periodic table of the elements one must admit that:

in any atom no more than two electrons may have the same quantum numbers $(n\,l\,m)$;

or equivalently:

in any atom no more than two electrons can occupy the same orbital.

The above statement is a particular case of a general principle in Nature, known as ***Pauli principle***, that we will enunciate in a more precise form and discuss in the next chapter.

As a consequence, in the ground state of helium both electrons occupy the orbital $n = 1, l = m = 0$; in lithium ($Z = 3$) two electrons occupy the orbital $n = 1, l = m = 0$ and the third electron occupies the next orbital $n = 2, l = m = 0$. In the notation (14.3) the ground state of He is

$$|(1\,0\,0), (1\,0\,0)\rangle . \tag{14.5}$$

It should be evident that, particularly for atoms with many electrons, the notation (14.3) is rather cumbersome: not only, it also is incorrect inasmuch as it presupposes that the electrons can be numbered, i.e. that they can be distinguished from one another; for this reason one prefers to give the ***electronic configuration***, i.e. the distribution of the electrons in the various orbitals, by representing the orbitals by the symbols $1s$, $2s$, $2p$, $3s$, $\cdots$, in which the number indicates the value of n and the letters s, p, d, f, $\cdots$ the values of l $(0, 1, 2, 3, \cdots)$; one then writes as the exponent of such symbols the number of electrons that occupy the corresponding orbitals.

So, for example, the ground-state configurations of the atoms from helium to carbon ($Z = 6$) are

$$\text{He}: (1s)^2; \quad \text{Li}: (1s)^2 2s; \quad \text{Be}: (1s)^2(2s)^2;$$
$$\text{B}: (1s)^2(2s)^2 2p; \quad \text{C}: (1s)^2(2s)^2 2p^2 \tag{14.6}$$

and more complicated ones will be met in the sequel. When all the $2l + 1$ orbitals corresponding to given n, l are occupied (2 electrons per orbital), one says that the ***shell*** (n, l) is complete and it is put between parentheses: in all the cases considered in (14.6) we see that the shell $1s$ is always complete and the shell $2s$ is complete starting from beryllium on. In order to fill the shell $2p$, $3 \times 2 = 6$ electrons are needed and this occurs for the first time in the case of neon that has $Z = 10$, so $\text{Ne}: (1s)^2(2s)^2(2p)^6$.

In order to write the ground-state configurations of the various atoms, it is necessary to know in which order the orbitals must be filled, i.e. the sequence in energy of the orbitals themselves as it results by taking into account the Coulombic repulsion among the electrons. The filling order of the orbitals is the following:

$$1s \quad 2s \quad 2p \quad 3s \quad 3p \quad 4s \quad 3d \quad 4p \quad 5s \quad \cdots . \tag{14.7}$$

Note that the sequence in energy is not the natural one corresponding to increasing n. Already between $4s$ and $3d$ an inversion occurs: we will come back later in this chapter on the origin of this fact.

The independent-electron approximation should be good if the term $\sum e^2/r_{\alpha\beta}$ were responsible of only small corrections to the energy levels

(14.4). Things do not go this way, as we can realize by considering the helium atom. Since in the independent-electron approximation the ground state of He is that in which both electrons occupy the orbital $1s$, the first-ionization energy E_I (namely the energy necessary to extract one electron) should be $Z^2 \times 13.6 \simeq 54\,\text{eV}$. Experimentally instead the first-ionization energy is $24.6\,\text{eV}$: the effect of the repulsion between the two electrons is that of reducing the first-ionization energy by a sizeable $30\,\text{eV}$.

In Sect. 14.3 we will explicitly calculate the first-order perturbative correction due to the term e^2/r_{12} and we will find a value of $20.4\,\text{eV}$ for the first-ionization energy, a value that starts being acceptable.

It is then clear, a fortiori in the case of atoms with many electrons, that the term $\sum e^2/r_{\alpha\beta}$ cannot be neglected: we will examine in due time its effect on the structure of atomic levels, but for the time being – at least in the first approximation – one can imagine to take it into account for a perturbative calculation in which $\sum_\alpha H_\alpha^0$ is the unperturbed Hamiltonian: in such a case, as we have learned in perturbation theory (and explicitly seen in the discussion of the Stark effect), the approximate eigenvectors of the Hamiltonian corresponding to the eigenvalues approximated to the first order still are eigenvectors of the unperturbed Hamiltonian. In the present case, in general, the approximate eigenvectors are not tout-court the vectors (14.3), because the eigenvalues of the unperturbed Hamiltonian are degenerate: even ignoring the degeneracy on l, to the 0-th order there still is (for each electron) the degeneracy on m and the so called *exchange degeneracy* that, as the name says, corresponds to the fact that, if the quantum numbers of two electrons are exchanged, the energy is unchanged: in any event the approximate eigenvectors are *linear combinations* of the vectors (14.3) corresponding to the same configuration: actually, the electronic configuration pinpoints an unperturbed energy level without either attributing a value of m to the single electrons, or saying which electron is in which orbital.

Therefore the concept of electronic configuration 'survives', *at least* up to the first order, the effect of the Coulombic repulsion among the electrons: this observation partly reassesses the vectors (14.3), namely the classification of the atomic states in the independent-electron scheme, inasmuch as it is from it that the electronic configuration is determined.

We may examine the importance of the concept of electronic configuration also in a different way: let us assume we know the energy levels of the Hamiltonian (14.1) and let us also imagine to adiabatically 'switch off' the Coulombic repulsion among the electrons: in other terms, let us multiply the term $\sum e^2/r_{\alpha\beta}$ by a parameter λ and ask ourselves what happens to the energy levels when λ is varied with continuity from 1 to 0. Since the effect of the Coulombic repulsion is that of removing, at least partially, the degeneracy present in the independent-electron approximation, some groups of levels, separated for $\lambda \neq 0$, will merge for $\lambda = 0$ in a unique level corresponding to a well determined electronic configuration: so, on top of the genealogical tree of each energy level of (14.1), there is the electronic configuration from which it arises.

Viceversa, given the electronic configuration, we will learn how to determine the number of branches into which the genealogical tree develops, i.e. the number of energy levels that originate from it: Fig. 14.1, that we now report without explaining its details, illustrates the idea in the case of the first configurations

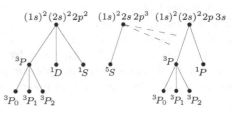

Fig. 14.1

of carbon: the letters S, P, D (as said in Sect. 10.4) indicate the values $L = 0, 1, 2$ of the angular momentum of the levels (we indeed know that the energy levels of the Hamiltonian (14.1) are classified by the total angular momentum L); the second generation levels (the bottom ones) derive from interactions relativistic in nature, that we have not yet taken into account.

So, in conclusion, to summarize the situation in a few words, one can say that the independent-electron approximation is unsuited for the calculation of energy levels, but it plays an essential role in the classification of the states and, therefore, in the determination of the number (and the quantum numbers) of the levels that stem from any electronic configuration.

14.2 The Variational Method

The variational method allows one to find an upper limit for the energy of the ground state for a (sometimes even complex) system: from the conceptual point of view it is a very simple method and, what is more, it is very effective: for these reasons it is widely used in atomic as well as in molecular, nuclear physics and so on.

The starting point is the following: if H is a Hamiltonian and E_0 its lowest (unknown) eigenvalue, for any vector $|A\rangle$ one has

$$\langle A \,|\, H \,|\, A \rangle \geq E_0 \,, \qquad \langle A \,|\, A \rangle = 1 \qquad (14.8)$$

in which the equality sign holds only if $|A\rangle$ is an eigenvector corresponding to the eigenvalue E_0. The inequality (14.8) is obvious: the mean value of measurements of H in the state $|A\rangle$ certainly is greater or equal to the minimum eigenvalue. Formally, if $|E_n\rangle$ is a basis of eigenvectors of H,

$$|A\rangle = \sum_{n=0}^{\infty} a_n |E_n\rangle \;\Rightarrow\; \langle A \,|\, H \,|\, A \rangle = \sum_{n=0}^{\infty} E_n |a_n|^2 \geq \sum_{n=0}^{\infty} E_0 |a_n|^2 = E_0 \,.$$

The variational method consists in choosing a set of vectors (if we are in the Schrödinger representation, such vectors are called "trial functions") $|A; \lambda_1, \lambda_2, \cdots, \lambda_n\rangle$ indexed by the parameters $\lambda_1, \lambda_2, \cdots, \lambda_n$ and calculating the minimum of the function

$$F(\lambda_1, \lambda_2, \cdots \lambda_n) \equiv \frac{\langle A; \lambda_1, \lambda_2, \cdots \lambda_n \,|\, H \,|\, A; \lambda_1, \lambda_2, \cdots \lambda_n \rangle}{\langle A; \lambda_1, \lambda_2, \cdots \lambda_n \,|\, A; \lambda_1, \lambda_2, \cdots \lambda_n \rangle}$$

(in the case the vectors $| A ; \lambda_1 , \lambda_2 , \cdots , \lambda_n \rangle$ are normalized the denominator is not needed); the result is an overestimate of E_0: of course, the smarter the choice of the family $| A ; \lambda_1 , \lambda_2 , \cdots , \lambda_n \rangle$, the better the estimate of E_0.

The estimate can be improved by increasing the 'degrees of freedom' of the family of vectors: for example, let $\{| k \rangle\}$ be a (either finite or infinite) set of *independent* vectors (not necessarily orthogonal to one another); the family

$$| A ; \lambda_1 , \lambda_2 , \cdots , \lambda_n \rangle = \sum_{k=1}^{n} \lambda_k | k \rangle$$

is the n-dimensional manifold $\mathcal{V}_n \subset \mathcal{H}$ generated by the first n vectors of the set and, being $\mathcal{V}_{n+1} \supset \mathcal{V}_n$, the estimate can only improve with increasing n.

The variational method can be also used to obtain an overestimate of the first excited level, provided the family $| A ; \lambda_1 , \lambda_2 , \cdots , \lambda_n \rangle$ is chosen *orthogonal* to the (even unknown) ground state: this can be achieved if, for example, it is a priori known that the ground state is even (under spatial inversion) and the first excited state is odd (as it happens in one-dimensional problems) or that they, in any event, belong to different eigenvalues of an operator whose eigenvectors can be easily characterized (as those of the spatial inversion).

We conclude the present section with the demonstration of the quantum version of the **virial theorem**, that we will use in the next section to estimate, by means of the variational method, the energy of the ground state of the helium atom.

For a system with n degrees of freedom, let

$$H = \sum_{i=1}^{n} \frac{p_i^2}{2m_i} + V(q_1, \cdots, q_n) \equiv T + V$$

where $V(q_1, \cdots, q_n)$ is a homogeneous function of degree k:

$$\sum_{i=1}^{n} x_i \frac{\partial V}{\partial x_i} = k V(x_1, \cdots, x_n) .$$

Thanks to the properties of commutators (4.48) and (4.54) one has

$$\sum_{i=1}^{n} [p_i q_i , H] = \sum_{i=1}^{n} \left(p_i [q_i , T] + [p_i , V] q_i \right) = i \hbar \sum_{i=1}^{n} \left(\frac{p_i^2}{m_i} - q_i \frac{\partial V}{\partial q_i} \right)$$
$$= i \hbar (2T - k V) .$$

If $| E \rangle$ is a generic (proper) eigenvector of H, one has

$$\sum_{i=1}^{n} \langle E | [p_i q_i , H] | E \rangle = 0 = 2\langle E | T | E \rangle - k \langle E | V | E \rangle \equiv 2\overline{T} - k\overline{V}$$

i.e. $2\overline{T} = k\overline{V}$ and, owing to $\overline{T} + \overline{V} = E$, finally:

$$\overline{T} = \frac{k}{2+k}\,E\,, \qquad\qquad \overline{V} = \frac{2}{2+k}\,E\,.$$

The Coulomb potential is homogeneous of degree -1, therefore $\overline{V} = 2E$.

In the case of the harmonic oscillator $k = 2$, whence $\overline{T} = \overline{V} = \frac{1}{2}E_n$ for all the energy levels, so (5.7), that refers to the ground state, only is a particular case.

14.3 The Lowest Energy Level of Helium

We now will estimate the energy of the ground state of the helium atom: we will start with a perturbative calculation to the first order, then we will make a variational calculation and compare the two results. It is convenient, not only for the sake of generality but also in view of the variational calculation, to consider atoms with two electrons: the hydrogen ion H$^-$ (hydrogen with two electrons), helium, the lithium (Li$^+$) ionized once, then Be^{++} etc.

The Hamiltonian of an atom with two electrons in the central field of the nucleus with charge $Z\,e$ is

$$H = \frac{\vec{p}_1^{\,2}}{2m} - \frac{Z\,e^2}{r_1} + \frac{\vec{p}_2^{\,2}}{2m} - \frac{Z\,e^2}{r_2} + \frac{e^2}{r_{12}}\,, \tag{14.9}$$

in which we will consider the last term as the perturbation:

$$H = H_0 + H'\,, \qquad\qquad H' = \frac{e^2}{r_{12}}\,. \tag{14.10}$$

The unperturbed ground state is given by (14.5) and, since it is nondegenerate, the first-order perturbative correction is

$$\begin{aligned}
\delta E_0 &= \langle\,(1\,0\,0),(1\,0\,0)\mid \frac{e^2}{r_{12}}\mid (1\,0\,0),(1\,0\,0)\,\rangle \\
&= \iint \psi_{100}(\vec{r}_2)^*\,\psi_{100}(\vec{r}_1)^*\,\frac{e^2}{r_{12}}\,\psi_{100}(\vec{r}_1)\,\psi_{100}(\vec{r}_2)\,\mathrm{d}V_1\,\mathrm{d}V_2 \\
&= \iint |\psi_{100}(\vec{r}_2)|^2\,\frac{e^2}{r_{12}}\,|\psi_{100}(\vec{r}_1)|^2\,\mathrm{d}V_1\,\mathrm{d}V_2
\end{aligned}$$

where, thanks to (11.8), the first of (10.42) and the first of (11.41)

$$\psi_{100}(r,\theta,\phi) = \sqrt{\frac{1}{4\pi}}\left(\frac{Z}{a_{\mathrm{B}}}\right)^{3/2} 2\,\mathrm{e}^{-Z\,r/a_{\mathrm{B}}} \tag{14.11}$$

and therefore

$$\delta E_0 = \left(\frac{e\,Z^3}{\pi\,a_{\mathrm{B}}^3}\right)^2 \iint \mathrm{e}^{-2Z\,r_2/a_{\mathrm{B}}}\,\frac{1}{r_{12}}\,\mathrm{e}^{-2Z\,r_1/a_{\mathrm{B}}}\,\mathrm{d}V_1\,\mathrm{d}V_2\,. \tag{14.12}$$

The integral in (14.12) is the electrostatic energy of the charge distribution

$$\rho(r) = -e\,|\psi_{100}(r,\theta,\phi)|^2 = \frac{-e\,Z^3}{\pi\,a_B^3}\,e^{-2Z\,r/a_B} \equiv -\frac{e\,\kappa^3}{8\pi}\,e^{-\kappa\,r}, \quad \kappa = \frac{2Z}{a_B}$$

in the field of a second distribution identical with the first one; so it can be calculated by the methods of electrostatics: the field $E(r)$ generated by the distribution $\rho(r)$ is calculated by means of Gauss theorem (the integrals of the type $\int r^n \exp(-\kappa\,r)\,\mathrm{d}r$ are evaluated as $(-1)^n\,\mathrm{d}^n/\mathrm{d}\kappa^n \int \exp(-\kappa\,r)\,\mathrm{d}r$)

$$E(r) = \frac{1}{r^2}\int_0^r \rho(r')\,\mathrm{d}V' = e\left(\frac{e^{-\kappa\,r}-1}{r^2} + \frac{\kappa}{r}e^{-\kappa\,r} + \frac{\kappa^2}{2}e^{-\kappa\,r}\right)$$

and the potential is (use that $\int(1/r^2 - \kappa/r)\exp(-\kappa\,r)\,\mathrm{d}r = -\exp(-\kappa\,r)/r$)

$$\varphi(r) = e\left(\frac{e^{-\kappa\,r}-1}{r} + \frac{\kappa}{2}e^{-\kappa\,r}\right) \tag{14.13}$$

and finally (also in the last step only integrals of the type $\int r^n \exp(-\kappa\,r)\,\mathrm{d}r$ occur)

$$\delta E_0 = \int \rho(r)\,\varphi(r)\,\mathrm{d}V = -\frac{e^2\,\kappa^3}{8\pi}\times 4\pi\times\left(-\frac{5}{8\kappa^2}\right) = \frac{5}{8}\frac{Z\,e^2}{a_B} \tag{14.14}$$

and in conclusion the energy of the ground state of the atoms with two electrons is:

$$E_0^{(1)} = E_0 + \delta E_0 = -2\times\frac{Z\,e^2}{2a_B} + \frac{5}{8}\frac{Z\,e^2}{a_B} = -\left(Z^2 - \frac{5}{8}Z\right)\frac{e^2}{a_B}. \tag{14.15}$$

The latter is the opposite of the energy necessary to *completely* ionize the atom. The first ionization is therefore:

$$E_I = -E_0^{(1)} - Z^2\frac{e^2}{2a_B} = \left(\frac{1}{2}Z^2 - \frac{5}{8}Z\right)\frac{e^2}{a_B} \tag{14.16}$$

where $e^2/a_B = 2\times 13.6 = 27.2\,\mathrm{eV}$.

Equation (14.16) exhibits that the first-ionization energy of He to the first order is as anticipated, i.e. 20.4 eV. It also exhibits that the experimentally established existence of the ion H^- (that possesses only one bound state) is not predicted by the perturbative theory: indeed, for $Z = 1$ the first-ionization energy turns out to be negative ($-3.4\,\mathrm{eV}$), which means that, if one stops at the first order of perturbation theory, the hydrogen atom cannot bind a second electron.

We now make the variational calculation. First of all, we have to decide which trial functions should be considered: in this choice we follow an idea that is physical in character: each of the two electrons partially screens the charge of the nucleus, so on the average the other electron is subject to the (attractive) action of a charge $Z' < Z$, and this symmetrically holds for both electrons. The simplest way to implement this idea consists in taking as trial function the wavefunction of two electrons in the orbital $1s$ of an hydrogen-like atom endowed with a nuclear charge $Z'e$, and in considering Z' as the variational parameter with respect to which the mean value of the Hamiltonian (14.9) is to be minimized:

$$| A; Z' \rangle \rightarrow \Psi(\vec{r}_1, \vec{r}_2; Z') = \frac{1}{4\pi} \left(\frac{Z'}{a_{\mathrm{B}}} \right)^3 4\,e^{-Z' r_1/a_{\mathrm{B}}}\, e^{-Z' r_2/a_{\mathrm{B}}} \quad (14.17)$$

i.e. $\Psi(\vec{r}_1, \vec{r}_2; Z')$ is the product of two functions of the type (14.11). In order to calculate the mean value of (14.9) we proceed in the following way: first of all, the mean value of the Coulombic repulsion term is given by (14.14) with Z' replacing Z; then we write

$$H_0 = \left(\frac{\vec{p}_1^2}{2m} - \frac{Z'\,e^2}{r_1} + \frac{\vec{p}_2^2}{2m} - \frac{Z'\,e^2}{r_2} \right) - \left(\frac{(Z - Z')\,e^2}{r_1} + \frac{(Z - Z')\,e^2}{r_2} \right) . \quad (14.18)$$

The trial function (14.17) is an eigenfunction of the first term of (14.18) corresponding to the eigenvalue $E_0^{Z'} = 2 \times (-Z'^2\,e^2/2a_{\mathrm{B}})$ that, as a consequence, coincides with the mean value; in order to calculate the mean value of the second term of (14.18) we exploit the virial theorem according to which the mean value of the potential energy

$$-\frac{Z'\,e^2}{r_1} - \frac{Z'\,e^2}{r_2}$$

is twice the value of the (total) energy $E_0^{Z'}$: so the mean value of the second term of (14.18) is given by

$$\frac{2\,E_0^{Z'}}{Z'}\,(Z - Z') = -Z'\,(Z - Z')\frac{2e^2}{a_{\mathrm{B}}} .$$

Collecting all the contributions:

$$\langle A; Z' \mid H \mid A; Z' \rangle = -Z'^2 \frac{e^2}{a_{\mathrm{B}}} - 2Z'\,(Z - Z')\frac{e^2}{a_{\mathrm{B}}} + \frac{5}{8}Z'\frac{e^2}{a_{\mathrm{B}}}$$

$$= \left(Z'^2 - 2Z\,Z' + \frac{5}{8}\,Z' \right)\frac{e^2}{a_{\mathrm{B}}}$$

that has its minimum for $Z' = Z - 5/16$; therefore

$$E_0 \leq -\left(Z - \frac{5}{16} \right)^2 \frac{e^2}{a_{\mathrm{B}}} \quad (14.19)$$

and for the first-ionization energy one has

$$E_{\mathrm{I}} \geq \left(\frac{1}{2}Z^2 - \frac{5}{8}Z + \frac{25}{256} \right)\frac{e^2}{a_{\mathrm{B}}} . \quad (14.20)$$

In the case of He, E_{I} increases from 20.4 to 23.1 eV, whereas the H^- ion carries on not being bounded ($E_{\mathrm{I}} = -0.74$ eV against the experimental result, about $+0.5$ eV).

14.4 The Alkali Atoms

The alkali atoms are:

$$\mathrm{Li}\,(Z=3), \quad \mathrm{Na}\,(Z=11), \quad \mathrm{K}\,(Z=18), \quad \mathrm{Rb}\,(Z=37), \quad \mathrm{Cs}\,(Z=55)$$

and the ground-state configurations have the form: (complete shells) ns. For example:

$$\mathrm{Na}:\; (1s)^2(2s)^2(2p)^6\,3s\;; \qquad \mathrm{K}:\; (1s)^2(2s)^2(2p)^6(3s)^2(3p)^6\,4s\;. \qquad (14.21)$$

Note that in the case of potassium (K), in agreement with the sequence (14.7), once the $3p$ shell has been filled, the last electron goes in the $4p$ shell instead of the $3d$.

The alkali atoms have one electron more than the noble gases He, Ne, Ar, Kr, Xe, and this fact explains many properties, like the fact that one only electron – the 'extra' one – determines both its physical and chemical properties. From the chemical point of view the extra electron is the valence electron (the alkali atoms have valence 1); it is also called the **optical electron** inasmuch as it is responsible for optical transitions, whereas the remaining $Z-1$ electrons, that make up what is called the **atomic core**, do not take part in optical transitions: indeed they occupy a particularly stable configuration, that of the noble gas with $Z-1$ electrons, but even more stable than the latter because of the higher nuclear charge: for example, while the first-ionization energy of He is 24.6 eV, from (14.20) it appears that the one relative to $\mathrm{Li^+}$ is ≥ 74 eV (the experimental value is 75.6 eV).

The first–ionization energies of the alkali atoms are the lowest among all the elements: they decrease from the 5.4 eV of Li to the 3.9 eV of Cs.

We now intend to investigate, at least qualitatively, the structure of the levels of alkali atoms and, therefore, how the emission and absorption spectra of vapours of alkali atoms should appear: thanks to the stability of the atomic core (namely an energy of the order of some ten electronvolt is necessary to excite one of its electrons), in all the excited states of an alkali atom undergoing an optical transition only the optical electron intervenes and it is exactly such levels – in which only the optical electron is excited – that we now have in mind to investigate.

We have already seen that, in general, the problem of determining – even by a simple perturbative calculation – the energy levels of an atom is a problem in which all the Z electrons must be considered along with the Coulombic repulsion term: the simplicity of alkali atoms precisely consists in the fact that for them the problem can be brought back, at least to some extent, to that of one only electron – the optical one – in a central force field. The approximation consists in considering the atomic core as a rigid distribution of electric charge spherically symmetric around the nucleus: the electron moves in the field generated by the nucleus and by such a distribution while the effect that the electron can have on the distribution, i.e. any effect of deformation (or polarization) of the atomic core, is neglected. In other words: the atomic core influences the motion of the optical electron, but is not influenced by the latter.

For the optical electron we will have the Hamiltonian

$$H = \frac{\vec{p}^{\,2}}{2m} - \frac{Z\,e^2}{r} + V(r) \tag{14.22}$$

where $V(r)$ is the central potential generated by the atomic core. In the next section we will show how and within which approximations (14.22) can be derived from (14.1), giving a rigorous formal justification to the arguments that have led us to (14.22).

In order to discuss the consequences of (14.22) it is necessary to know $V(r)$, i.e. the charge distribution $\rho(r)$ of the atomic core; as we will see in the next section, $\rho(r)$ itself is known provided the wavefunction of the $Z-1$ electrons of the atomic core is known: but obviously the latter is not known. We will therefore limit ourselves to draw from (14.22) some conclusions that only are qualitative in character, based on the information we have about $\rho(r)$ and, therefore, on $V(r)$. We indeed know that:

i) $\int \rho(r)\,\mathrm{d}\vec{r} = -(Z-1)\,e$;

ii) the density $\rho(r)$ is mainly concentrated within a sphere of radius R_0, of the order of one ångström, that corresponds to the size of the ionized alkali atom; for $r \gtrsim R_0$ we expect $\rho(r)$ to be very small and to decrease exponentially (as the radial probability densities of Fig. 11.3).

It then follows from Gauss theorem that

$$U(r) \equiv -\frac{Z\,e^2}{r} + V(r) \approx \begin{cases} -\dfrac{e^2}{r} & \text{for } r \gtrsim R_0 \\[2mm] -\dfrac{Z\,e^2}{r} & \text{for } r \to 0 \,. \end{cases} \tag{14.23}$$

Given the form of $U(r)$ (Fig. 14.2 refers to the case $Z=3$), let us see what can be expected, on intuitive grounds, about the levels of the optical electron.

Let us reason à la Bohr: if the electron went around orbits external to the atomic core, by Gauss theorem it always would be in the Coulombic field of a point-like charge e and its levels would exactly be those of the hydrogen atom, with the relative degeneracy on l coming about. Deviations from such a

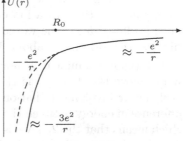

Fig. 14.2

situation must be expected for small values of the quantum numbers n ad l: indeed, for a given n, the orbits with a lesser l are more eccentric ellipses that, therefore, penetrate the atomic core to a larger extent ('penetrating orbits'); as n is decreased, the major axis in the ellipses decreases, the orbits become smaller and smaller and, as a consequence, the electron 'spends more time' within the atomic core. But, on the other side, it is clear that the more the electron penetrates the atomic core, the higher is the charge it sees: no longer the charge e it sees when it is outside the core, but the charge $Z e$ of the nucleus decreased by the charge relative to the part of the atomic core that is at a distance from the nucleus smaller than the distance of the optical electron itself (again Gauss theorem). The effect of this is that, always in comparison with the case of an electron in presence of a point-like charge e, the optical electron experiences a higher attractive force, i.e. it is more bound, i.e. its energy is smaller. And the more the orbit penetrates and the longer the time the electron spends within the atomic core, the higher the effect. So it must happen that, for a given n, the states with different l (from $l = 0$ to $l = n-1$) are no longer degenerate in energy, but those with smaller l (more penetrating orbits) have lesser energy: therefore all the levels of an alkali atom with a given value of n have an energy smaller than the level of hydrogen with the same n. Such differences in energy among levels with the same n and different l, as well as the differences with the corresponding levels in hydrogen, become smaller and smaller as n increases.

For now we will content ourselves with this picture à la Bohr, postponing to the next section a truly quantum treatment of the problem, that will lead to the same results.

By comparing with the experimental results on the energies of alkali atoms we can calculate how much the energy levels are lowered as a consequence of the penetration of the atomic core. Let us consider, for example, the lithium atom: the ionization energy is 5.4 eV, so the energy of the lowest energy level is −5.4 eV: if the effect of penetration is neglected, the energy of the lowest energy level of Li should be that of the $n = 2$ level of H (we recall that Pauli principle demands that the optical electron of Li has $n = 2$), namely $−13.6/4 = −3.4$ eV. So, due to penetration, the lowest energy level of Li is lowered by about 2 eV.

The effect is even more considerable in Na, due to the higher nuclear charge: indeed, the ionization energy of Na is 5.14 eV, to be compared with the $13.6/9 = 1.5$ eV one would have by neglecting penetration; therefore in this case the lowering is about 3.6 eV.

Let us now examine how much is the effect on the excited levels with $l \neq 0$. In Li, above the 2s there is the 2p level: the transitions between these two levels occur involving radiation with wavelength $\lambda = 6707$ Å, therefore the difference in energy between the 2p and the 2s levels is $12400/6707 \simeq 1.85$ eV, which means that the 2p level is lowered only by about 0.15 eV. But in Na the difference between the 3p (the first excited level) and the 3s (lowest energy level) is about 2.1 eV, so the lowering of the 3p level is about 1.5 eV, about ten times the result found for the 2p in Li! The main reason for this fact is that 2p is a circular orbit, since it has the maximum value for l compatible with $n = 2$, whereas 3p is an elliptical, i.e. more penetrating, orbit. Indeed, if we consider the 3d level in Na, corresponding to a circular orbit, one finds that the effect of penetration is smaller than one tenth of electronvolt. In this respect it is interesting to note that already in Na the $4s$ level is lower than the $3d$: since it is a level with $l = 0$, the effect of penetration is sufficient, in spite of the higher value of n, to make it overstep the $3d$ level. It is exactly this effect due to penetration that determines the energy sequence of orbitals reported in (14.7) and, in particular, the inversion between $4s$ and $3d$.

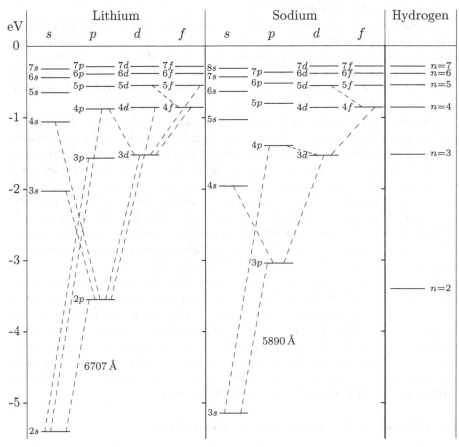

Fig. 14.3

In Fig. 14.3 the diagrams of the energy levels of Li and Na are reported, side to side with those of H, for the sake of comparison.

The transversal dashed lines represent the possible and observed transitions: the transitions from the p level to the s lowest energy level ($2s$ for Li, $3s$ for Na) form the **principal series** (p-series), those from the s levels to the lowest p level ($2p$ for Li, $3p$ for Na) form the **sharp series** (s-series), the transitions $s \to d$ give rise to the **diffuse series** (d-series), finally those $f \to d$ form the **fundamental series** (f-series): this is the terminology of spectroscopists that gave rise to the use of representing by the letters $s, p, d, f, \cdots$ the states with angular momentum $l = 0, 1, 2, 3, \cdots$.

Note that all the transitions reported in Fig. 14.3 occur only between levels belonging to adjacent columns: even if only some of the observed transitions are reported, it is experimentally established that in the vapours of the alkali metals the transitions take place only between s and p, p and d-and-s, d and p-and-f, i.e. only between states that differ in the value of l by one unit. This fact is in agreement with the selection rule (13.51) $\Delta l = \pm 1$ that holds for systems with only one electron.

Another important aspect of the spectra of vapours of the alkali metals is the fact that, if they are analyzed by a spectroscope of a sufficiently high resolution power, the observation shows that a **fine structure** is displayed by the lines of such spectra: namely one sees that the above lines are not simple lines, but the consist of two or three lines very close to each other. This effect is not very evident in Li, but it is so in the other alkali metals.

So, for example, in Na the transition from $3p$ to $3s$ that, according to Fig. 14.3, should give rise to a single line with $\lambda \simeq 5890\,\text{Å}$ (the well known yellow light of sodium lamps), actually gives rise to two lines with $\lambda_1 \simeq 5890\,\text{Å}$ and $\lambda_2 \simeq 5896\,\text{Å}$: the resolution power necessary to distinguish this **doublet** must be $\Delta\nu/\nu = \Delta\lambda/\lambda \simeq 6/6000 = 10^{-3}$. Things go in such a way as if the level $3p$ were not a single level (as it comes out from the theory so far developed), but a doublet of levels distant about $2\times10^{-3}\,\text{eV}$ ($\Delta E/E = \Delta\lambda/\lambda$) from each other. Experiment shows that all the lines of the p and s series are doublets whereas the other ones (d and f series) are triple lines; however in the latter case the fine structure is somewhat harder to detect.

In concluding this section we may state that the theory enables us to understand at least the 'gross' features of the spectra of the vapours of alkali metals; experiment then shows the existence of a fine structure, that the theory so far developed does not even qualitatively account for: in order to explain the fine structure one should admit that the alkali atoms possess more energy levels, i.e. in turn more states, than those predicted by the theory. As already announced, this is one of the problems that will be solved by attributing further degrees of freedom to the electron: the spin.

14.5 Quantum Treatment of the Alkali Atoms

The aim of this section is to translate in a formal way the treatment we have given in the previous section, on intuitive grounds, of the alkali atoms.

It is certainly true that we will find by a more complicated way some results that can be considered as already achieved, basing on the essential physical aspects of the problem; but translating the problem in formal terms opens the possibility of a quantitative, albeit approximate, treatment. Furthermore this subject allows us to understand the meaning and the limits of the concept of atomic core that, at least in the first approximation, is applied not only to the noble-gas-like configurations, but also to complete-shell configurations (as, for instance, the electrons of the shells $(1s)^2(2s)^2$ in carbon).

Let us start from the Hamiltonian (14.2). A perturbative calculation in which the term $\sum e^2/r_{\alpha\beta}$ is considered as the perturbation cannot be very meaningful, at least for atoms with a somewhat large value of Z (Na, K, ...): indeed, the number of terms in the summation is $Z(Z-1)/2$, i.e. it grows fast with Z and each term carries a contribution that can be estimated as e^2/a_B, i.e. about $20\,\text{eV}$. Therefore the first-order corrections are too large to consider the perturbative method as reliable.

Therefore we will proceed in a different way, trying to implement the atomic-core approximation of the previous section. Le us rewrite (14.2) in the following way:

$$H = H_1^0 + H' + \sum_{\alpha=2}^{Z} \frac{e^2}{r_{1\alpha}} \qquad (14.24)$$

where the variables with index 1 refer to the optical electron and H' is the Hamiltonian of the $Z-1$ electrons (those belonging to the atomic core). The term $\sum e^2/r_{1\alpha}$ represents the interaction between the optical electron and the electrons of the atomic core and, since it contains the variables of all the electrons in a non separable way, it generates correlations between the optical electron and the other ones: so it is only possible to talk about states of the entire system and not about states of the optical electron alone. In other words, the exact eigenfunctions of H cannot be factorized into the product of a wavefunction relative to the optical electron times a wavefunction relative to the electrons of the atomic core.

Let now $\Phi(\vec{x}_2, \cdots, \vec{x}_Z)$ be the eigenfunction of H' corresponding to the lowest energy E':

$$H'\Phi = E'\Phi . \qquad (14.25)$$

The wavefunction Φ is therefore the wavefunction of the ground state of the ionized alkali atom, i.e. of the atomic core (as we will see in the next chapter, the ground state of the ionized alkali atom is a state with angular momentum $L=0$, therefore the eigenvalue E' is nondegenerate).

Let us introduce the quantity

$$V(\vec{x}_1) = \sum_{\alpha=2}^{Z} e^2 \int \frac{|\Phi(\vec{x}_2, \cdots, \vec{x}_Z)|^2}{|\vec{x}_1 - \vec{x}_\alpha|} \, d\vec{x}_2 \cdots d\vec{x}_Z \qquad (14.26)$$

and examine its physical meaning: $|\Phi(\vec{x}_2, \cdots, \vec{x}_Z)|^2$ integrated over all the variables but the α-th is the position probability density $\rho^{(\alpha)}(\vec{x}_\alpha)$ for the α-th electron: as a consequence, $-e\,\rho^{(\alpha)}(\vec{x})$ has the meaning of charge density in the point $\vec{x}$ due to the α-th electron and

$$-e\,\rho(\vec{x}) = -e \sum_{\alpha=2}^{Z} \rho^{(\alpha)}(\vec{x})$$

is the charge distribution in the point $\vec{x}$ due to all the electrons of the atomic core. Equation (14.26) can be written as

$$V(\vec{x}_1) = \sum_{\alpha=2}^{Z} e^2 \int \frac{\rho^{(\alpha)}(\vec{x}_\alpha)}{|\vec{x}_1 - \vec{x}_\alpha|} \, d\vec{x}_\alpha = e^2 \int \frac{\rho(\vec{x})}{|\vec{x}_1 - \vec{x}|} \, d\vec{x} \qquad (14.27)$$

and is the potential energy of the optical electron in the field produced by the charge distribution of the atomic core assumed as rigid, i.e. not deformed by the addition of the optical electron.

Since the ground state of the ionized alkali atom has angular momentum $L = 0$, the wavefunction $\Phi(\vec{x}_2, \cdots, \vec{x}_Z)$ is invariant under rotations: this means that if $\vec{x}_2', \cdots, \vec{x}_Z'$ are obtained by operating the same rotation on all the coordinates $\vec{x}_2, \cdots, \vec{x}_Z$, one has

$$\Phi(\vec{x}_2', \cdots, \vec{x}_Z') = \Phi(\vec{x}_2, \cdots, \vec{x}_Z) \ . \tag{14.28}$$

This entails that any $\rho^{(\alpha)}(\vec{x})$ is rotationally invariant, i.e it only depends on r. Let us indeed consider $\rho^{(2)}(\vec{x})$. Owing to (14.28) and $\mathrm{d}\vec{x}_3 \cdots \mathrm{d}\vec{x}_Z = \mathrm{d}\vec{x}_3' \cdots \mathrm{d}\vec{x}_Z'$ (recall that the Jacobian of the transformation from $\vec{x}_3, \cdots, \vec{x}_Z$ to $x_3', \cdots, \vec{x}_Z'$ is 1), one has

$$\rho^{(2)}(\vec{x}) = \int \left| \Phi(\vec{x}, \vec{x}_3, \cdots, \vec{x}_Z) \right|^2 \mathrm{d}\vec{x}_3 \cdots \mathrm{d}\vec{x}_Z = \int \left| \Phi(\vec{x}', \vec{x}_3', \cdots, \vec{x}_Z') \right|^2 \mathrm{d}\vec{x}_3 \cdots \mathrm{d}\vec{x}_Z$$

$$= \int \left| \Phi(\vec{x}', \vec{x}_3', \cdots, \vec{x}_Z') \right|^2 \mathrm{d}\vec{x}_3' \cdots \mathrm{d}\vec{x}_Z' = \rho^{(2)}(\vec{x}') \ .$$

So also $\rho(\vec{x})$ is spherically symmetric and as a consequence $V(\vec{x}) = V(r)$. It is now convenient to rewrite (14.24) in the following way:

$$H = H' + H_1^0 + V(r_1) + \left(-V(r_1) + \sum_{\alpha=2}^{Z} \frac{e^2}{r_{1\alpha}} \right) \ . \tag{14.29}$$

Equation (14.29) just is (14.2) written in a different way: so far we have made no approximation. Now we intend to consider the term

$$-V(r_1) + \sum_{\alpha=2}^{Z} \frac{e^2}{r_{1\alpha}} \equiv V'(\vec{x}_1, \cdots, \vec{x}_Z) \tag{14.30}$$

as the perturbation. As a consequence, the unperturbed Hamiltonian is

$$H^0 = H' + H_1^0 + V(r_1) \tag{14.31}$$

that is one displaying separate variables: $H_1^0 + V(r_1)$ only depends on the variables of the optical electron, H' on the variables of the other electrons. Therefore a complete set of eigenfunctions of H^0 consists in the products of eigenfunctions of $H_1^0 + V(r_1)$ times the eigenfunctions of H'. In this approximation, in which the V' given by (14.30) is neglected, one is allowed to speak about the state of the optical electron and the state of the atomic core as of two uncorrelated things. Furthermore, if we are interested in the states of the atom in which the atomic core is not excited – and is therefore described by the eigenfunction $\Phi(\vec{x}_2, \vec{x}_3, \cdots, \vec{x}_Z)$ – we must only find the eigenfunctions of $H_1^0 + V(r_1)$, which exactly is the Hamiltonian (14.22) of the previous section.

What we have gained is that we now know what we have neglected in writing (14.22), so we are also able to calculate its effect. Assuming that we know both $\Phi(\vec{x}_2, \vec{x}_3, \cdots, \vec{x}_Z)$ and the eigenfunctions $\psi(\vec{x}_1)$ of $H_1^0 + V(r_1)$, let us calculate the effect of $V'(\vec{x}_1, \cdots, \vec{x}_Z)$ by means of perturbation theory.

Let us first consider the matrix elements of V' between eigenstates of H^0 of the type

$$\Psi_1(\vec{x}_1, \cdots, \vec{x}_Z) = \psi_1(\vec{x}_1)\, \Phi(\vec{x}_1, \cdots, \vec{x}_Z)$$
$$\Psi_2(\vec{x}_1, \cdots, \vec{x}_Z) = \psi_2(\vec{x}_1)\, \Phi(\vec{x}_1, \cdots, \vec{x}_Z) \tag{14.32}$$

where ψ_1 and ψ_2 are two eigenfunctions of $H_1^0 + V(r_1)$, possibly equal to each other. One has

$$\int \Psi_1^*\, V(\vec{x}_1)\, \Psi_2\, \mathrm{d}\vec{x}_1 \cdots \mathrm{d}\vec{x}_Z = \int \psi_1^*(\vec{x}_1) V(r_1)\, \psi_2(\vec{x}_1)\, \mathrm{d}\vec{x}_1$$

$$= \sum_{\alpha=2}^{Z} e^2 \int \psi_1^*(\vec{x}_1) \frac{\left|\Phi(\vec{x}_2, \cdots, \vec{x}_Z)\right|^2}{|\vec{x}_1 - \vec{x}_\alpha|}\, \psi_2(\vec{x}_1)\, \mathrm{d}\vec{x}_1 \cdots \mathrm{d}\vec{x}_Z$$

$$= \sum_{\alpha=2}^{Z} \int \Psi_1^*(\vec{x}_1, \cdots, \vec{x}_Z) \frac{e^2}{|\vec{x}_1 - \vec{x}_\alpha|}\, \Psi_2(\vec{x}_1, \cdots, \vec{x}_Z)\, \mathrm{d}\vec{x}_1 \cdots \mathrm{d}\vec{x}_Z$$

whence

$$\int \Psi_1^* \left(\sum_{\alpha=2}^{Z} \frac{e^2}{r_{1\alpha}} - V(r_1) \right) \Psi_2\, \mathrm{d}\vec{x}_1 \cdots \mathrm{d}\vec{x}_Z \equiv \int \Psi_1^*\, V'\, \Psi_2\, \mathrm{d}\vec{x}_1 \cdots \mathrm{d}\vec{x}_Z = 0$$

for any Ψ_1, Ψ_2 with the form (14.32). As a consequence, only the matrix elements of V' between eigenfunctions of H^0 that differ in the eigenfunction of H' – i.e. in the eigenfunction of the atomic core – can be nonvanishing. So not only the first-order perturbative contribution of V' is vanishing, but the matrix elements of V' that may contribute to higher orders are divided by large energy differences (see (12.26)), inasmuch as excitation energies of the atomic core.

Once the validity of (14.22) is established, let us see how quantum mechanics allows us to find again the results of the previous section about the energy levels of the alkali atoms. To this end we rewrite the Hamiltonian (14.22): $H = \vec{p}^2/2m - Z\, e^2/r + V(r)$, as

$$H = \frac{\vec{p}^2}{2m} - \frac{e^2}{r} + U'(r), \qquad U'(r) \equiv V(r) - \frac{(Z-1)\, e^2}{r}. \tag{14.33}$$

$U'(r)$ is the difference between the potential energy of the optical electron in the field of the atomic core (charge $+(Z-1)\, e$) and that in the field of a point-like charge $\left(-(Z-1)\, e\right)$ in the center of the distribution: so $U'(r) \simeq 0$ for $r \gtrsim R_0$ and is nonvanishing for $r \lesssim R_0$ only because the charge of the atomic core has a finite extension; it therefore accounts for the penetration of the optical electron inside the atomic core. The qualitative behaviour of

$U'(r)$ can be deduced from Fig. 14.2 as the difference between the solid curve and the dashed one and is reported in Fig. 14.4.

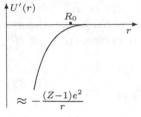

Fig. 14.4

Let us now treat $U'(r)$ as a perturbation: we must calculate the effect of $U'(r)$ on the levels of a hydrogen atom. To the first order, since the matrix elements of $U'(r)$ between states with different values of l and/or m are vanishing, we must calculate the mean values

$$\Delta E_{nl} = \langle\, n,\, l,\, m \mid U'(r) \mid n,\, l,\, m \,\rangle \qquad (14.34)$$

that immediately provide us with the first-order corrections to the energy levels. Note that such corrections ΔE_{nl} may depend on n and l, but not on m because $U'(r)$, being spherically symmetric, cannot remove the degeneracy on m.

In the Schrödinger representation the expression for the mean values (14.34) is

$$\Delta E_{nl} = \int_0^\infty r^2 \left| R_{nl}(r) \right|^2 U'(r)\, dr = \int_0^\infty \left| u_{nl}(r) \right|^2 U'(r)\, dr \qquad (14.35)$$

where the $u_{nl}(r)$ are the reduced radial functions defined by (11.10). Since $U'(r)$ practically is nonvanishing only for $r < R_0$, the integral in (14.35) is determined by the behaviour of $|u_{nl}(r)|^2$ in this interval and, in addition, given that $U'(r)$ diverges for $r \to 0$, , the behaviour of $|u_{nl}(r)|^2$ near the origin gains in relevance.

We know from Sect. 11.3 that $|u_{nl}(r)|^2 \simeq r^{(2l+2)}$ for $r \to 0$ and therefore the higher the value of l, the faster its decrease: this entails that ΔE_{nl}, that is negative definite, decreases in absolute value as l increases. It is likewise clear that $|\Delta E_{nl}|$ decreases as n increases because, as n increases, the probability distribution $|u_{nl}(r)|^2$ is more and more concentrated around higher values of r, as it appears in Fig. 14.5 (in the next page) in which we have reported the radial probability densities $|u_{nl}(r)|^2$ for $n = 2$ and $n = 3$: all the curves are drawn at the same scale, in order to facilitate the comparison. It helps to have in mind that in Li the size R_0 of the atomic core is about $0.5\,\text{Å}$ whereas in Na $R_0 \simeq 0.8\,\text{Å}$: the comparison should be done among curves $|u_{nl}(r)|^2$ with different values of l and fixed n, as well as among those with $n = 2$ and $n = 3$ for a given l, paying attention only to the region $r < R_0$. Note in particular that where $|u_{20}|^2$ has its first maximum (the only one relevant for (14.34)), $|u_{21}|^2$ still is practically vanishing; observe, in addition, that the maximum of $|u_{30}|^2$ is lower then that of $|u_{20}|^2$.

The translation of Bohr language into that of quantum mechanics should by now be clear: for fixed major axis, i.e. for given n, the more penetrating orbits correspond to the states for which the probability of finding the optical electron close to the nucleus is higher (behaviour r^{2l} of $|\psi|^2$, occurrence of the maximum close to the origin); for fixed value of the angular momentum l

the orbits with higher n correspond to the states for which the probability of finding the electron far from the nucleus is higher, i.e. to the wavefunctions that, even having the same behaviour at the origin, give rise to lower maxima of $|\psi|^2$ close to the origin.

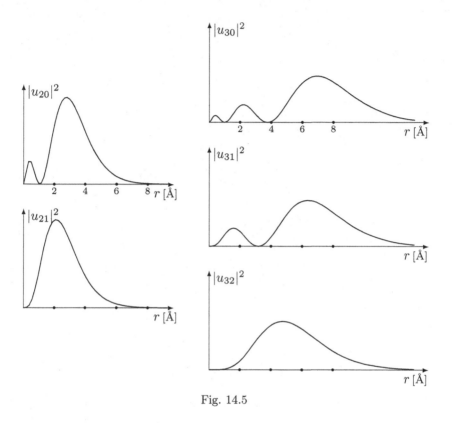

Fig. 14.5

Chapter 15

Composite Systems. Pauli Principle

15.1 The Hilbert Space for a System with Many Particles: Tensor Product of Hilbert Spaces

What we are going to discuss now only is a problem of language and notation.

Let 1 and 2 be two particles. The dynamical variables of particle 1 are q_1, p_1 and their functions $f(q_1, p_1)$, those of particle 2 are q_2, p_2 and $f(q_2, p_2)$. If we only have one of the particles, we know how to characterize its states: there are several ways for doing it, for example a state is characterized by being an eigenstate of one or more observables, or by its representatives in some representation, or also by the set of the mean values of all the observables (see Sect. 4.8). Things are unchanged if the system consists of both particles, only the set of observables changes, now being the set of all functions of q_1, p_1 and q_2, p_2: indeed we have found no particular problems in going from one-dimensional to three-dimensional systems, and here we are facing the same situation. It is however convenient, when dealing with a system consisting of two (or more) particles, to use terminology and notation that emphasize the fact that the system is made out of two (or more) subsystems. One reason for doing so is that, under certain circumstances, one is not interested in one of the two systems and can bring the study back to that of just one particle: this is what we have several times expressed by saying that the system is a 'separate variable' one (see Sects. 11.2, 13.4 and 14.5); one more reason is that there exist states of the composite system in which each of the two particles is in a well determined state and therefore, in such cases, one is allowed to talk about the state of a subsystem, while in general only talking about states of the entire system is meaningful.

The problem is the following: let $\mathcal{H}_1$ be the state space of particle 1, $\mathcal{H}_2$ that of particle 2 and $\mathcal{H}$ the state space of the composite system. Which is the relationship among $\mathcal{H}_1$, $\mathcal{H}_2$ and $\mathcal{H}$? We will face this problem in the Schrödinger representation, instead of doing it in the abstract, because it is simpler, but all the conclusions apply to all 'composite' systems for which the observables can be expressed as functions of two sets of observables such that those of a set commute with all those belonging to the other set, much as

© Springer International Publishing Switzerland 2016
L.E. Picasso, *Lectures in Quantum Mechanics*, UNITEXT for Physics,
DOI 10.1007/978-3-319-22632-3_15

either q_1, p_1 and q_2, p_2 for the system of two particles, to which we will now refer, or q, p and the spin variables, as we will see in the next section.

In the Schrödinger representation $\mathcal{H}_1$ and $\mathcal{H}_2$ are the space of square summable functions $\phi(x_1)$ and $\chi(x_2)$, $\mathcal{H}$ is the space of the square summable functions $\Psi(x_1, x_2)$. If $\phi_m(x_1)$ is an orthonormal basis in $\mathcal{H}_1$ and $\chi_n(x_2)$ is an orthonormal basis in $\mathcal{H}_2$, then $\phi_m(x_1) \chi_n(x_2)$ is an orthonormal basis in $\mathcal{H}$, i.e. for any $\Psi(x_1, x_2) \in \mathcal{H}$

$$\Psi(x_1, x_2) = \sum_{m\,n} c_{m\,n}\, \phi_m(x_1)\, \chi_n(x_2) \ . \tag{15.1}$$

Equation (15.1) expresses the fact that $\mathcal{H}$ is the **tensor product** of the factor spaces $\mathcal{H}_1$ and $\mathcal{H}_2$:

$$\mathcal{H} = \mathcal{H}_1 \otimes \mathcal{H}_2 \ .$$

In general the tensor product of two spaces $\mathcal{H}_1$ and $\mathcal{H}_2$ is defined in the following way: if $|m\rangle_1$ is a basis in $\mathcal{H}_1$ and $|n\rangle_2$ is a basis in $\mathcal{H}_2$, $\mathcal{H}$ is that space in which a basis is provided by the pairs $\{|m\rangle_1, |n\rangle_2\}$ and the scalar product is that induced in a 'natural' way by the scalar products in $\mathcal{H}_1$ and $\mathcal{H}_2$, as a consequence of the following notation: the vectors $\{|m\rangle_1, |n\rangle_2\}$ are denoted by $|m\,n\rangle$ or also $|m\rangle|n\rangle$ (the indices 1 and 2 are omitted, but the order is important: $|m\,n\rangle \neq |n\,m\rangle$) and the corresponding 'bras' by $\langle n\,m|$ or also $\langle n|\langle m|$; then the scalar product is defined by

$$\langle l\,k \mid m\,n \rangle \equiv \langle k \mid m \rangle \times \langle l \mid n \rangle = \delta_{km}\, \delta_{ln} \ .$$

The definition of $\mathcal{H}$ turns out to be independent (up to isomorphisms) of the choice of the bases in $\mathcal{H}_1$ and $\mathcal{H}_2$: if the bases one starts with are changed, the result is equivalent to a change of the basis in $\mathcal{H}$.

In $\mathcal{H}$ there are *both* vectors of the type

$$|X\rangle = \sum_{m\,n} a_m\, b_n\, |m\,n\rangle = \left(\sum_m a_m\,|m\rangle\right) \times \left(\sum_n b_n\,|n\rangle\right)$$
$$= |A\rangle\,|B\rangle \equiv |A\,B\rangle \tag{15.2}$$

that correspond to states in which the subsystem 1 is in the state $|A\rangle$ and the subsystem 2 is in the state $|B\rangle$ (such states are called **factorized states**), *and* vectors

$$|X\rangle = \sum_{m\,n} c_{m\,n}\,|m\,n\rangle \ , \qquad c_{m\,n} \neq a_m \times b_n \tag{15.3}$$

i.e. $|X\rangle = |A\,B\rangle + |C\,D\rangle + \cdots$, in which case one can only talk of the state of the whole system and not of states of the single subsystems: the latter are called **entangled states**.

If $\xi^{(1)}$ is a dynamical variable relative to subsystem 1 (e.g. q_1, or p_1) it also is a dynamical variable of the composite system: the operator on $\mathcal{H}$ associated with it acts only on the component m of the basis $|m\,n\rangle$ and, in the case of the factorized states $|X\rangle \equiv |A\rangle|B\rangle$, only on $|A\rangle$; formally, the operator on $\mathcal{H} = \mathcal{H}_1 \otimes \mathcal{H}_2$ associated with $\xi^{(1)}$ should be written as $\xi^{(1)} \otimes \mathbb{1}^{(2)}$, where $\xi^{(1)}$ acts on $\mathcal{H}_1$ and $\mathbb{1}^{(2)}$ is the identity operator on $\mathcal{H}_2$; likewise for the variables relative to subsystem 2. However, in order to keep the notation as 'light' as possible, we will simply write $\xi^{(1)}$ and $\eta^{(2)}$ (or even ξ_1, η_2). The notation $|A\rangle|B\rangle$, instead of $|AB\rangle$, will be used especially in connection with operators that act only on one subsystem ('single particle' dynamical variables or 'one body' operators).

15.2 The Spin of the Electron

Many experimental facts, apart from those already cited, lead in a more or less direct way to the necessity of attributing to the electron an **intrinsic magnetic moment** $\vec{\mu}_s$, in addition to the orbital magnetic moment (13.58). This means that the electron is not fully described by the variables $\vec{q}$ and $\vec{p}$, but other variables must be introduced in addition to these. Which are these new variables? To answer this question we assume that, in analogy with (13.58), the intrinsic magnetic moment $\vec{\mu}_s$ is connected with an **intrinsic angular momentum** or **spin** by means of the relation

$$\vec{\mu}_s = -g\, \frac{e}{2mc}\, \vec{s}\,, \qquad\qquad \vec{s}^{\,\dagger} = \vec{s} \tag{15.4}$$

where g is a numerical factor to be determined experimentally. Again in analogy with (13.58) we should be tempted to put $g = 1$. We will instead see that, in order to explain the anomalous Zeeman effect, we will have to assume $g \neq 1$. Furthermore, the formulation of quantum mechanics that embodies the relativity principle (Dirac equation) predicts that electron is endowed with the spin and, correspondingly, with a magnetic moment related to the spin by (15.4) with $g = 2$, in excellent agreement with the experimental facts $\big(g = 2.002\,319\,304\,3737(82)$: the so called "$g - 2$" is explained by quantum electrodynamics$\big)$.

Let us take $\vec{s}$ as the new dynamical variables for the electron. Since they represent an angular momentum, the following commutation relations must be satisfied:

$$[\,s_i\,,\,s_j\,] = \mathrm{i}\,\hbar\,\epsilon_{ijk}\,s_k\,. \tag{15.5}$$

We assume, in addition, that the s_i commute with $\vec{q}$ and $\vec{p}$:

$$[\,s_i\,,\,q_j\,] = 0\,, \qquad\qquad [\,s_i\,,\,p_j\,] = 0\,. \tag{15.6}$$

So the s_i are dynamical variables independent of the 'orbital' variables $\vec{q}$ and $\vec{p}$: it is for this reason that they are said to represent an intrinsic or *spin* angular momentum of the electron.

In order to explain the experimental facts we must admit that the components of $\vec{s}$ only have *two* different eigenvalues: such an assumption, we will shortly see, is directly connected with other "two" we have already met: the doublets of the alkali atoms and the two that appears in the Pauli principle formulation we have given in Sect. 14.1.

If the components of $\vec{s}$, which are components of an angular momentum, have only two eigenvalues, these must necessarily be $+\frac{1}{2}\hbar$ and $-\frac{1}{2}\hbar$: this entails that each s_i^2 only has the eigenvalue $\frac{1}{4}\hbar^2$ and is, as a consequence, a multiple of the identity. Therefore also the operator

$$\vec{s}^{\,2} = s_x^2 + s_y^2 + s_z^2 = \frac{3}{4}\hbar^2 = s(s+1)\hbar^2$$

is a c-number and $s = \frac{1}{2}$. For this reason the electron is said to be a particle endowed with spin $1/2$. It is not the only one: also the proton, the neutron, the μ-mesons etc. are particles with spin $1/2$. The difference between the orbital angular momentum $\vec{L}^{\,2}$ and the spin angular momentum is in the fact that $\vec{L}^{\,2}$ has infinitely many eigenvalues, $\vec{s}^{\,2}$ has just one eigenvalue. Another important difference is that the eigenvalues of the s_i are half-integers whereas those of the L_i are integers. The *total* angular momentum of the electron is therefore

$$\vec{J} = \vec{L} + \vec{s}$$

namely the sum of the orbital and the spin angular momenta and the rotation operators of both the orbital and the spin variables are given by $\big($see (10.29)$\big)$

$$U(\vec{n}, \phi) = e^{-i\,\vec{J}\cdot\vec{n}\,\phi/\hbar} = e^{-i\,(\vec{L}+\vec{s})\cdot\vec{n}\,\phi/\hbar} = e^{-i\,\vec{L}\cdot\vec{n}\,\phi/\hbar}\,e^{-i\,\vec{s}\cdot\vec{n}\,\phi/\hbar}$$

(the last step is legitimate *only* because the addenda $\vec{L}\cdot\vec{n}$ and $\vec{s}\cdot\vec{n}$ in the exponent *commute* with each other). The eigenvalues of the components of $\vec{J}$ are half-integers and, if we denote by $j(j+1)\hbar^2$ the eigenvalues of $\vec{J}^{\,2}$, the same holds also for the values of j: $j = \frac{1}{2},\,\frac{3}{2},\,\frac{5}{2},\,\cdots$.

We will now look for a representation of the operators s_i. The s_i are operators on the space $\mathcal{H}$ of the state vectors and, as a consequence, they should be represented by infinite dimensional matrices; however, since the spin variables commute with the orbital variables, $\mathcal{H} = \mathcal{H}_{\text{orb}} \otimes \mathcal{H}_{\text{spin}}$, and the representation we are looking for is a representation on $\mathcal{H}_{\text{spin}}$ whose dimension is 2: the s_i are therefore represented by 2×2 matrices.

We choose a representation in which s_z is diagonal and we will denote by $|+\rangle$ and $|-\rangle$ the two vector of the basis corresponding to the eigenvalues $+\frac{1}{2}\hbar$ and $-\frac{1}{2}\hbar$ of s_z:

$$s_z|+\rangle = +\frac{1}{2}\hbar|+\rangle, \qquad s_z|-\rangle = -\frac{1}{2}\hbar|-\rangle. \tag{15.7}$$

We know that (15.7) do not univocally determine the basis, i.e. the phases of the vectors $|+\rangle$ and $|-\rangle$ (see the discussion in Sect. 6.1). We will shortly realize this fact and we will also see how this arbitrariness can be eliminated. The representation of s_z obviously is

$$s_z \to \frac{1}{2}\hbar \begin{pmatrix} 1 & 0 \\ 0 & -1 \end{pmatrix}. \tag{15.8}$$

The x component s_x will be represented by a matrix like

$$s_x \rightarrow \begin{pmatrix} a & b \\ b^* & c \end{pmatrix} . \tag{15.9}$$

In order to determine a, b and c we proceed in the following way: since s_y^2 is a multiple of the identity, one has

$$0 = [\, s_y^2 \,,\, s_z \,] = s_y \,[\, s_y \,,\, s_z \,] + [\, s_y \,,\, s_z \,]\, s_y = i\,\hbar \,(s_y \, s_x + s_x \, s_y)$$

and the analogous relations that are obtained by permuting x, y and z, i.e.

$$\begin{aligned} \{\, s_x \,,\, s_y \,\} &\equiv s_x \, s_y + s_y \, s_x = 0 \\ \{\, s_y \,,\, s_z \,\} &\equiv s_y \, s_z + s_z \, s_y = 0 \\ \{\, s_z \,,\, s_x \,\} &\equiv s_z \, s_x + s_x \, s_z = 0 \end{aligned} \tag{15.10}$$

(the expression $\{\, \xi \,,\, \eta \,\}$ is called the **anticommutator** of ξ and η).

If we demand that the last of (15.10) be satisfied by the representatives (15.8) and (15.9) of s_z and s_x, one obtains $a = c = 0$, therefore

$$s_x \rightarrow \begin{pmatrix} 0 & b \\ b^* & 0 \end{pmatrix} .$$

Furthermore $s_x^2 = \frac{1}{4}\hbar^2$ must hold, which means that $b = \frac{1}{2}\hbar\,e^{i\varphi}$, φ being a real arbitrary phase. It should be now clear that, basing only on the relations that define s_x, it is not possible to specify any further its representation, i.e. it is not possible to determine the phase of b: indeed $b = \langle + \mid s_x \mid - \rangle$ exactly reflects the arbitrariness that (15.7) leave on the phases of the vectors $\mid + \rangle$ and $\mid - \rangle$. Viceversa, it is clear that fixing the relative phase of the vectors $\mid + \rangle$ and $\mid - \rangle$ is equivalent to fixing the phase of b; so if we chose $\varphi = 0$, i.e. $b = \frac{1}{2}\hbar$, the representation is completely determined. In conclusion:

$$s_x \rightarrow \frac{1}{2}\hbar \begin{pmatrix} 0 & 1 \\ 1 & 0 \end{pmatrix} . \tag{15.11}$$

The representation of s_y can be obtained, for example, by $[\, s_z \,,\, s_x \,] = i\,\hbar\,s_y$. One has

$$s_y \rightarrow \frac{1}{2}\hbar \begin{pmatrix} 0 & -i \\ +i & 0 \end{pmatrix} . \tag{15.12}$$

We have thus determined a representation of the operators s_i: the one in which s_z is diagonal and the phases are determined by the representation of s_x. Usually (15.8), (15.11) and (15.12) are written as

$$\begin{cases} s_i \rightarrow \frac{1}{2}\hbar\,\sigma_i \\ \sigma_x = \begin{pmatrix} 0 & 1 \\ 1 & 0 \end{pmatrix} , \quad \sigma_y = \begin{pmatrix} 0 & -i \\ +i & 0 \end{pmatrix} , \quad \sigma_z = \begin{pmatrix} 1 & 0 \\ 0 & -1 \end{pmatrix} . \end{cases} \tag{15.13}$$

The matrices σ_i are known as **Pauli matrices** and the representation we have found is called **Pauli representation**.

Let us now take the orbital degrees of freedom of the electron into account: if, ignoring the spin, a representation is determined by a *complete* set of compatible observables relative *only* to the orbital degrees of freedom, in order to obtain a representation that takes into account also the existence of the spin, it will suffice to add, to the above set of observables, one component of the spin, e.g. s_z. So, for example, if one starts with the Schrödinger representation in which the q_i are diagonal, and if to the q_i we add s_z, we obtain a representation in which the vectors of the basis are the simultaneous eigenvectors of the q_i and of s_z, namely

$$| \vec{x}, + \rangle \quad \text{and} \quad | \vec{x}, - \rangle , \qquad \vec{x} \in \mathbb{R}^3 .$$

The representatives of the state $| A \rangle$ then are

$$\psi_{A+}(\vec{x}) = \langle \vec{x}, + \, | \, A \rangle , \qquad \psi_{A-}(\vec{x}) = \langle \vec{x}, - \, | \, A \rangle \qquad (15.14)$$

i.e. a pair of wavefunction that we can represent as a two-component vector:

$$| A \rangle \rightarrow \begin{pmatrix} \psi_{A+}(\vec{x}) \\ \psi_{A-}(\vec{x}) \end{pmatrix} . \qquad (15.15)$$

In the representation (15.15) the operators s_i operate as 2×2 matrices, for example those given by (15.13). This happens for all the representations pinpointed by {orbital observables + one of the s_i}.

We will see that also representations in which observables involving both orbital and spin variables are diagonal are possible and useful: one example will be provided by $\vec{J}^2 = (\vec{L} + \vec{s})^2$ and/or J_z. In this and in similar cases the vectors of the basis are entangled vectors of the space $\mathcal{H} = \mathcal{H}_{\text{orb}} \otimes \mathcal{H}_{\text{spin}}$.

15.3 Composition of Angular Momenta

The law for the composition of vectors in classical physics is well known: if $\vec{M}_1$ and $\vec{M}_2$ are two (independent) angular momenta whose moduli M_1 and M_2 are known, we know that the modulus M of the resulting angular momentum $\vec{M} = \vec{M}_1 + \vec{M}_2$ may have any value between $|M_1 - M_2|$ and $M_1 + M_2$.

The same problem, whose relevance is clarified by the following examples, shows up in quantum mechanics.

Let us consider an atom with two electrons: we know that, in the independent-electron approximation, each electron is characterized by a well determined value of the angular momentum: let l_1 and l_2 be the orbital angular momenta of the two electrons. Which are the possible values L of the total angular momentum?

Also: the optical electron of an alkali metal is, for example, in a p ($l = 1$) state; the electron also possesses the spin ($s = 1/2$): which are the possible values of the total angular momentum j of the electron?

More to it: let us consider the total spin operators $\vec{S} = \vec{s}_1 + \vec{s}_2$ for two electrons and let us denote by $S(S+1)\,\hbar^2$ the eigenvalues of $\vec{S}^2$. Which are the possible values of S, i.e. of the total spin of the two electrons?

Let us try to formulate the problem in a more precise way and, just to be concrete, let us consider the first example cited above, two electrons of orbital angular momenta l_1 and l_2. Let us introduce the total orbital angular momentum $\vec{L} = \vec{L}_1 + \vec{L}_2$ and denote, as usual, by $L(L+1)\,\hbar^2$ the eigenvalues of $\vec{L}^2$. For given values of l_1 and l_2, the number of *independent* states of the system consisting of two electrons is $(2l_1 + 1) \times (2l_2 + 1)$ (we think we have fixed also the principal quantum numbers n_1 and n_2): indeed this is the number of states of the type

$$| l_1\, m_1\,,\ l_2\, m_2 \rangle\,, \qquad -l_1 \leq m_1 \leq +l_1\,, \quad -l_2 \leq m_2 \leq +l_2\,. \qquad (15.16)$$

We should not expect that any state of the type (15.16) also is an eigenstate of $\vec{L}^2$: indeed it is true that $\vec{L}^2$ commutes with both $\vec{L}_1^2$ and $\vec{L}_2^2$, inasmuch as any component of $\vec{L}$ commutes with $\vec{L}_1^2$ and $\vec{L}_2^2$ because the latter are scalar operators:

$$[\vec{L}\,,\vec{L}_1^2\,] = 0 = [\vec{L}\,,\vec{L}_2^2\,] \qquad \Rightarrow \qquad [\vec{L}^2\,,\vec{L}_1^2\,] = 0 = [\vec{L}^2\,,\vec{L}_2^2\,]$$

but

$$[L_{1z}\,,\vec{L}^2\,] \neq 0\,, \qquad\qquad [L_{2z}\,,\vec{L}^2\,] \neq 0$$

because

$$\vec{L}^2 = (\vec{L}_1 + \vec{L}_2)^2 = \vec{L}_1^2 + \vec{L}_2^2 + 2(L_{1x} L_{2x} + L_{1y} L_{2y} + L_{1z} L_{2z})$$

(L_{1z} commutes with $\vec{L}_1^2$, with $\vec{L}_2^2$ but does not commute with the terms containing L_{1x} and L_{1y}) so that, if any state of the type (15.16) were an eigenstate of $\vec{L}^2$, it would follow that in any subspace consisting of vectors with given l_1 and l_2 – and in turn in the whole $\mathcal{H}$ – the (non-commuting) operators $\vec{L}^2$, L_{1z} and L_{2z} would have a complete set of simultaneous eigenvectors. However, since both $\vec{L}^2$ and L_z commute with $\vec{L}_1^2$ and $\vec{L}_2^2$, we can search, in the subspace of dimension $(2l_1+1)\times(2l_2+1)$ of all the states with given l_1 and l_2, for those states that are also eigenstates of $\vec{L}^2$ and L_z: such states, that we will denote by

$$| l_1\, l_2\, L\, M \rangle\,, \qquad (15.17)$$

necessarily will be linear combinations of the states (15.16).

To summarize: we have the possibility of classifying the states by means of either the eigenvalues of

$$\vec{L}_1^2\ L_{1z} \qquad \vec{L}_2^2\ L_{2z} \qquad (15.18)$$

or the eigenvalues of

$$\vec{L}_1^2 \, \vec{L}_2^2 \qquad L^2 \, L_z \, . \tag{15.19}$$

In the first case the states (15.16) are obtained, whereas in the second the states (15.17) are found.

The problem is that of finding which values can L assume in (15.17) for given values of l_1 and l_2. The solution of this problem is based on the obvious consideration that, no matter how the states are classified, the number of independent states with given l_1 and l_2 must be always the same, namely $(2l_1 + 1) \times (2l_2 + 1)$.

Another consideration that plays a crucial role in the solution of the problem is that if there exists, for given l_1 and l_2, a state with some value of L and of M, then necessarily there exist $2L + 1$ independent states with that value of L: according to the classification (15.17) these are the states with all the values of M between $-L$ and $+L$. This happens because the operators $\vec{L}_1^2$, $\vec{L}_2^2$ and $\vec{L}^2$ commute with all the components of $\vec{L}$ and therefore their eigenvalues are degenerate with respect to M; or, equivalently, if a state with given L and M is found, all the $2L+1$ states $|\, l_1 \, l_2 \, L \, M \,\rangle$ with $-L \le M \le +L$ can be found by means of the raising and lowering operators L_+ and L_-.

The solution of the problem of finding the possible values for L goes in the following way.

1. The states (15.16) are also eigenstates of L_z corresponding to the eigenvalues $M = m_1 + m_2$; the maximum value for M is obtained for $m_1 = l_1$ and $m_2 = l_2$: as a consequence, there exists one only state with $M = l_1 + l_2$ that, owing to the lemma of p. 87 (Sect. 4.10), must be en eigenstate also of $\vec{L}^2$: indeed the triple l_1, l_2, $M = l_1 + l_2$ of eigenvalues of $\vec{L}_1^2$, $\vec{L}_2^2$ and L_z is nondegenerate and $\vec{L}^2$ commutes with these operators; furthermore in this case L must necessarily have the value $l_1 + l_2$: certainly not less, for $L \ge M$; certainly not more because otherwise, due to the observation made above, there should exist states with $M > l_1 + l_2$.

2. Among the states (15.16) there exist two independent states with the value $M = l_1 + l_2 - 1$: they are

$$|\, l_1 \, m_1 = l_1 \,,\, l_2 \, m_2 = l_2 - 1 \,\rangle \,, \qquad |\, l_1 \, m_1 = l_1 - 1 \,,\, l_2 \, m_2 = l_2 \,\rangle \,. \tag{15.20}$$

   A suitable linear combination of the states (15.20) must be the state with $L = l_1 + l_2$ and $M = l_1 + l_2 - 1$ (that must exists); setting this state aside, there remains one only state with $M = l_1 + l_2 - 1$, i.e. the linear combination of the states (15.20) orthogonal to the linear combination which gives the state $|\, l_1 \, l_2 \,,\, L = l_1 + l_2 , M = l_1 + l_2 - 1 \,\rangle$: by the same reasoning of point 1 above, such a state must be an eigenstate of $\vec{L}^2$ and the value of L must be $l_1 + l_2 - 1$.

3. Among the states (15.16) there exist three independent states with $M = l_1 + l_2 - 2$; two suitable linear combinations of them provide the two states with $L = l_1 + l_2$, $M = l_1 + l_2 - 2$ and $L = l_1 + l_2 - 1$, $M = l_1 + l_2 - 2$; the third combination orthogonal to the two above has, therefore, $L = l_1 + l_2 - 2 \cdots$. Where shall we stop? Let $l_1 \le l_2$. It is very easy to

realize that the number of independent states with a given value of M is *constant* – and equal to $2l_1 + 1$ – when M takes a value between $l_2 - l_1$ and $-(l_2 - l_1)$, i.e. it no longer increases as M decreases: let indeed $M = m_1 + m_2$ such that $-(l_2 - l_1) \le M \le l_2 - l_1$; this value of M can be obtained in $2l_1 + 1$ different ways, by taking $m_1 = -l_1, -l_1 + 1, \cdots, +l_1$ and correspondingly $m_2 = M - m_1$, which covers all the values permitted for m_2: $m_2 \le M + l_1 \le l_2$ and $m_2 \ge M - l_1 \ge -l_2$.

It is clear that the procedure stops when the number of states with a given value of M stops increasing as M itself decreases, i.e. it stops with $L = l_2 - l_1$. Indeed it is straightforward to check that

$$\sum_{L=l_2-l_1}^{l_2+l_1} (2L + 1) = (2l_1 + 1) \times (2l_2 + 1)$$

as it must be, because the number of independent vectors does not change.

The following table, relative to the particular case $l_1 = 1$, $l_2 = 2$ may be useful to summarize the procedure discussed above.

M	3	2	1	0	-1	-2	-3
$m_1 + m_2$	$1+2$	$1+1$	$1+0$	$1-1$	$1-2$	$0-2$	$-1-2$
		$0+2$	$0+1$	$0+0$	$0-1$	$-1-1$	
			$-1+2$	$-1+1$	$-1+0$		
	$\downarrow$	$\downarrow$	$\downarrow$	$\downarrow$	$\downarrow$	$\downarrow$	$\downarrow$
	$L=3$	$L=3$	$L=3$	$L=3$	$L=3$	$L=3$	$L=3$
		$L=2$	$L=2$	$L=2$	$L=2$	$L=2$	
			$L=1$	$L=1$	$L=1$		

We have thus solved the problem of adding two angular momenta: the value L can assume, once l_1 and l_2 are given, are

$$l_1 + l_2, \qquad l_1 + l_2 - 1, \qquad \cdots \qquad , |l_1 - l_2|$$

so that, according to the classification by means of the eigenvalues of (15.19) one has:

$$|l_1 l_2 L M\rangle, \qquad -L \le M \le +L \qquad |l_1 - l_2| \le L \le l_1 + l_2. \qquad (15.21)$$

Of course all the above applies to any two independent angular momenta.

The previous discussion enabled us only to determine which are the values of L in the classification (15.17), but not to determine how the vectors (15.17) are expressed as linear combinations of the vectors (15.16):

$$|l_1 l_2 L M\rangle = \sum_{m_1 m_2} C_{l_1 m_1 \; l_2 m_2}^{L \, M} |l_1 m_1, l_2 m_2\rangle. \qquad (15.22)$$

The coefficients $C_{l_1 m_1 \, l_2 m_2}^{L \; M}$ are called **Clebsch-Gordan coefficients**; the
– very important – problem of their determination will not be faced here; we
will limit ourselves to discuss it in a particular case that will have a remarkable
interest for us in the sequel: it is the case, already cited in the introduction of
the present section, of the composition of the spins of two electrons.

The spin states of two electrons can be classified by means of the eigenvalues of s_{1z} and s_{2z} (these are the operators (15.18), where one should note
that $\vec{s}_1^{\,2}$ and $\vec{s}_2^{\,2}$ – the analogues of $\vec{L}_1^2$ and $\vec{L}_2^2$ – are omitted inasmuch as
multiples of the identity) and in this case four states are obtained

$$|\, s'_{1z} \, s'_{2z} \,\rangle , \qquad\qquad s'_{1z} = \pm\frac{1}{2}\hbar , \qquad\qquad s'_{2z} = \pm\frac{1}{2}\hbar$$

(i.e. states of the type (15.16)) that we will write with obvious notation (see
(15.7))

$$|++\rangle , \qquad |+-\rangle , \qquad |-+\rangle , \qquad |--\rangle . \qquad\qquad (15.23)$$

We can alternatively classify the spin states of the two electrons by using the
eigenvalues of $\vec{S}^{\,2} = (\vec{s}_1 + \vec{s}_2)^2$ and of S_z (the operators in (15.19)) and one
obtains the states

$$|\, S \, S'_z \,\rangle , \qquad S = 0, 1 , \qquad -S \leq S'_z \leq +S$$

namely

$$|1 +1\rangle , \qquad |1 \quad 0\rangle , \qquad |1 -1\rangle , \qquad |0 \quad 0\rangle . \qquad\qquad (15.24)$$

The spin 1 states ($S = 1$) are called the **triplet states**, whereas the spin 0
state ($S = 0$) is called the **singlet state**. We want now to determine how
the states (15.24) are expressed as linear combinations of the states (15.23).
First of all, it is evident that

$$|1 +1\rangle = |++\rangle , \qquad\qquad |1 -1\rangle = |--\rangle .$$

The third triplet state, that with $S'_z = 0$, will be a linear combination of
$|+-\rangle$ and $|-+\rangle$; it could be obtained by applying the lowering operator
$S_- = S_x - \mathrm{i}\, S_y$ to the state $|1 +1\rangle$; actually the calculation is unnecessary, as
it is sufficient to observe that the vector $|1 +1\rangle = |++\rangle$ is **symmetric** under
the **permutation** of the quantum numbers s'_{1z}, s'_{2z} of the two electrons,
the operator $S_- = s_{1-} + s_{2-}$ is symmetric under the exchange of the two
electrons ($1 \leftrightarrow 2$), whence also the vector $S_-|++\rangle$ will be symmetric
under the permutation of the quantum numbers of the two electrons: there
is only one symmetric linear combination of $|-+\rangle$ and $|+-\rangle$, that is
$|-+\rangle + |+-\rangle$, therefore

$$|1 \quad 0\rangle = \frac{1}{\sqrt{2}} \left(|-+\rangle + |+-\rangle \right) . \qquad\qquad (15.25)$$

Now the singlet state is the only state orthogonal to the state (15.25), i.e.

$$|0 \quad 0\rangle = \frac{1}{\sqrt{2}} \left(|-+\rangle - |+-\rangle \right)$$

that turns out to be **antisymmetric**. Summarizing the results:

$$\left. \begin{array}{l} |1 +1\rangle = |++\rangle \\[2mm] |1 \quad 0\rangle = \dfrac{1}{\sqrt{2}} \left(|-+\rangle + |+-\rangle \right) \\[2mm] |1 -1\rangle = |--\rangle \end{array} \right\} \begin{array}{l} \text{symmetric} \\ \text{triplet states} \end{array}$$

$$\left. |0 \quad 0\rangle = \frac{1}{\sqrt{2}} \left(|-+\rangle - |+-\rangle \right) \right\} \begin{array}{l} \text{antisymmetric} \\ \text{singlet state.} \end{array} \tag{15.26}$$

15.4 Pauli Exclusion Principle

The Pauli principle plays a fundamental role in the understanding of the structure of matter and, in particular, of atoms: it applies to all the cases in which one has to deal with two or more electrons and its roots are in the fact that electrons are particles indistinguishable from one another. Let us try to understand what does it mean that electrons are indistinguishable from one another and what does this fact entail.

Let us consider a system consisting of two electrons: the variables suitable for the description of this system are

$$\vec{q}_1, \ \vec{p}_1, \ \vec{s}_1 \ ; \qquad \vec{q}_2, \ \vec{p}_2, \ \vec{s}_2 \ . \tag{15.27}$$

It is clear that, if we maintained that, for example, the $\vec{q}_1$ were observables, we should give a meaning to the expression "the electron 1", i.e. we should be able to distinguish one electron from the other: so, for example, if q_{1x} were an observable, the instrument associated with it should provide a result only when it 'sees' the electron 1, while it should not interact at all with the electron 2. This is absurd because this statement would be equivalent to admit that the two electrons had different interaction and properties therefore they would not be identical particles. As a consequence, it should be clear that not all the Hermitian functions of (15.27) can be considered as observables for the system consisting of the two electrons, but only those functions $f(\vec{q}_1, \vec{p}_1, \vec{s}_1; \vec{q}_2, \vec{p}_2, \vec{s}_2)$ that are **symmetric** under the permutation of the indices 1 and 2 (for example $q_{1x} + q_{2x}$, $|q_{1x} - q_{2x}|$ etc.).

This does not imply that two electrons occupying different (orthogonal) states cannot be distinguished: in other words phrases like either "an electron is in the state $|A\rangle$ and the other in the state $|B\rangle$" or "the electron that is in the state $|A\rangle$"... are meaningful. What is meaningless is to state that "electron 1 is in the state $|A\rangle$ and electron 2 is in the state $|B\rangle$". One could be tempted to call "electron 1" the one which is in the state $|A\rangle$ and electron 2 the one in the state $|B\rangle$: in such a way the electrons are considered distinguishable on the basis of the fact that they are in different states; such distinguishability is meaningful until the electrons remain in orthogonal states, which can fail due both to time evolution and to measurements made on the system.

It is legitimate to ask oneself if all the above considerations, and the conclusions we will draw from them, apply as well to macroscopic objects like either billiard balls or brand-new coins: it is however clear that the identity in such cases exclusively refers to macroscopic properties (shape, size, mass, density, ...) whereas it certainly fails on the level of microscopic properties (number of atoms the objects are made out of, states of the single atoms, ...). This means that two macroscopic, so-called identical, objects certainly are not identical with each other or that, at least in principle, they are distinguishable. One can also state that two macroscopic 'identical' objects are and always will be distinguishable for the fact that they are (and will always be) in different states (not only as far as microscopic states are concerned, but also in different states of position). We will come back later on the subject.

Let us now introduce, for a system of two electrons, the **exchange operator** Π defined by

$$\Pi\,\vec{q}_1\,\Pi^{-1} = \vec{q}_2\,, \quad \Pi\,\vec{p}_1\,\Pi^{-1} = \vec{p}_2\,, \quad \Pi\,\vec{s}_1\,\Pi^{-1} = \vec{s}_2$$

$$\Pi\,\vec{q}_2\,\Pi^{-1} = \vec{q}_1\,, \quad \Pi\,\vec{p}_2\,\Pi^{-1} = \vec{p}_1\,, \quad \Pi\,\vec{s}_2\,\Pi^{-1} = \vec{s}_1\,. \tag{15.28}$$

Therefore the indistinguishability of the two electrons is translated into the statement that all the observables of the system commute with Π:

$$\Pi\,\xi\,\Pi^{-1} = \xi\,. \tag{15.29}$$

The operator Π exists and is unitary, thanks to the von Neumann theorem cited in Sect. 6.3 (the transformation $(\vec{q}_1, \vec{p}_1, \vec{s}_1) \leftrightarrow (\vec{q}_2, \vec{p}_2, \vec{s}_2)$ leaves the canonical commutation relations unchanged and maps self-adjoint operators in self-adjoint operators). However (15.28) define Π only up to a phase factor: from (15.28) it follows that, if $|\,A\,B\,\rangle$ is a 'factorized' vector in $\mathcal{H} = \mathcal{H}_1 \otimes \mathcal{H}_2$ (see (15.2))

$$\Pi\,|\,A\,B\,\rangle = \mathrm{e}^{\mathrm{i}\varphi}\,|\,B\,A\,\rangle$$

with the phase φ independent of the (factorized) vector [indeed:

$$\langle\,A\mid C\,\rangle\langle\,B\mid D\,\rangle = \langle\,B\,A\mid C\,D\,\rangle = \langle\,B\,A\mid \Pi^\dagger\,\Pi\mid C\,D\,\rangle$$
$$= \mathrm{e}^{\mathrm{i}\,(\varphi_{CD}-\varphi_{AB})}\langle\,A\,B\mid D\,C\,\rangle = \mathrm{e}^{\mathrm{i}\,(\varphi_{CD}-\varphi_{AB})}\langle\,A\mid C\,\rangle\langle\,B\mid D\,\rangle$$

hence $\varphi_{CD} = \varphi_{AB}$]. Therefore Π can be (re)defined $(\Pi \to \mathrm{e}^{-\mathrm{i}\varphi}\,\Pi)$ in such a way that

$$\Pi\,|\,A\,B\,\rangle = |\,B\,A\,\rangle\,. \tag{15.30}$$

As a consequence of the definition (15.30) one has

$$\Pi^2 = \mathbb{1}\,, \qquad\qquad \Pi^{-1} = \Pi^\dagger = \Pi$$

whence Π, much as the space-inversion operator I, only has the eigenvalues ± 1. The vectors corresponding to the eigenvalue $+1$ are said **symmetric**

$(|\mathcal{S}\rangle)$ whereas those corresponding to the eigenvalue -1 are said **antisymmetric** $(|\mathcal{A}\rangle)$ under the exchange of the two electrons:

$$\Pi\,|\,\mathcal{S}\rangle = +|\,\mathcal{S}\rangle \qquad\qquad \Pi\,|\,\mathcal{A}\rangle = -|\,\mathcal{A}\rangle$$

and any vector can be expressed as the superposition of a vector of type $|\,\mathcal{S}\,\rangle$ and a vector of type $|\,\mathcal{A}\,\rangle$. In other word, the Hilbert space $\mathcal{H}$ is the direct sum of the subspaces $\mathcal{H}_S$ and $\mathcal{H}_A$ consisting respectively of the symmetric and the antisymmetric vectors:

$$\mathcal{H} = \mathcal{H}_S \oplus \mathcal{H}_A = (\mathcal{H}_1 \otimes \mathcal{H}_2)_S \oplus (\mathcal{H}_1 \otimes \mathcal{H}_2)_A \ .$$

If at a given time the state of the system consisting of the two electrons is represented by a vector in $\mathcal{H}_S$, since *all* the observables commute with Π, the state of the system will always remain in $\mathcal{H}_S$, as a consequence of either time evolution or of the perturbation due to any measurement processes. For the same reason, if initially the state is represented by a vector in $\mathcal{H}_A$, it will always remain in $\mathcal{H}_A$, no matter what will happen to the system. Moreover, if the system were in a superposition $|\,\mathcal{S}\rangle + |\,\mathcal{A}\rangle$, any observation (i.e. any measurement of a nondegenerate observable) would force the system to end up either in $\mathcal{H}_S$ or in $\mathcal{H}_A$ and there would remain forever.

Given that things go in this way, it could happen that for any system consisting of two electrons the only possible states are only either the symmetric states or the antisymmetric ones: this is possible but, a priori, one cannot say this is true: we are then left with nothing else to do but "questioning Nature".

Let us consider an atom of helium: the configuration with minimum energy is $(1s)^2$, i.e. the electrons possess the same orbital quantum numbers. The two electrons may, in addition, have total spin either $S = 0$ or $S = 1$. We thus have the two possibilities

$$|\,(1s)^2, S = 0\,\rangle\,, \qquad\qquad |\,(1s)^2, S = 1\,\rangle\,.$$

Under the action of the operator Π, that exchanges both the spin and the orbital quantum numbers of the two electrons, the first is an antisymmetric vector: it is symmetric with respect to the permutation of the orbital quantum numbers only, antisymmetric under the permutation of the spin quantum numbers $\big($see (15.26)$\big)$; the second vector is symmetric. The spectroscopic analysis of He says that in Nature only the singlet $(S = 0)$ $(1s)^2$ state exists, and that there exists no $(1s)^2$ states with $S = 1$. This not only confirms that not all the vectors in $\mathcal{H}$ correspond to physical states, but also that the only possible states are those represented by antisymmetric vectors. We can now enunciate the

Pauli Principle: *The only possible states for a system of two electrons are those represented by vectors antisymmetric under the exchange of the two electrons. For a system consisting of more than two electrons, the only possible states are those represented by vectors antisymmetric under the exchange of any pair of electrons.*

As perhaps already happened, we will often say "antisymmetric states" instead of "states represented by antisymmetric vectors", consistently with the warning given at the end of Sect. 4.1; note however that, even if the vectors are antisymmetric, the states they represent are symmetric under the exchange of the electrons inasmuch as $|\mathcal{A}\rangle$ and $-|\mathcal{A}\rangle$ represent the same state.

The Pauli principle, in the form we have enunciated it, applies to all the systems consisting of identical particles endowed with half-integer spin (protons, neutrons etc.) whereas for the particles with integer spins $(0, 1, \cdots)$ – as e.g. the π-mesons etc. – only the symmetric states are possible. The particles of the first type are called **fermions**, those of the second type **bosons**.

It should be clear that, in comparison with a system consisting of two distinguishable particles, for a system of two identical fermions one has, as a consequence of the Pauli principle, a lesser number of states: for distinguishable particles, as e.g. a positron and an electron, if one of them is in the state $|A\rangle$ and the other one in the state $|B\rangle$, the possible independent states are two, precisely $|AB\rangle$ and $|BA\rangle$, while if both particles are electrons (or positrons) only the state

$$|AB\rangle - |BA\rangle$$

is permitted: in this case one cannot say which electron is in the state $|A\rangle$ and which in $|B\rangle$, but only that one is in the state $|A\rangle$ and one in $|B\rangle$.

Furthermore, two electrons cannot be in the same state $|A\rangle$, because such a state would be represented by the symmetric vector $|AA\rangle$: for this reason the Pauli principle is an exclusion principle.

We wish now to show that the statement made in Sect. 14.1, according to which no more than two electrons in an atom can have the same orbital quantum numbers $n\,l\,m$, indeed is a consequence of the Pauli principle.

First of all, it should be clear that, as in the case of the configuration $(1s)^2$ of He, two electrons can occupy the same orbital only if their total spin is $S = 0$: only in this way the state, that is symmetric with respect to the permutation of only the orbital quantum numbers, turns out to be antisymmetric under the permutation of all the (both orbital and spin) quantum numbers of the two electrons. But one cannot have more than two electrons in the same orbital since, existing only two independent spin states for each electron, it is impossible to construct a spin state for three or more electrons that is antisymmetric under the permutation of the spin quantum numbers of any pair of electrons. For example $|+++\rangle$ and $|---\rangle$ are symmetric under any permutation; $|+-\rangle, |+-+\rangle, |-++\rangle$ etc. are symmetric with respect to some permutation, so none of their linear combinations may be completely antisymmetric: the eight states of spin of three electrons split up into four states with total spin $S = 3/2$, symmetric, and two pairs of states with total spin $S = 1/2$, that are neither symmetric nor antisymmetric.

Let us now go back to the problem of the distinguishability of indistinguishable particles: Pauli principle substantially says that never a system consisting of two electrons is one with 'separate variables' and never the states are factorized (see Sect. 15.1): states of the type $|AB\rangle$ are not permitted, only states of the type $|AB\rangle - |BA\rangle$ are possible; more to it: it is not one with separate variables the system that consists of *all* the electrons in the Universe!

But then, what does it mean to study a system, for example, with one only electron, like the hydrogen atom? Should we antisymmetrize the state of 'our' electron with that of all the other electrons, albeit far apart? Luckily: no. Let us consider a system consisting of two electrons or, more in general, of either two identical fermions or bosons, in two states $|A\rangle$ and $|B\rangle$ orthogonal to each other (for example, very far from each other and sufficiently localized) and let us examine under which circumstances the behaviour of such a state is different from that of the state $|AB\rangle$ of two distinguishable particles. We have seen in Sect. 4.8 that all the information about the properties of a state is written in the collection of the mean values of the observables. So let us examine which are the differences between the mean values of the observables in the state

$$|X\rangle = |A_1 \ B_2\rangle \tag{15.31}$$

of the distinguishable particles and in the (normalized) states

$$|Y_\pm\rangle = \frac{1}{\sqrt{2}}\Big(|A_1 \ B_2\rangle \pm |B_1 \ A_2\rangle\Big) \tag{15.32}$$

of two identical particles (the $+$ or $-$, according to whether one is dealing with either bosons or fermions; the indices 1 and 2, albeit superfluous, are written for the sake of clarity).

Let ξ be an observable. One has

$$\langle X \mid \xi \mid X\rangle = \langle B_2 \ A_1 \mid \xi \mid A_1 \ B_2\rangle \tag{15.33}$$

$$\langle Y_\pm \mid \xi \mid Y_\pm\rangle = \frac{1}{2}\Big(\langle B_2 \ A_1 \mid \pm \langle A_2 \ B_1 \mid\Big)\, \xi\,\Big(|A_1 \ B_2\rangle \pm |B_1 \ A_2\rangle\Big)$$

$$= \frac{1}{2}\Big(\langle B_2 \ A_1 \mid \xi \mid A_1 \ B_2\rangle + \langle A_2 \ B_1 \mid \xi \mid B_1 \ A_2\rangle$$

$$\pm\, \langle B_2 \ A_1 \mid \xi \mid B_1 \ A_2\rangle \pm \langle A_2 \ B_1 \mid \xi \mid A_1 \ B_2\rangle\Big) \ .$$

Thanks to (15.29), $\Pi^\dagger = \Pi^{-1}$ and $\Pi\,|A\ B\rangle = |B\ A\rangle$,

$$\langle A_2 \ B_1 \mid \xi \mid B_1 \ A_2\rangle = \langle A_2 \ B_1 \mid \Pi^\dagger\, \xi\, \Pi \mid B_1 \ A_2\rangle = \langle B_2 \ A_1 \mid \xi \mid A_1 \ B_2\rangle$$
$$\tag{15.34}$$
$$\langle A_2 \ B_1 \mid \xi \mid A_1 \ B_2\rangle = \langle A_2 \ B_1 \mid \Pi^\dagger\, \xi\, \Pi \mid A_1 \ B_2\rangle = \langle B_2 \ A_1 \mid \xi \mid B_1 \ A_2\rangle$$

whence

$$\langle Y_\pm \mid \xi \mid Y_\pm\rangle = \langle B_2 \ A_1 \mid \xi \mid A_1 \ B_2\rangle \pm \langle B_2 \ A_1 \mid \xi \mid B_1 \ A_2\rangle \ . \tag{15.35}$$

The difference between the cases of bosons and fermions is given by the sign of the second term in the right hand side of (15.35) ("interference term"), which is absent in (15.33), i.e. in the case of distinguishable particles. So, in all cases in which the interference term are vanishing, a system consisting of two electrons is not different from a system consisting of either two bosons or of two distinguishable particles.

For example, if (as we said in the beginning) the states $|A\rangle$ and $|B\rangle$ are far apart from each other and sufficiently localized (i.e. the wavefunctions ψ_A and ψ_B have disjoint supports), for any observable that only depends on the position variables $f(\vec{q}_1, \vec{q}_2)$ (e.g. e^2/r_{12}), the interference term (in this case called *exchange integral*) is vanishing:

$$\langle B_2\, A_1 \mid \xi \mid B_1\, A_2 \rangle = \iint \psi_B^*(\vec{x}_2)\psi_A^*(\vec{x}_1)\, f(\vec{x}_1,\vec{x}_2)\, \psi_B(\vec{x}_1)\psi_A(\vec{x}_2)\, dV_1\, dV_2 = 0$$

because, in the case we are considering, $\psi_A^*(\vec{x}_1)\psi_B(\vec{x}_1) = 0 = \psi_B^*(\vec{x}_2)\psi_A(\vec{x}_2)$

Clearly also the matrix elements of observables that are polynomials in $\vec{p}_1$ and $\vec{p}_2$ are vanishing, because the supports of the derivatives coincide with the supports of the functions.

Another important case in which the interference terms vanish is that of the 'one particle' observables, i.e. of observables of the type

$$\xi = \xi_1 + \xi_2 = f(\vec{q}_1, \vec{p}_1, \vec{s}_1) + f(\vec{q}_2, \vec{p}_2, \vec{s}_2)\ .$$

Indeed:

$$\langle B_2\, A_1 \mid (\xi_1 + \xi_2) \mid B_1\, A_2 \rangle =$$
$$= \langle A_1 \mid \xi_1 \mid B_1 \rangle \times \langle B_2 \mid A_2 \rangle + \langle A_1 \mid B_1 \rangle \times \langle B_2 \mid \xi_2 \mid A_2 \rangle = 0$$

since $\langle B \mid A \rangle = \langle A \mid B \rangle = 0$.

In conclusion, in all the cases when one expects that the two identical particles could be considered as distinguishable, the theory helps us and confirms our expectations.

Chapter 16

Many-Electron Atoms

16.1 The Energy Levels of the Helium Atom

The helium atom has two electrons, therefore it is very suitable to illustrate the consequences that Pauli principle has in atomic physics.

The Hamiltonian of the He atom, in the nonrelativistic approximation and assuming – for the sake of simplicity – the nucleus fixed at the origin, is

$$ H = \frac{\vec{p}_1^{\,2}}{2m} + \frac{\vec{p}_2^{\,2}}{2m} - \frac{2e^2}{r_1} - \frac{2e^2}{r_2} + \frac{e^2}{r_{1\,2}} \,. \tag{16.1} $$

In the independent-electron approximation the ground-state configuration is $(1s)^2$; the generic excited configuration is $(n\,l)(n'\,l')$. However we will consider the energy levels of the atom corresponding to configurations in which only one electron is excited, i.e. configurations of the type $(1s)(n\,l)$: the reason for doing so is in the fact – to be discussed in Sect. 16.5 – that, to a good approximation, such configurations are the only ones that can be radiatively excited starting from the ground state.

Similarly to Sect. 15.4, in which we have introduced the operator Π that exchanges the index 1 with the index 2 of all the (orbital and spin) variables of the two electrons, we now introduce the operators Π_o and Π_s that respectively exchange with each other the indices 1 and 2 relative either to the only orbital and the only spin variables:

$$ \Pi_o\,\vec{q}_{1,2}\,\Pi_o^{-1} = \vec{q}_{2,1} \,, \quad \Pi_o\,\vec{p}_{1,2}\,\Pi_o^{-1} = \vec{p}_{2,1} \,, \quad \Pi_o\,\vec{s}_{1,2}\,\Pi_o^{-1} = \vec{s}_{1,2} $$
$$ \tag{16.2} $$
$$ \Pi_s\,\vec{s}_{1,2}\,\Pi_s^{-1} = \vec{s}_{2,1} \,, \quad \Pi_s\,\vec{q}_{1,2}\,\Pi_s^{-1} = \vec{q}_{1,2} \,, \quad \Pi_s\,\vec{p}_{1,2}\,\Pi_s^{-1} = \vec{p}_{1,2} \,. $$

Exactly as Π, also Π_o and Π_s are defined by (16.2) up to a phase factor: clearly it is possible to choose the arbitrary phase factors in such a way that

$$ \Pi_o\,\Pi_s = \Pi_s\,\Pi_o = \Pi \,, \qquad \Pi_o^2 = \Pi_s^2 = \mathbb{1} $$

so that both Π_o and Π_s only have the eigenvalues ± 1. Π_s has the value $+1$ on the triplet states and -1 on the singlet states $\big($see (15.26)$\big)$; the states

© Springer International Publishing Switzerland 2016
L.E. Picasso, *Lectures in Quantum Mechanics*, UNITEXT for Physics,
DOI 10.1007/978-3-319-22632-3_16

on which Π_o has the value $+1$ are the states that are symmetric under the exchange of the only orbital variables, those on which it has the value -1 are the antisymmetric ones: let us denote the former by $|\,(\text{orb. symm.}), \cdots \rangle$, the latter by $|\,(\text{orb. antisymm.}), \cdots \rangle$. Particular antisymmetric states $|\mathcal{A}\rangle$ (i.e. those with $\Pi = -1$) are the states:

$$|\,(\text{orb. symm.}), S = 0 \rangle, \qquad\qquad |\,(\text{orb. antisymm.}), S = 1 \rangle \qquad (16.3)$$

but obviously also their linear combinations are possible.

In order to study the levels of He, let us temporarily forget about the fact that electrons possess spin (this is legitimate since in the Hamiltonian (16.1) the spin variables are not present) and, as a consequence, let us also forget about the Pauli principle. The exchange operator Π_o commutes with the Hamiltonian (16.1), so we can classify the levels of the He atom by means of the eigenvalue of Π_o: for example, the ground state $(1s)^2$ corresponds to the eigenvalue $+1$ of Π_o. Let us now consider the excited configuration $1s\,2s$: two states correspond to this configuration, precisely

$$|\,1s\,2s\,\rangle, \qquad\qquad |\,2s\,1s\,\rangle \qquad\qquad (16.4)$$

that are not eigenstates of Π_o: indeed

$$\Pi_o\,|\,1s\,2s\,\rangle = |\,2s\,1s\,\rangle, \qquad\qquad \Pi_o\,|\,2s\,1s\,\rangle = |\,1s\,2s\,\rangle\,.$$

In the independent-electron approximation the two states (16.4) have the same energy: such a degeneracy is called *exchange degeneracy*. If we now consider the term e^2/r_{12} as a perturbation, the exchange degeneracy is removed: we must indeed diagonalize the 2×2 matrix of the perturbation $H' = e^2/r_{12}$ among the states (16.4):

$$\begin{pmatrix} \langle 2s\,1s\,|\,H'\,|\,1s\,2s\,\rangle & \langle 2s\,1s\,|\,H'\,|\,2s\,1s\,\rangle \\[2mm] \langle 1s\,2s\,|\,H'\,|\,1s\,2s\,\rangle & \langle 1s\,2s\,|\,H'\,|\,2s\,1s\,\rangle \end{pmatrix} \qquad (16.5)$$

(following the notation we have established in Sect. 15.1, we denote by $\langle B\,A\,|$ the "bra" corresponding to the "ket" $|\,A\,B\,\rangle$). The diagonal elements of (16.5) (much like the off-diagonal elements) are equal to each other, thanks to

$$\Pi_o\,H'\,\Pi_o^{-1} = H'\,, \qquad\qquad \Pi_o^{\dagger} = \Pi_o^{-1}\,.$$

The demonstration is the same as that reported in (15.34). As a consequence, the matrix (16.5) has the form:

$$\begin{pmatrix} a & b \\ b & a \end{pmatrix} \qquad\qquad (16.6)$$

(independently of the fact that the wavefunctions of the states $1s$ and $2s$ are real, b is a real number, as we shall see shortly). The eigenvalues and the corresponding eigenvectors of (16.5) are

$$E_+ = a + b\,, \qquad |E_+\rangle = \frac{1}{\sqrt{2}}\left(|\,1s\,2s\,\rangle + |\,2s\,1s\,\rangle\right)$$

$$E_- = a - b\,, \qquad |E_-\rangle = \frac{1}{\sqrt{2}}\left(|\,1s\,2s\,\rangle - |\,2s\,1s\,\rangle\right).$$

$$(16.7)$$

Both $|E_+\rangle$ and $|E_-\rangle$ are eigenvectors of Π_o corresponding respectively to the eigenvalues $+1$ and -1: this particular aspect of the result has a more general validity than that attributable to the perturbative calculation, because we know that the eigenstates of the Hamiltonian (16.1), if nondegenerate, must have a well-defined orbital symmetry. Another aspect of the result (16.7) is the following: which of the two levels E_+ and E_- has a higher energy depends on the sign of b; in the Schrödinger representation b is given by the following *exchange integral*:

$$b = \int \psi_{1s}^*(\vec{r}_1)\,\psi_{2s}^*(\vec{r}_2)\,\frac{e^2}{r_{12}}\,\psi_{1s}(\vec{r}_2)\,\psi_{2s}(\vec{r}_1)\,\mathrm{d}V_1\mathrm{d}V_2$$

$$= \int \Phi^*(\vec{r}_1)\,\frac{e^2}{r_{12}}\,\Phi(\vec{r}_2)\,\mathrm{d}V_1\mathrm{d}V_2\,, \qquad \Phi(\vec{r}) \equiv \psi_{1s}(\vec{r})\,\psi_{2s}^*(\vec{r})\,.$$

The exchange integral is formally identical with twice the electrostatic energy of the (complex) charge distribution $\rho(\vec{r}) = e\,\Phi(\vec{r})$, therefore it is (real and) positive: $b > 0$.

In conclusion, the perturbative calculation says that the degeneracy is removed ($b \neq 0$) and that $E_+ > E_-$. This result has a very simple physical interpretation, that provides a strong argument that leads us to think that it stays true also for the exact eigenvalues: since the difference in energy between the symmetric (under Π_o) state $|E_+\rangle$ and the antisymmetric state $|E_-\rangle$ is due to the term e^2/r_{12}, it is reasonable to expect that the state with lower energy is the one in which the two electrons have a lesser probability to be close to each other, inasmuch as the Coulombic repulsion – whose effect is to increase the energy – is less relevant. The antisymmetric state has an antisymmetric wavefunction:

$$\Psi(\vec{r}_1, \vec{r}_2) = \langle \vec{r}_1\,\vec{r}_2 \mid E_-\rangle = -\langle \vec{r}_1\,\vec{r}_2 \mid \Pi_o \mid E_-\rangle = -\langle \vec{r}_2\,\vec{r}_1 \mid E_-\rangle = -\Psi(\vec{r}_2, \vec{r}_1)$$

and if we put $\vec{r}_1 = \vec{r}_2 = \vec{r}$, one has

$$\Psi(\vec{r}, \vec{r}) = -\Psi(\vec{r}, \vec{r}) = 0$$

so, the probability-density of finding the two electrons in the same point being zero, in the state $|E_-\rangle$ the probability of finding the two electrons close to each other is relatively small. This leads to conclude that $E_- < E_+$, in agreement with what has been found by perturbation theory: this result is known as the first **Hund rule**.

If we now consider the excited configurations, for example $1s\,2p$, the reasoning goes on in the same way: in absence of the Coulombic repulsion there occurs the exchange degeneracy between the states (the quantum number m, that takes the values 0, ± 1 is omitted for the sake of brevity)

$$| 1s\, 2p \rangle\,, \qquad\qquad | 2p\, 1s \rangle\,.$$

The repulsion removes this degeneracy and the approximate eigenstates are

$$\frac{1}{\sqrt{2}} \left(| 1s\, 2p \rangle - | 2p\, 1s \rangle \right) \qquad\qquad \text{antisymmetric}$$

$$\frac{1}{\sqrt{2}} \left(| 1s\, 2p \rangle + | 2p\, 1s \rangle \right) \qquad\qquad \text{symmetric}\,.$$

The same argument presented for the states originated from the configuration $1s\,2s$ allows us to conclude that the former, the antisymmetric ones, have an energy lower than the latter's. And so on for the other configurations.

Let us now take the electron spins into considerations. The two electrons may have total spin either $S = 0$ or $S = 1$; if the Pauli principle were not there, to any orbital state of the atom we should associate both the values of S: any energy level would be degenerate on S since, the spin variables not appearing in the Hamiltonian (16.1), the energy only depends on the orbital variables and not on the value of S. But the Pauli principle demands that if a state has a defined orbital (i.e. with respect to Π_o) symmetry, then it must have the opposite spin (i.e. with respect to Π_s) symmetry – see (16.3) – so all the levels corresponding to symmetric orbital states must have $S = 0$ and the orbital antisymmetric must have $S = 1$. In particular the lowest energy level must have, as we already know, $S = 0$:

$$| 1s\, 1s, S = 0 \rangle\,.$$

It is then clear that the He levels do not display any degeneracy with respect to S: on the one hand, because the Pauli principle admits only one value of S for any state with a well defined orbital symmetry; on the other hand, because the states with opposite symmetry originated by the same electronic configuration exhibit the exchange degeneracy only in the independent-electron approximation: but the latter degeneracy is removed by the Coulombic repulsion between the two electrons.

To summarize: the He atom exhibits two series of energy levels, the singlet ($S = 0$) levels and the triplet ($S = 1$) levels or, equivalently those corresponding to orbitally symmetric and orbitally antisymmetric states. Any electronic configuration of the type $(1s)(n\,l)$ – with the exception of the fundamental one $(1s)^2$ – gives rise to both a singlet and a triplet level: the singlet state has an energy slightly higher than the corresponding triplet state, owing to the different effect the inter-electronic Coulombic repulsion has on states with opposite orbital symmetry. The difference in energy between the singlet and triplet states originated by the configuration $1s\,2s$ is about 0.8 eV, whereas that between the states originated by the configuration $1s\,2p$ is about 0.25 eV. We report in Fig. 16.1 the diagram of the lowest levels of He, as an illustration of what we have said so far.

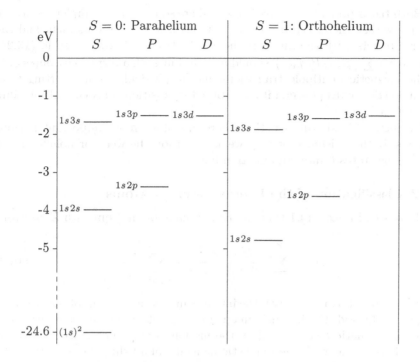

Fig. 16.1

The terminology **parahelium** for the set of levels with $S = 0$ and **ortho-helium** for the set of levels with $S = 1$ originates from the fact that the electromagnetic transitions between the two types of levels are extremely rare (very weak spectral lines), so that for a long time it was held that the two series belonged not to a unique substance, but that, as a mater of fact, He consisted of a mixture of two different chemical species, called exactly parahelium and orthohelium. The dilemma was that one did not succeed in isolating the two chemical species from each other.

To conclude, let us show that there exists a selection rule that prohibits the transitions between the levels with $S = 0$ and those with $S = 1$: indeed, the electromagnetic transition probability (13.24) is determined by the matrix element of an operator that only depends on the orbital variables, and therefore commutes with all the components of $\vec{S}$ and in particular with $\vec{S}^2$: as a consequence, the **selection rule** $\Delta S = 0$ has to be satisfied in electromagnetic transitions, which exactly excludes transitions between states with total spin $S = 0$ and $S = 1$.

Another way to realize that the above transitions are forbidden and that, thanks to the Pauli principle, is equivalent to that we have just presented, is based on the observation that the operator $\sum_\alpha \vec{p}_\alpha \cdot \vec{A}(\vec{q}_\alpha, t)$ commutes with Π_o and therefore cannot induce transitions between states with different orbital symmetries.

Both the arguments we have presented are somewhat incomplete, since we now know that the electron possesses an intrinsic magnetic moment and the latter interacts with the magnetic field $\vec{B}(\vec{x}, t) = \nabla \wedge \vec{A}(\vec{x}, t)$, so in (13.24) also a term $\sum_\alpha \vec{\mu}_{s\,\alpha} \cdot \vec{B}(\vec{q}_\alpha, t)$ should appear; but such a term contributes only to the magnetic multipole transitions and, in particular, only starting from the magnetic quadrupole can it contribute to transitions between orthohelium and parahelium. ,

In Sect. 16.5 we will see that there are other more important reasons that justify the existence of very weak transitions between orthohelium and parahelium states ("intercombination lines").

16.2 Classification of the Energy Levels of Atoms

We have seen in Sect. 14.1 that the Hamiltonian of an atom with Z electrons is

$$H_0 = \sum_{\alpha=1}^{Z} \left(\frac{\vec{p}_\alpha^2}{2m} - \frac{Z\,e^2}{r_\alpha} \right) + \sum_{\alpha > \beta} \frac{e^2}{r_{\alpha\beta}} \,. \tag{16.8}$$

However (16.8) is correct only if the intrinsic magnetic moment of the electrons is ignored. Indeed, the intrinsic magnetic moment of each electron interacts with the magnetic field created by the motion of all the charges the atom is made of, as well as with the magnetic moments of all the other electrons. All this implies the existence of further terms to be added to (16.8). Nevertheless these further terms, inasmuch as relativistic corrections to (16.8), only bring – at least in the case of light atoms (if Z is small, the velocities of the electrons are much smaller than c) – small corrections to the levels of (16.8), so it will be justified to treat them as a perturbation.

In order to make a perturbative calculation it is necessary to know in advance how to classify the eigenstates of the unperturbed Hamiltonian H_0 and which are the degeneracies of its eigenvalues. This problem has already been faced in Sect. 13.6 (see (13.42)): however the classification of the levels of H_0 discussed under those circumstances only had a provisional character, since it did not take into account either the electron spin or the Pauli principle. It is for this reason that, in the present section, we reconsider the problem of classifying the energy levels of H_0.

We must find the higher possible number of constants of motion (i.e. of observables that commute with H_0): this will allow us both to extract – from among the latter observables – a (possibly overabundant) set of compatible observables, by means of which we will classify the eigenstates of H_0 and to obtain precise information – by exploiting the degeneracy theorem of Sect. 7.2 – about the degree of degeneracy of the eigenvalues of H_0.

All the components of the total orbital angular momentum are constants of motion:

$$[\vec{L}, H_0] = 0 \,, \qquad\qquad \vec{L} = \sum_{\alpha=1}^{Z} \vec{L}_\alpha \tag{16.9}$$

whereas, on the contrary, the single $\vec{L}_\alpha$ are *not* constant of motion: indeed, the Hamiltonian (16.8) is a scalar if the coordinates $\vec{q}_\alpha$ and $\vec{p}_\alpha$ of *all* the electrons are simultaneously rotated; instead H_0 does not stay unchanged if the coordinates of a single electron are rotated (it is the term $\sum e^2/r_{\alpha\beta}$ that is not invariant, the single $\vec{L}_\alpha$ are constants of motion only in the independent-electron approximation).

Since in H_0 the spin variables do not appear, each $\vec{s}_\alpha$ commutes with H_0 and as a consequence, in particular,

$$[\vec{S},\,H_0] = 0\,, \qquad \vec{S} = \sum_{\alpha=1}^{Z} \vec{s}_\alpha\,. \tag{16.10}$$

Furthermore, also the space-inversion operator I commutes with H_0. Therefore a set of compatible observables consists of

$$H_0 \quad I \quad \vec{L}^2 \quad L_z \quad \vec{S}^2 \quad S_z \tag{16.11}$$

and the eigenstates of H_0 can be classified by means of the eigenvalues of such observables:

$$|\,E^0 \quad w \quad L \quad M \quad S \quad S_z'\,\rangle\,. \tag{16.12}$$

Since *all* the components of $\vec{L}$ and $\vec{S}$ commute with H_0 (and with I), each energy level is degenerate with respect to both $M\ (-L \le M \le +L)$ and to $S_z'\ (-S \le S_z' \le +S)$ and has a degree of degeneracy equal to

$$g_{E^0} = (2L+1)\times(2S+1)\,. \tag{16.13}$$

The last statement deserves a particular comment. In arriving at the conclusion that degeneracy of each level exactly is that given by (16.13) we have excluded that the levels may be degenerate on L and/or S: indeed we know that the degeneracy theorem is not able to establish that the degree of degeneracy is so much, but only that it is *at least* so much – i.e. it is able only to give a lower bound. So we cannot show that the eigenvalues of H_0 are not degenerate on L and/or S: we can only produce plausibility arguments.

As far as L is concerned, we have no good reason to suspect such a degeneracy: as we will see in the next section – when discussing the carbon atom as an example – a degeneracy on L may occur only in the independent-electron approximation, but such a degeneracy is removed by the Coulombic repulsion among the electrons (one could say that in Physics the 'compulsory government' principle holds: *all what is not mandatory is forbidden*; in the present case the degeneracy on L is not demanded by symmetry reasons, as that on M is, so it is forbidden: the physical basis for such a 'principle' is that Nature is sufficiently complicated that there always exists some interaction that removes the degeneracies that can be removed).

As far as S is concerned, the argument is more complex. Indeed, as already observed, not only the components of $\vec{S}$ are constants of motion, but also all the single $\vec{s}_\alpha$: this would entail a spin degeneracy equal to 2 for each single electron, i.e. 2^Z for each energy level, instead of $2S + 1$. We have already examined in the case of the helium atom the mechanism by which the Pauli principle prevents, in Nature, such a degeneracy: the number of permitted spin states is less than 2^Z and, in addition, the different values of S correspond to different orbital states having different energies.

In conclusion the energies of the states (16.12) depend both on L and S:

$$E^0 = E_{LS}$$

and their degree of degeneracy is given by (16.13). As either L and/or S are varied, different energy levels are obtained, the differences in energy being of the order of the eV.

The classification in terms of (16.11) is the analogue of the classification in terms of (15.18) of Sect. 15.3. It is then clear that, (16.9) and (16.10) implying that

$$[\vec{J}, H_0] = 0, \qquad \vec{J} = \vec{L} + \vec{S} \qquad (16.14)$$

it is also possible to classify the eigenstates of H_0 by means of the eigenvalues of

$$H_0 \quad I \quad \vec{L}^2 \quad \vec{S}^2 \quad \vec{J}^2 \quad J_z \qquad (16.15)$$

analogous to (15.19):

$$| E^0 \quad w \quad L \quad S \quad J \quad J_z' \rangle . \qquad (16.16)$$

According to the latter classification, the degeneracy of the energy levels, whose degree always is $(2L + 1) \times (2S + 1)$, is looked upon as the degeneracy on J, that takes the values between $|L - S|$ and $L + S$ and, for each value of J, on J_z', that ranges from $-J$ to $+J$.

In the sequel we will often adopt the spectroscopic notation to indicate the quantum numbers of an energy level: a capital letter S, P, D, $\cdots$ to indicate the value of L ($= 0, 1, 2, \cdots$) with an upper left character to indicate the spin multiplicity $2S + 1$ (sometimes also a small o is exposed as an upper right character if the level has the space-inversion parity -1: o stands for 'odd'); such a symbol is named spectroscopic *term* (not to be confused with the spectroscopic terms introduced in Sect. 2.4). For example, for the levels of helium, discussed in the previous section, we have the following terms:

$$(1s)^2 \, {}^1S , \qquad 1s\,2s \, {}^1S \qquad 1s\,2s \, {}^3S \qquad 1s\,2p \, {}^1P^o , \qquad 1s\,2p \, {}^3P^o .$$

In the above case we have indicated the configuration that generates each spectroscopic term, so the use of the ^{o} to indicate that the level has odd parity is superfluous because the parity can be read directly in the configuration:

$$w = (-1)^{l_1 + \cdots + l_z} \qquad (\neq (-1)^L \, !) . \qquad (16.17)$$

16.3 Relationship between Electronic Configuration and Spectroscopic Terms: the Carbon Atom

In the previous section we have examined how the energy levels of an atom can be classified: the classifications we have discussed – albeit rigorous in the approximation in which the interactions involving the spin magnetic moments of the electrons are neglected – are too general and provide little information: in particular they do not provide any information about the sequence of the energy level of an atom and the corresponding quantum numbers. On the other hand, in Sect. 14.1 we have introduced the independent-electron classification: the latter gives all the information one may wish about the energy levels of an atom, not only the information concerning energies and quantum numbers, but it even allows one to write the wavefunction of any atomic state in terms of the hydrogen-like wavefunctions. Unfortunately, however, the independent-electron approximation is too gross so that it is not a priori evident how much such large quantity of information is reliable.

Our aim is now to show that it is possible to extract, from the independent-electron approximation, both some qualitative information about the position of the energy levels of an atom, and rigorous information about the quantum numbers relative to such levels.

Instead of discussing the problem in general, we rather prefer to illustrate the concepts and the methodology in a particular case: the carbon atom.

The problem we want to face is the following: which are the lowest energy levels of carbon? The carbon atom has six electrons; the ground-state configuration is therefore

$$(1s)^2 (2s)^2 2p^2 .$$

First of all, let us try to estimate the first-ionization energy of carbon: for this problem we can consider the four electrons of the complete shells $1s$ and $2s$ as the atomic core. However, in the present case, the concept of atomic core is much less meaningful than it is in the case of the alkali atoms: this is due to the fact that the two $2p$ 'external' electrons have the same n as two of the electrons of the atomic core. So saying the two $2p$ electrons are external just is a way of saying: differently from the alkali atoms, here it happens that the first excited configuration is $(1s)^2 2s 2p^3$ (see Fig. 14.1), i.e. it is exactly one electron of the atomic core that changes its state.

If we neglected the repulsion between the $2p$ electrons, the ionization energy should be $4 \times 13.6/4 = 13.6\,\text{eV}$ (the nuclear-plus-atomic-core net charge is 2), but this probably is un upper bound. If instead we considered one of the two $2p$ electrons as part of the atomic core (and in this way we would overestimate the repulsion between the external electrons), the energy necessary to extract the other $2p$ electron should be $1 \times 13.6/4 = 3.4\,\text{eV}$, which certainly is a lower bound. It seems legitimate to expect a value closer to $13.6\,\text{eV}$ than to $3.4\,\text{eV}$, also in view of a possible penetration in the atomic core $(1s)^2 (2s)^2$: indeed, the experimental value for the first-ionization energy is $11.3\,\text{eV}$.

Which are the quantum numbers L and S of the 'atomic core'? Since we are dealing with four electrons, all with $l = 0$, obviously also $L = 0$. As far as S is concerned, we know that, due to the Pauli principle, two electrons in the same orbital always have $S = 0$. So, adding the spin of the two $(1s)^2$ electrons – which is zero – with that of the two $(2s)^2$ electrons – which is zero as well – we obtain $S = 0$. So the atomic core $(1s)^2 (2s)^2$ has $L = 0$ and $S = 0$, i.e. it does not contribute to values of L and S of the atom.

It is true in general – always as a consequence of the Pauli principle – that any complete shell (as e.g. the $(2p)^6$ shell of Na) has $L = 0$ and $S = 0$: this is obvious for what regards S, because any pair of electrons in the same orbital is in a spin singlet state. As far as L is concerned, let us consider as an example just a p^6 shell: it is clear that $M = 0$, because there are two electrons with $m = +1$, two with $m = 0$ and two with $m = -1$; but this statement stays true even if we decide to classify the states of the single electrons by means of the eigenvalues of either the x or the y components of the $\vec{L}_\alpha$: so, for a complete shell, the eigenvalues of all the components of $\vec{L}$ are vanishing, therefore $L = 0$.

If one needed to explicitly write the state of the six electrons in the independent-electron approximation – that must be antisymmetric under the permutation of the quantum numbers of any pair of electrons – the **Slater determinant** could be used: calling (for the sake of brevity) $a, b, c, \cdots, f$ the six independent states $|n; l = 1; m = 0, \pm 1; s' = \pm\frac{1}{2}\rangle$, one has (the indices $1, \cdots, 6$ refer to the six electrons):

$$|p^6\rangle = \det \begin{vmatrix} a_1 & b_1 & c_1 & \cdots & f_1 \\ a_2 & b_2 & c_2 & \cdots & f_2 \\ \vdots & \vdots & \vdots & \ddots & \vdots \\ a_6 & b_6 & c_6 & \cdots & f_6 \end{vmatrix}.$$

Use of the properties of the determinant allows one to explicitly check that the result, that is an antisymmetric linear combination of 6!=720 vectors of the type (14.3), is independent of the chosen basis (e.g. the states could exactly be classified by either the x or the y component of the angular momentum).

The fact that a complete shell has $L = 0$ justifies the statement, made in Sect. 14.5, that the wavefunction $\Phi(\vec{x}_2, \cdots, \vec{x}_Z)$ of the atomic core of an alkali atom has spherical symmetry.

For what regards the states originated by the ground-state configuration of the carbon atom, the quantum numbers L and S are determined only by the two $2p$ electrons: they have $l_1 = 1$ and $l_2 = 1$, so the possible values for L are $L = 0$ (S), $L = 1$ (P) and $L = 2$ (D).

If we provisionally ignore the electron spins, we have nine independent states:

$$|n_1 = 2,\ l_1 = 1,\ m_1;\ n_2 = 2,\ l_2 = 1,\ m_2\rangle, \qquad m_1, m_2 = 0, \pm 1 \quad (16.18)$$

or also:

$$|n_1 = n_2 = 2,\ l_1 = l_2 = 1;\ L M\rangle, \qquad L = 0, 1, 2;\ -L \le M \le +L \quad (16.19)$$

that, in the independent-electron approximation, all have the same energy. Note that in the present case in which $n_1 = n_2$ and $l_1 = l_2$ – called the case of **equivalent electrons** – the exchange degeneracy ($m_1 \leftrightarrow m_2$) is contained in the degeneracy on m_1 and m_2, so it gives nothing new.

Let us now imagine to take into account the Coulombic repulsion between the two external electrons and to treat it as a perturbation to the first order: we have to diagonalize the 9×9 matrix consisting of the matrix elements of $H' = e^2/r_{12}$ among either the states (16.18) or (16.19). Since H' commutes with $\vec{L}^2$ and L_z, it is convenient to use the basis (16.19): in this basis, indeed, the only nonvanishing matrix elements of H' are the diagonal ones, i.e. those with the same L and the same M; furthermore such matrix elements – that are the first-order corrections to the energy level – depend only on L but not on M, owing to the fact that H' is a scalar with respect to $\vec{L}$ and, as a consequence, the energy levels are degenerate with respect to M.

The states (16.19), but not the states (16.18), are therefore the approximate eigenstates of the Hamiltonian: the Coulombic repulsion has removed the degeneracy on L, so the configuration $2p^2$ gives rise to three levels respectively with $L = 0, 1, 2$; each of these levels is degenerate on M with degeneracy degrees respectively equal to $1, 3, 5$.

Note that, *to this level of approximation*, the stationary states of the atom – i.e. the states (16.19) – still are classified by the quantum numbers n_α and l_α (but not by the m_α) relative to the single electrons: therefore these are **approximately good** quantum numbers (i.e. good up to the first order in H'): also in this sense, assigning the electronic configuration (namely the n_α and the l_α of all the electrons) is something better than the independent-electron approximation.

Before we take spin into consideration, we want to understand which of the results, obtained by means of a perturbative treatment of the Coulombic repulsion, have a general validity and which ones do not.

For instance, one of the results we have found is that, starting from the $(1s)^2 (2s)^2 2p^2$ configuration and 'turning on' the interaction between the two $2p$ electrons, three levels with quantum numbers $L = 0, 1, 2$ are obtained, and all of them with parity $w = +1$ (see (16.17)): the problem is whether these are the exact quantum numbers or if they are correct only to the first order in H'. The same problem arises for the quantum numbers (L, parity w and orbital symmetry) we found in Sect. 16.1 for the levels of the helium atom. It is clear, therefore, that this is a problem of general nature.

In order to discuss this problem, we imagine to multiply the term representing the Coulombic repulsion among the electrons by a parameter λ that varies from 0 to 1: in such a way we obtain the (fictitious) Hamiltonian

$$H(\lambda) = H_0 + \lambda \sum_{\alpha > \beta} \frac{e^2}{r_{\alpha\beta}}$$

that for $\lambda = 0$ reduces to H_0 – the independent-electron approximation – and for $\lambda = 1$ is the 'true' Hamiltonian of the atom.

We expect that, as λ is continuously varied from 0 to 1, both the eigenstates of $H(\lambda)$ and the quantum numbers that classify them continuously depend on λ: it is then clear that, if such quantum numbers only take discrete values – as L and w – the only way in which they can be continuous functions of λ is to be independent of λ, i.e. to be constants. Therefore the quantum numbers (L, w etc.) that are deduced from a given electronic configuration (i.e. for $\lambda = 0$) exactly are the quantum numbers of those eigenstates of $H(\lambda = 1)$ that are originated by the given electronic configuration (**adiabatic theorem**). This argument shows how it is indeed possible (as anticipated in the discussion at the end of Sect. 14.1) to extract rigorous information on the quantum numbers of the energy levels from the independent-electron approximation.

Note however that the adiabatic theorem discussed above only applies to the discrete eigenvalues of those observables that commute with *any* $H(\lambda)$ – as $\vec{L}^2$, I, Π_o – and not, for example, to the n_α and l_α by means of which it is not possible to classify the *exact* eigenstates of H.

Let us now take the electron spin into consideration. The two $2p$ electrons may have total spin either $S = 0$ or $S = 1$: in order to know which value of S is compatible with each of the three levels S, P, D, the orbital symmetry (the eigenvalue of Π_o) of such levels must be known. It is clear that the state with $m_1 = m_2 = 1$ has $L = 2$ and is symmetric ($n_1 = n_2$, $l_1 = l_2$, $m_1 = m_2$); furthermore, since Π_o commutes with all the components of $\vec{L}$, and in particular with the lowering operator L_-, all the states with $L = 2$, $M = 2, \cdots, -2$ are symmetric. With the two independent states with $M = 1$ (we omit, for the sake of brevity, $n_1 = n_2 = 2$, $l_1 = l_2 = 1$)

$$| m_1 = 1 \; m_2 = 0 \rangle \qquad\qquad | m_1 = 0 \; m_2 = 1 \rangle$$

the symmetric linear combination

$$| m_1 = 1 \; m_2 = 0 \rangle + | m_1 = 0 \; m_2 = 1 \rangle$$

and the antisymmetric linear combination

$$| m_1 = 1 \; m_2 = 0 \rangle - | m_1 = 0 \; m_2 = 1 \rangle$$

can be formed. The first, being symmetric, must have $L = 2$; therefore the second must have $L = 1$. As a consequence all the states with $L = 1$ are antisymmetric (degeneracy on M of the eigenvectors of Π_o). With the three states with $M = 0$

$$| m_1 = 1 \; m_2 = -1 \rangle \qquad | m_1 = -1 \; m_2 = 1 \rangle \qquad | m_1 = 0 \; m_2 = 0 \rangle$$

only one combination can be antisymmetric: this is so because the third state $| m_1 = 0 \; m_2 = 0 \rangle$ is symmetric, therefore the three states give rise to two independent symmetric combinations. The antisymmetric one must have $L = 1$, a particular symmetric combination has $L = 2$ and, in conclusion, the orthogonal one – the last one we are left with – has $L = 0$.

Summarizing:

$$
\begin{array}{lll}
(2p)^2\ S\ , & \Pi_o = +1 & \text{(symmetric)} \\
(2p)^2\ P\ , & \Pi_o = -1 & \text{(antisymmetric)} \\
(2p)^2\ D\ , & \Pi_o = +1 & \text{(symmetric)}\ .
\end{array}
$$

Recalling that, because of the Pauli principle, if a state has $\Pi_o = 1$ it must have (total) spin $S = 0$, whereas if it has $\Pi_o = -1$ it must have $S = 1$, we have for the three levels of the carbon atom the following terms:

$$
^1S \qquad ^3P \qquad ^1D\ .
$$

In order to decide which of these three levels should be the lowest in energy, we can repeat what we said about the levels of helium originating by the same configuration: the level ^3P, that is antisymmetric under the exchange of the orbital quantum numbers, must have an energy lower than the other two, in accordance with observation: this is an instance of application of the already mentioned first Hund rule according to which the term with the highest spin has the lowest energy. Above the ^3P level, the second Hund rule requires that, among the ^1D and the ^1S that have the same multiplicity (i.e. the same spin), the one with the higher orbital angular momentum has the lowest energy. Also the rationale of the second Hund rule is based on the effect of the Coulombic repulsion: according to a classical picture, in the states with higher orbital angular momentum the electrons are orbiting in the same direction, therefore they mostly stay apart from each other and the Coulombic repulsion has lesser effect than when orbiting in opposite direction.

Summing-up, above the ^3P level, at a distance of 1.2 eV, there is the ^1D, finally there is the ^1S level that is 2.75 eV above the lowest energy level: in this way one sees that, much as in the case of helium, the levels originated by the same configurations, but differing in either L and/or S, have energies whose differences are of the order of the electronvolt.

16.4 Spin-orbit Interaction. Fine Structure of the Energy Levels

Let us now take into consideration the relativistic corrections to the Hamiltonian (16.8). Among the several relativistic corrections to be added to (16.8), the most relevant is the so called **spin-orbit interaction**.

To understand the physical origin of such interaction let us consider, for the sake of simplicity, an hydrogen-like atom with nuclear charge $Z\,e$. If we are in a frame in which the electron is at rest, it is the nucleus that rotates around the electron: as a consequence, the electron 'feels' the magnetic field generated by the motion of the nucleus. Such a field is given by

$$
\vec{B} = -\frac{\vec{v}}{c} \wedge \vec{E} = \frac{\vec{v}}{c} \wedge \frac{\vec{r}}{r}\frac{\mathrm{d}\varphi(r)}{\mathrm{d}r} = -\frac{1}{mc}\frac{1}{r}\frac{\mathrm{d}\varphi(r)}{\mathrm{d}r}\,\vec{L}\ , \qquad\qquad \varphi = \frac{e}{r}
$$

and gives rise to the interaction (15.4) with the intrinsic magnetic moment of the electron:

$$H' = (g-1) \times \frac{-e}{2\,m^2\,c^2} \frac{1}{r} \frac{d\varphi}{dr} \vec{L} \cdot \vec{s} = \frac{1}{2m^2c^2} \frac{Z\,e^2}{r^3} \vec{L} \cdot \vec{s}. \qquad (16.20)$$

The term $g-1$ instead of g in (16.20) is not justifiable within the proposed calculation ($H' = -\vec{\mu}_s \cdot \vec{B}$): the discrepancy originates from the fact that the motion of the electron is not rectilinear and uniform and, in passing to the frames in which the electron is instantaneously at rest ("tangent frames") the electron spin undergoes a rotation, called **Thomas precession**. A correct treatment, based on the relativistic Dirac equation, provides, besides the correct value of g ($= 2$), also the term -1 appearing in (16.20). The last term in (16.20) follows therefore from having put $g-1 = 2-1 = 1$.

The interaction (16.20) is called "spin-orbit interaction" because it couples the orbital and the spin angular momenta: however it should be clear that the origin of such interaction is the interaction between the spin *magnetic* moment and the *magnetic field* 'felt' by the electron.

Let us estimate the order of magnitude of the effects ΔE_{so} we must expect: if in (16.20) we replace r by $n^2\,a_B/Z$ (see (2.10)) and $\vec{L} \cdot \vec{s}$ by $\hbar^2$, we have

$$\Delta E_{so} \simeq \frac{\hbar^2}{2m^2\,c^2} \frac{Z^4\,e^2}{n^6\,a_B{}^3} = \frac{Z^4}{2n^6} \alpha^2 \frac{e^2}{a_B}. \qquad (16.21)$$

In (16.21) we note the factor Z^4/n^6: indeed, since $\overline{1/r^3}$ depends on l, the dependence on n varies from n^{-3} to n^{-6} according to the value of l; $\alpha \simeq 1/137$ is the fine-structure constant defined in (13.40) and $e^2/a_B = 27.2\,\mathrm{eV}$ is the atomic energy unit. Later on we shall see what are the typical values for ΔE_{so}.

For an atom having Z electrons, the spin-orbit interaction we must add to (16.8) is

$$\sum_{\alpha=1}^{Z} \xi(r_\alpha)\,\vec{L}_\alpha \cdot \vec{s}_\alpha \qquad (16.22)$$

where $\xi(r_\alpha)$ is related to the nuclear potential as in (16.20).

In addition to the interaction (16.22) there will be also other terms that we symbolically indicate as

$$\sum_{\alpha\beta} \cdots \vec{L}_\alpha \cdot \vec{s}_\beta, \qquad \sum_{\alpha\beta} \cdots \vec{s}_\alpha \cdot \vec{s}_\beta, \qquad \cdots$$

the first of which, called spin-other-orbit interaction, has an origin analogous to (16.22); the second is due to the interaction among the magnetic moments of the electrons; the last $\cdots$ stand for other corrections relativistic in nature.

In conclusion, the complete atomic Hamiltonian will be

$$H_{at} = H_0 + \sum_{\alpha=1}^{Z} \xi(r_\alpha)\,\vec{L}_\alpha \cdot \vec{s}_\alpha + \cdots = H_0 + H' \qquad (16.23)$$

with H_0 given by (16.8).

The idea is to consider H' as a perturbation and to see which effects does it produce on the atomic levels. Of course, since H' is very complicated, nor is it completely known, we cannot expect to calculate quantitatively the effect of H' on the levels of an atom – the latter not being exactly known as well – but we can only obtain qualitative information about the structure of the levels and general information about their degree of degeneracy, 'good' or 'approximately good' quantum numbers etc.

For a first-order treatment of the effect of H' we must only take into account the matrix elements of H' among states with the same energy E_{LS} and neglect all the other ones (***Russell–Saunders*** approximation): this is justified by the fact that, as we already know, in any atom the differences in energy due to the only electrostatic interactions – the distances among the E_{LS} levels – are, at least for the lowest lying levels, of the order of the electronvolt whereas, in view of the estimate (16.21), we know that the effects of (16.21) are, at least for the lighter atoms, of the order of $10^{-4} \div 10^{-2}\,\mathrm{eV}$ (the more excited levels are closer to each other, but on them – on the other hand – the effect of H' is smaller).

We must then consider the matrix relative to the perturbation H' among the $(2L+1)\times(2S+1)$ states (16.12) or (16.16), with given values of L and S: to decide which of the two bases is more convenient to be used, let us note the following: no matter how complicated H_{at} given by (16.23) may be, inasmuch as the Hamiltonian of an *isolated* atom (i.e. no external fields are present) certainly it commutes with all the components of the total angular momentum $\vec{J}$ (all the interactions – spin-orbit, spin-spin, $\dots$, – as well as H_0, are scalars under the simultaneous rotations of all the orbital and spin variables of all the electrons), so thanks to (16.14) also H' commutes with $\vec{J}$ and, therefore, in particular with J_z and $\vec{J}^2$, whereas H' does not commute with either L_z or S_z – indeed with any component of either $\vec{L}$ or $\vec{S}$ (the spin-orbit terms do not separately commute with $\vec{L}$ or $\vec{S}$). This entails that the matrix that represents H' is diagonal in the basis (16.16) (fixed values of L and S), whereas it is not such in the basis (16.12). There is no doubt that the choice of the basis (16.16) is more convenient for the perturbative calculation. Furthermore – and note that this argument is analogous to that concerning the levels of the carbon atom in the previous section – the (only nonvanishing) matrix elements of H'

$$\langle\, E_{LS}\ L\ S\ J\ J'_z \,|\, H' \,|\, E_{LS}\ L\ S\ J\ J'_z \,\rangle$$

only depend on J, that ranges from $|L-S|$ to $L+S$ (as well as, of course, on the level E_{LS} under consideration), but do not depend on J'_z, that takes the values between $-J$ and $+J$, because H' commutes with all the components of $\vec{J}$ and cannot, as a consequence, remove the degeneracy on J'_z.

So we have

$$\langle\, E_{LS}\ L\ S\ J\ J'_z \,|\, H' \,|\, E_{LS}\ L\ S\ J\ J'_z \,\rangle = \Delta E_{LSJ} \qquad (16.24)$$

and the ΔE_{LSJ} are the first-order corrections produced by the perturbation H' to the levels E_{LS} .

To summarize: the approximated eigenvalues of the Hamiltonian (16.23) are of the form:

$$E_{LSJ} = E_{LS} + \Delta E_{LSJ} \qquad (16.25)$$

and the corresponding approximated eigenvectors are the vectors (16.16)

$$|E_{LS}\, L\, S\, J\, J'_z\rangle\, . \qquad (16.26)$$

Equation (16.25) expresses the fact that, owing to the relativistic corrections H' that couple the orbital with the spin degrees of freedom, the energy levels no longer only depend on L and S, but also on J; this means that H' has partially removed the degeneracy $(2L+1) \times (2S+1)$ of the levels of H_0: it has completely removed the degeneracy on J (that it was not forced to keep by any invariance argument), leaving however the degeneracy on J'_z (that is protected by the invariance of H_{at} under rotations).

If in (16.25) J is varied while L and S are kept fixed, it is only the correction ΔE_{LSJ} that changes; since – at least in not too heavy atoms – these corrections are small ($\Delta E_{LSJ} \simeq 10^{-4} \div 10^{-2}\,$eV), this means that the levels of an atom must be structured in the form of **multiplets** (i.e. groups) of levels very close to one another (those with L and S fixed and J varying from $|L-S|$ to $L+S$). These multiplets are separated from each other by about $1\,$eV, since in going from a multiplet to the nearest one either L and/or S must change.

Such multiplets are called **fine-structure multiplets** or **fine-structure levels**; they are still represented by means of the spectroscopic terms by adding the value of J as a lower right index: for example ${}^3P_0$, ${}^3P_1$, ${}^3P_2$ (see Fig. 14.1). The reason for indicating – in the spectroscopic terms – the spin multiplicity $2S+1$ instead of the value S of the spin lies in the fact that often (precisely when $L \geq S$), $2S+1$ is just the number of levels the fine-structure multiplet consists of.

Let us now consider some examples of fine-structure levels.

Alkali atoms

The lowest energy level of all the alkali atoms is a ${}^2S$ level: $L = 0$, $s = \frac{1}{2}$ and therefore the only value for J is $J = \frac{1}{2}$: ${}^2S_{\frac{1}{2}}$. Therefore there is no fine structure, i.e. there is only a single twice degenerate level. The same applies to all the excited levels with $L = 0$. If $L > 0$, two values of J are possible: $J = L + \frac{1}{2}$ and $J = L - \frac{1}{2}$. One therefore has fine-structure doublets, for example ${}^2P_{\frac{1}{2}}$ and ${}^2P_{\frac{3}{2}}$, ${}^2D_{\frac{3}{2}}$ and ${}^2D_{\frac{5}{2}}$ etc.

The separation in energy between these doublets decreases with the increase of the principal quantum number n and is more sizeable for atoms with higher Z: in Li the separation between $2\,{}^2P_{\frac{1}{2}}$ and $2\,{}^2P_{\frac{3}{2}}$ (the first 2 is the value of n) is only $0.4 \times 10^{-4}\,$eV, whereas in Na the separation between $3\,{}^2P_{\frac{1}{2}}$ and $3\,{}^2P_{\frac{3}{2}}$ is about $2 \times 10^{-3}\,$eV (see the discussion at the end of Sect. 14.4).

Helium

For the singlet levels (parahelium) $S = 0$, whence $J = L$ and there is no fine structure. For the triplet levels (orthohelium) $S = 1$ so in all cases except that of levels with $L = 0$ one has fine-structure triplets: $J = L - 1$, $J = L$, $J = L + 1$. The levels with $L = 0$ have $J = 1$ and no fine structure. Anyway the separations within a triplet are very small ($\simeq 10^{-4}\,\text{eV}$) due to the low value of Z ($Z = 2$) and to the fact that only the excited levels $(1s)(n\,l)$ ($n \geq 2$, $l \geq 1$) may give rise to fine-structure triplets.

Carbon

From among the three levels considered in the previous section and originated by the configuration of minimum energy $(1s)^2 (2s)^2\, 2p^2$, only the ^3P gives rise to a fine-structure multiplet, the triplet $^3P_0$, $^3P_1$, $^3P_2$: the higher J, the higher the energy and the separations are about $2 \times 10^{-3}\,\text{eV}$ between the $^3P_1$ and the $^3P_0$, and about $5.4 \times 10^{-3}\,\text{eV}$ between $^3P_2$ and the $^3P_0$.

We show in Fig. 16.2 the diagram of the levels of the carbon atom corresponding to the configuration $(1s)^2 (2s)^2\, 2p^2$.

From left to right: in the independent-electron approximation, in presence of the only Coulombic repulsion and, finally, in presence of the relativistic corrections. Fig. 16.2, that re-proposes Fig. 14.1, is not in scale because of the great difference, of about three orders of magnitude, between the separation of the levels inside the multiplet ^3P and the distance from the ^1D, ^1S levels.

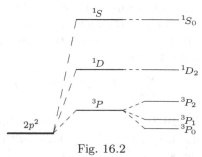

Fig. 16.2

As we said in the beginning of the present section, from among all the relativistic corrections that are included in H' the spin-orbit interaction (16.22) usually is the most relevant. If we take into consideration only the spin-orbit and the spin-other-orbit interactions and neglect in H' the spin-spin interaction, we can obtain a more detailed information about the structure of the multiplets, thanks to the following identity (the notation is the same as in Sect. 10.2):

$$\langle\, \varSigma''\, j\, m'' \mid V_i \mid \varSigma'\, j\, m'\,\rangle = \langle\, \varSigma''\, j\, m'' \mid \frac{\vec{M}\cdot\vec{V}}{\vec{M}^2}\, M_i \mid \varSigma'\, j\, m'\,\rangle \qquad (16.27)$$

where $\vec{M}$ is a generic angular momentum (either orbital or spin or total …) and $\vec{V}$ is a vector operator, i.e. an operator for which the commutation rules (10.9) hold: we will say $\vec{V}$ is a vector under $\vec{M}$; j and m refer to the angular momentum $\vec{M}$.

The demonstration of (16.27) is based on the identity (13.48) we rewrite here:

$$\left[\vec{M}^2,\, [\vec{M}^2,\, V_i]\right] = 2\hbar^2\left(\vec{M}^2 V_i + V_i\, \vec{M}^2\right) - 4\hbar^2\left(\vec{M}\cdot\vec{V}\right) M_i\,.$$

If we take the matrix elements of both sides between states with the same j, the left hand side gives a vanishing contribution:

$$\langle j \cdots | \, [\, \vec{M}^{\,2}, [\, \vec{M}^{\,2}, V_i]] \, | \, j \cdots \rangle =$$

$$\hbar^2 \Big(j(j+1) \langle j \cdots | \, [\, \vec{M}^{\,2}, V_i] \, | \, j \cdots \rangle - \langle j \cdots | \, [\, \vec{M}^{\,2}, V_i] \, | \, j \cdots \rangle j(j+1) \Big) = 0$$

whence

$$j(j+1) \, \hbar^2 \, \langle \, \Sigma'' \, j \, m'' \, | \, V_i \, | \, \Sigma' \, j \, m' \, \rangle = \langle \, \Sigma'' \, j \, m'' \, | \, (\vec{M} \cdot \vec{V}) \, M_i \, | \, \Sigma' \, j \, m' \, \rangle$$

from which (16.27) follows.

Equation (16.27) is known as **Wigner–Eckart theorem**, even if the way in which the theorem is usually formulated is slightly different and has a greater generality inasmuch as it applies not only to vector operators, but to tensor operators of any rank.

Let us apply (16.27) to the calculation of the first-order effect of spin-orbit like terms. In order to keep the notation as light as possible, we will refer only to the spin-orbit interaction given by (16.22). For the perturbative calculation we will use the basis (16.12) instead of (16.16): we then have the matrix elements

$$\sum_{\alpha} \langle E_{LS} \, L \, M'' \, S \, S'' \, | \, \xi(r_\alpha) \, \vec{L}_\alpha \cdot \vec{s}_\alpha \, | \, E_{LS} \, L \, M' \, S \, S' \, \rangle \, . \tag{16.28}$$

In (16.28) $\xi(r_\alpha) \, \vec{L}_\alpha$ is a vector operator under the total orbital angular momentum $\vec{L}$ ($\xi(r_\alpha)$ commutes with $\vec{L}$). Let us take its matrix elements among the states $| \, E_{LS} \, L \, M'' \, S \, S'' \, \rangle$ and $\vec{s}_\alpha | E_{LS} \, L \, M' \, S \, S' \, \rangle$ that are eigenstates of $\vec{L}^2$ with the same eigenvalue L ($\vec{s}_\alpha$ commutes with $\vec{L}$): by (16.27) we can make in (16.28) the replacement

$$\xi(r_\alpha) \, \vec{L}_\alpha \to \Big(\frac{\xi(r_\alpha) \vec{L} \cdot \vec{L}_\alpha}{\vec{L}^2} \Big) \, \vec{L} \; \Rightarrow \; \xi(r_\alpha) \, \vec{L}_\alpha \cdot \vec{s}_\alpha \to \Big(\frac{\xi(r_\alpha) \vec{L} \cdot \vec{L}_\alpha}{\vec{L}^2} \Big) \, \vec{L} \cdot \vec{s}_\alpha \, .$$

Now the same argument can be applied to the operator $\vec{s}_\alpha$, that is a vector operator under the total spin $\vec{S}$, of which the matrix elements with the same S are taken, so that in (16.28)

$$\sum_{\alpha} \xi(r_\alpha) \vec{L}_\alpha \cdot \vec{s}_\alpha \to \Lambda \, \vec{L} \cdot \vec{S} \, , \qquad \Lambda \equiv \sum_{\alpha} \Big(\frac{\xi(r_\alpha) \vec{L} \cdot \vec{L}_\alpha}{\vec{L}^2} \Big) \Big(\frac{\vec{S} \cdot \vec{s}_\alpha}{\vec{S}^2} \Big) \, . \tag{16.29}$$

So in the space spanned by the $(2L+1) \times (2S+1)$ vectors $| \, E_{LS} \, L \, M' \, S \, S' \, \rangle$, (16.22) can be replaced by $\Lambda \, \vec{L} \cdot \vec{S}$.

At this point we go back to the basis (16.16) in which the matrix that represents $\Lambda \, \vec{L} \cdot \vec{S}$ is diagonal and, since

$$\vec{J} = \vec{L} + \vec{S} \qquad \Rightarrow \qquad \vec{L} \cdot \vec{S} = \frac{1}{2} \Big(\vec{J}^2 - \vec{L}^2 - \vec{S}^2 \Big)$$

one has

$$\langle E_{LS}\,L\,S\,J\,J_z' \mid \Lambda \vec{L}\cdot\vec{S} \mid E_{LS}\,L\,S\,J\,J_z' \rangle =$$
$$= \frac{\hbar^2}{2}\langle E_{LS}\,L\,S\,J\,J_z' \mid \Lambda \mid E_{LS}\,L\,S\,J\,J_z' \rangle \Big(J(J+1) - L(L+1) - S(S+1) \Big).$$

The operator Λ commutes with all the components of $\vec{L}$ and $\vec{S}$, so in the subspace we are considering (whose dimension is $(2L+1)\times(2S+1)$) it is a multiple of the identity matrix (just as H_0; see also the theorem shown in Sect. 13.9 p. 244), therefore

$$\frac{\hbar^2}{2}\langle E_{LS}\,L\,S\,J\,J_z' \mid \Lambda \mid E_{LS}\,L\,S\,J\,J_z' \rangle \equiv \frac{\hbar^2}{2}\,A_{LS}$$

with A_{LS} depending only on L and S (as well as on E_{LS}), but *not* on J.

In conclusion, if we take into consideration only spin-orbit like interactions up to the first order, one has

$$\Delta E_{LSJ} = \frac{\hbar^2}{2}\,A_{LS}\Big(J(J+1) - L(L+1) - S(S+1) \Big). \qquad (16.30)$$

Equation (16.30) does not allow one to calculate the separations in energy among the fine-structure levels because A_{LS} is not known. However one knows that the contribution to A_{LS} due to the spin-orbit interaction alone (i.e. omitting the spin-other-orbit terms) is positive in the case all the shells are complete except the last one filled by less than a half, as e.g. in carbon: this is already sufficient to guarantee that in such cases the energy of the fine-structure levels increases as J increases (***normal*** multiplets, as e.g in carbon), while if $A_{LS} < 0$ one says that the multiplets are ***inverted***: the levels with higher J have a lower energy, as e.g. in the case of the triplet $^3P_{0,1,2}$ of oxygen generated by the lowest-energy configuration $(1s)^2\,(2s)^2\,2p^4$: $\ E(^3P_2) < E(^3P_1) < E(^3P_0)$.

From (16.30) the ***Landé interval rule*** follows:

$$E_{LSJ} - E_{L\,S\,J-1} = \hbar^2\,A_{LS}\,J \qquad (16.31)$$

and from (16.31), since the only unknown parameter is A_{LS}, when in a multiplet the number of levels is higher than two, it is possible to derive some relations involving the distances of the levels belonging to the multiplet: for example, in the case of the triplet $^3P_{0,1,2}$ of carbon and oxygen (16.31) entails that

$$E(^3P_2) - E(^3P_0) = 3\big(E(^3P_1) - E(^3P_0) \big).$$

In the case of carbon, by using the experimental values we have given for the distances among the fine-structure levels, the left hand side of (16.31) is about 5.4×10^{-3} eV whereas the right hand side is about 6.3×10^{-3} eV: the difference is approximately 15%.

Another result that can be derived from (16.30) is the following *sum rule*:

$$\sum_{J=|L-S|}^{L+S} (2J+1)\Delta E_{LSJ} = 0 \qquad (16.32)$$

expressing the fact that the 'center-of-energy' of the fine-structure levels (weighted by the degeneracy $2J+1$ of each of them) is not shifted.

In order to show (16.32) one can proceed by the direct calculation using the ΔE_{LSJ} given by (16.30), but – more easily – it suffices to observe that the left hand side of (16.32) is the sum of the eigenvalues, i.e. the trace, of the matrix

$$\langle\, E_{LS}\ L\ S\ J''\ J_z''\ |\ \Lambda\, \vec{L}\cdot\vec{S}\ |\ E_{LS}\ L\ S\ J'\ J_z'\,\rangle$$

and that the trace is independent of the representation, so we can calculate it in the basis (16.12):

$$\sum_{M\,S_z'}\langle\, E_{LS}\ L\ M\ S\ S_z'\ |\ \Lambda\, \vec{L}\cdot\vec{S}\ |\ E_{LS}\ L\ M\ S\ S_z'\,\rangle\,. \qquad (16.33)$$

Since both for L_x and L_y the selection rules $\Delta M = \pm 1$ apply (the same as (13.46) for D_x and D_y), (16.33) boils down to

$$\sum_{M\,S_z'}\langle\,\cdots\ M\ S_z'\ |\ \Lambda\, L_z\,S_z\ |\ \cdots\ M\ S_z'\,\rangle = \frac{\hbar^2}{2}\, A_{LS}\sum_{M\,S_z'} M\times S_z' = 0\,.$$

Equation (16.32) is useful to find the position E_{LS} of the unperturbed level from the experimental data about a fine-structure multiplet.

16.5 Hierarchy of Quantum Numbers. Selection Rules II

The discussion we have made about the structure of the atomic levels has required several steps: firstly we have considered the independent-electron approximation, then we have added the electrostatic repulsion among the electrons, finally we have taken into account the interactions that are relativistic in nature. Each of these steps is characterized by the existence of quantum numbers that are "good" within the considered approximation; in the next step some of such numbers stay "good", some are only "approximately good" (i.e. only up to the first order in the added interaction the eigenstates of the Hamiltonian are classified by means of these quantum numbers), finally other quantum numbers totally loose their value: this is a situation that shows up in many branches of physics, from nuclear physics to elementary particle physics.

This means that in the final classification of the atomic energy levels – i.e. the ones whose Hamiltonian is the complete Hamiltonian – there exists a hierarchy of quantum numbers: from the "good" ones to the "less good" ones and so on. It should be said that (16.23), that we have called the "complete Hamiltonian", still is an approximate Hamiltonian in which, for example, the interaction among the electrons and the nuclear magnetic moment – that is about 10^{-3} times the Bohr magneton (13.54) – is neglected: it gives rise to the **hyperfine structure** of the levels, ignoring which is legitimate only in the framework of low resolution power spectroscopy.

Once the limits of our discussion have been clarified, let us examine the hierarchy of the quantum numbers of an isolated atom: at the top we have J, J'_z and the space-inversion parity w; under such numbers, i.e. as a little bit "less good" numbers, there are L and S: indeed, $\vec{L}^2$ and $\vec{S}^2$ no longer are constants of motion in presence of spin-orbit interactions (except for one-electron atoms like hydrogen-like and alkali atoms), inasmuch as they do not commute with the single $\vec{L}_\alpha$ and $\vec{s}_\alpha$ in (16.23); however, up to the first order in the spin-orbit interaction, the eigenstates of the Hamiltonian still can be classified by means of L and S. This means that the "exact" eigenstates of the Hamiltonian will be linear combinations of states with the same J and J'_z (and parity) but with different values of L and/or S:

$$\alpha \,|\, L\,S\,J\,J'_z \rangle + \beta \,|\, L'\,S'\,J\,J'_z \rangle + \gamma \,|\, L''\,S''\,J\,J'_z \rangle + \cdots \qquad (16.34)$$

in which, if the first-order treatment of the spin-orbit interaction is reliable, one must have

$$|\alpha|^2 \gg |\beta|^2 + |\gamma|^2 + \cdots . \qquad (16.35)$$

Note that M and S'_z, that in absence of the spin-orbit interaction are good quantum numbers, loose their meaning in the very moment the spin-orbit is "switched on": this is due to the fact that, in presence of the only electro-static interactions, there is degeneracy on these quantum numbers and the introduction of the – no matter how small – spin-orbit interaction entails that the new eigenstates of the Hamiltonian, inasmuch as eigenstates of J^2, are linear combinations of states with different values of M and S'_z: such linear combinations, contrary to those of the type (16.34), are independent of the magnitude of the spin-orbit interaction and are univocally determined by the "geometric" rule of addition of angular momenta.

Making a step down in the hierarchy, we finally have the n_α and the l_α of the single electrons, i.e. the electronic configuration: also for these quantum numbers we could repeat what we have said for L and S, except the fact that the n_α and the l_α are only "approximately good" because of the inter-electronic repulsion, which is a rather strong interaction.

Which is the utility of this hierarchy of quantum numbers? In other words: which is the utility of the "approximately good" quantum numbers?

We have already seen the utility of giving the electronic configuration: from the one hand, the electronic configuration gives us a qualitative idea about the position of the energy levels, as e.g. the fact that – at least in the simplest cases – a more excited configuration gives rise to energy levels that, on the average, are higher than those originated by a less excited configuration; on the other hand, we have seen in Sect. 16.3 how a rigorous information about the quantum numbers of the levels can be obtained from the configuration. Moreover, the knowledge of the approximate quantum numbers L and S al-lows for the determination of the number of fine-structure sublevels within a multiplet and of the value of J for each of such sublevels.

But there is one more reason by which the quantum numbers – both the good and the less good ones – are important: the fact that, for each of them, we have a selection rule for the electromagnetic transitions: once the meaning of "approximate quantum numbers" has been clarified, as expressed by (16.34) and (16.35), it is evident that the more a given quantum number is "good", the better the selection rule relative to it will be verified, and viceversa.

We now complete this section with a list of the selection rules on the several quantum numbers we have already encountered, including for the sake of completeness also the already discussed selection rules and discussing for each of the validity and framework of applicability. Unless the contrary is stated, we will always understand *selection rules for electric-dipole transitions*.

1. Selection Rule on w, J, J_z:

$$w' \cdot w'' = -1 \tag{16.36}$$

$$\Delta J = \pm 1\,, 0 \qquad J = 0 \not\to J = 0 \quad \text{(rigorously prohibited)} \tag{16.37}$$

$$\Delta J_z = +1\,, -1\,, 0 \quad \text{(respectively for } D_+\,, D_-\,, D_z\text{)} \tag{16.38}$$

$$J_z = 0 \not\to J_z = 0 \qquad \text{if} \quad J' = J''\,. \tag{16.39}$$

Equation (16.36) has been discussed in Sect. 13.6. Equations (16.37) and (16.38) are formally identical with (13.47) and (13.44): indeed both are consequences of the fact that the electric dipole operator $\vec{D}$ has, both with $\vec{L}$ and $\vec{J}$, the commutation rules (10.9) of a vector. The selection rules (16.38) also are an immediate consequence of (13.44) and of the selection rule on S_z we will give in a while. The selection rule (16.39) has been written for the sake of completeness, but it will be neither shown nor utilized.

Thanks to the goodness of the quantum numbers they refer to, all the above selection rules are well verified: violations are to be attributed to either magnetic dipole or higher electric and magnetic multipole transitions.

2. Selection Rule on L and S:

$$\Delta L = \pm 1\,, 0 \qquad L = 0 \not\to L = 0 \quad \text{(rigorously prohibited)} \tag{16.40}$$

$$\Delta S = 0\,. \tag{16.41}$$

The selection rule (16.40) has been discussed in Sect. 13.6 while (16.41) has been introduced in Sect. 16.1 to explain the impossibility of transitions between orthohelium and parahelium.

The validity of the above selection rules is limited not so much by the electric dipole approximation but, rather, by the fact that L and S are approximate quantum numbers. The case of the alkaline earth atoms (as e.g. Mg, Ca, Hg, Ba, ...) is typical, in particular that of mercury Hg that has $Z = 80$; the ground-state configuration is

$$\text{(complete shells)}\,(6s)^2$$

and therefore, much as in He, the ground state is a state ^1S. The first excited configuration is (complete shells) $6s\,6p$

and gives rise, as in He, to states ^1P and ^3P. Due to the high value of Z, the spin-orbit interaction has non-negligible matrix elements between states with different energies (but with the same J), in particular between the states $^1P_1$ and $^3P_1$. The consequence of this fact is that, besides removing the degeneracy on J, the spin-orbit interaction appreciably mixes the states $^1P_1$ and $^3P_1$ with each other: therefore the two levels of Hg originated by the configuration $6s\,6p$ no longer have either a well defined spin or, as a consequence, a well defined orbital symmetry, but are linear combinations of states with $S = 0$ and $S = 1$ of the type (16.34):

$$|\,L = 1 \; S \simeq 0 \; J = 1 \; J_z\,\rangle = \alpha \; |\,^1P_1\,\rangle + \beta \; |\,^3P_1\,\rangle$$
$$|\,L = 1 \; S \simeq 1 \; J = 1 \; J_z\,\rangle = \alpha^* \; |\,^3P_1\,\rangle - \beta^* \; |\,^1P_1\,\rangle \;.$$

It is therefore clear that transitions to the ground state $^1S_0$ are possible from both levels, the probability transitions being proportional to $|\alpha|^2$ and $|\beta|^2$ respectively with, as in (16.35), $|\alpha|^2 \gg |\beta|^2$.

The lowest of the levels P_1 (that we have denoted as $|\,S \simeq 1\rangle$) was – improperly – classified by spectroscopists as a triplet level $^3P_1$ (in accordance with Hund's rule), so that one says that in this case the selection rule (16.41) is violated. The corresponding transition line ($\lambda = 2357\,\text{Å}$) is called an **intercombination line** (transition between states with 'different' spin). The same mechanism explains the very weak transitions between orthohelium and parahelium.

3. Selection Rule on L_z and S_z:

$$\Delta M = +1\,,\, -1\,,\, 0 \qquad \text{(respectively for } D_+\,,\, D_-\,,\, D_z\,) \qquad (16.42)$$
$$\Delta S_z = 0\,. \qquad\qquad\qquad\qquad\qquad\qquad\qquad\qquad\qquad (16.43)$$

Equation (16.42) is already known from Sect. 13.6; the selection rule expressed by (16.43), as already stated in Sect. 16.1, originates from the fact that the electric dipole operator $\vec{D}$ commutes with the components of $\vec{S}$ and with S_z in particular.

Such commutation rules are significant only when the stationary states of the atom can be classified by means of the eigenvalues of L_z and S_z and in presence of external fields that remove the degeneracy on these quantum numbers: we will see in the next section that both such conditions are met for an atom in a magnetic field so intense as to make the spin-orbit interaction negligible.

4. Selection Rule on the Electronic Configuration:
the only possible electromagnetic transitions are those occurring between states originating from configurations that differ in the quantum numbers $n,\,l$ of one only electron.

This selection rule does not presuppose the electric dipole approximation. If the latter holds, the further limitation (13.51) ($\Delta l = \pm 1$) on the variation of the l relative to the electron that changes its state applies.

The limits of validity of such selection rule are the same as for the classification by means of the electronic configuration, i.e. the first order in the Coulombic repulsion. However the selection rule on the configuration appears to be verified better than one would expect basing on the fact that the interaction among the electrons is a rather strong interaction: indeed the mixing (analogue to (16.34)) among different configurations due to the repulsion among the electrons is not very sizable (despite the term $\sum_{\alpha\beta} e^2/r_{\alpha\beta}$ may have nonvanishing matrix elements among states generated by different configurations) inasmuch as the differences among the unperturbed energies relative to different configurations usually are – at least for not too excited configurations – very large.

The proof of the selection rule 4 is based on the fact that the electromagnetic transition probability (13.24) is determined by the matrix element of an operator that is the sum of the operators $\xi_\alpha = \vec{p}_\alpha \cdot \vec{A}(\vec{x}_\alpha)$, each of which is a single-particle operator, i.e. it only contains the variables that refer to a single electron. It is therefore clear that, if the states $|E_a^0\rangle$ and $|E_b^0\rangle$ differ in the quantum numbers of two or more electrons, every matrix element $\langle E_b^0 \mid \xi_\alpha \mid E_a^0 \rangle$ is vanishing and, as a consequence, the transition probability between the states $|E_a^0\rangle$ and $|E_b^0\rangle$ is vanishing as well.

This selection rule is the reason, anticipated in Sect. 16.1, why we restricted our discussion on the energy levels of He to configurations where only one electron is excited.

16.6 Atoms in a Magnetic Field: the Anomalous Zeeman Effect

The introduction of the electron spin has allowed for the formulation of the Pauli principle as well as – due to the fact that the spin is associated with a magnetic moment – for the explanation of the fine structure of the energy levels. It is then clear that the spin must play a fundamental role also in the case of an atom in an external magnetic field, thanks to the interaction between the magnetic moment associated with it and the field. In this section we will examine which differences occur both in the treatment and in the results with respect to the treatment of Sect. 13.7, in which the existence of the spin was ignored.

The discussion of Sect. 13.7 had the Hamiltonian (13.56) as starting point; the changes, due to the spin of the electron, we must now make in (13.56) are two. First we must add the interaction among the magnetic moments of all the electrons and the magnetic field $\vec{B}$; this interaction is given by

$$-\sum_{\alpha=1}^{Z} \vec{\mu}_{s\,\alpha} \cdot \vec{B} = g\,\frac{e}{2m\,c}\,\vec{S} \cdot \vec{B}\ .$$

Secondly, while in (13.56) H_0 was the Hamiltonian of the isolated atom only inclusive of the electrostatic interactions, we must now consider the Hamiltonian (16.23) that includes also the spin-orbit interaction, as well as the other relativistic interactions.

The atomic Hamiltonian in a magnetic field then is (always neglecting the terms proportional to $\vec{B}^2$):

$$H = H_{at} + \frac{e}{2mc}\,(\vec{L} + g\,\vec{S})\cdot\vec{B}$$

or, choosing the z-axis parallel to the direction of $\vec{B}$:

$$H = H_{at} + \frac{eB}{2mc}\,(L_z + g\,S_z) \qquad (16.44)$$

where H_{at} is the *complete* Hamiltonian of the isolated atom, given by (16.23). Let us first show that necessarily $g \neq 1$. If in (16.44) we put $g = 1$,

$$H = H_{at} + \frac{eB}{2mc}\,J_z \qquad \text{(in the case } g = 1)$$

and, since J_z commutes with H_{at}, just as in the situation of Sect. 13.7, one has that the *exact* eigenvalues are

$$E^0 + \mu_B B\,J_z' \qquad \text{(in the case } g = 1)$$

E^0 being the eigenvalues of H_{at}. Also in this case, as in Sect. 13.7, due to the selection rules (16.38) $\Delta J_z' = \pm 1,\,0$ one has that, thanks to the effect of the magnetic field, each line splits up into three lines: one finds once more the normal Zeeman effect and the anomalous Zeeman effects remains unexplained.

So let us once for all take $g = 2$, whence (16.44) may be written in the equivalent forms

$$H = H_{at} + \frac{eB}{2mc}\,(L_z + 2S_z) \qquad (16.45)$$

and

$$H = H_{at} + \frac{eB}{2mc}\,(J_z + S_z)\ . \qquad (16.46)$$

In Sect. 13.7, as well as in the just discussed case $g = 1$, it has been possible to exactly calculate the effect of the magnetic field on the energy levels of the atom. Now, with either the Hamiltonian (16.45) or (16.46), this is no longer possible because the term that represents the interaction with the magnetic field

$$H' = \frac{eB}{2mc}\,(J_z + S_z)\ . \qquad (16.47)$$

does not commute with H_{at}: indeed it is S_z that does not commute with the spin-orbit interaction contained in H_{at}.

It is then clear that, if we could neglect the relativistic corrections, it would still be possible to exactly calculate the effect on the energy levels of the atom: from the physical point of view this is possible provided a very strong magnetic field is at one's disposal, such that the separations between the Zeeman sublevels turn out much larger than the separations among the fine-structure levels, to the point that the latter can be altogether neglected.

Since the Zeeman separations are of the order of $\mu_B B$, the condition that enables one to neglect the relativistic corrections in H_{at} is

$$\mu_B B \gg \Delta E_{LSJ} . \qquad (16.48)$$

For example in the case of Li, where the splitting between the $2\,^2P_{\frac{1}{2}}$ and $2\,^2P_{\frac{3}{2}}$ levels is about $0.4 \times 10^{-4}\,\text{eV}$, (16.48) is satisfied by a magnetic field B with intensity much higher than 10^4 gauss.

The situation is still less favourable in atoms with higher Z. In any event let us start the discussion with the case of the lithium atom that, inasmuch as exactly soluble, is the simplest:

Strong Field

As just said, strong field means that (16.48) is satisfied. It is convenient to take the Hamiltonian in the form (16.45) and – provisionally – neglect all the relativistic corrections in H_{at}. Then L_z and S_z commute with H_{at} and the eigenvectors (16.12) of H_{at} also are eigenvectors of H:

$$\left(H_{at} + \frac{B}{2mc}\,(L_z + 2S_z)\right)| E_{LS}\,L\,S\,S_z' \rangle = \left(E_{LS} + \mu_B B\,(M + 2S_z')\right)| E_{LS}\,L\,S\,S_z' \rangle .$$

The eigenvalues of H then are

$$E_{LS} + \mu_B B\,(M + 2S_z') , \qquad L \le M \le +L , \qquad -S \le S_z' \le +S . \qquad (16.49)$$

In the present approximation the distance between two adjacent Zeeman sub-levels still is $\mu_B B$ for all the levels with $L \ne 0$. It is instead $2\mu_B B$ for the levels with $L = 0$ and $S \ne 0$. The number of components into which, owing to the field, any spectral line splits up still is three, due to the selection rules $\Delta M = 0 \pm 1,\,0$ and $\Delta S_z' = 0$ listed in point 3 of the previous section. Therefore, again, there is a central π line and two side σ lines: the case we have discussed presents all the features of the normal Zeeman effect.

Since, by neglecting the relativistic corrections, we have exactly solved the problem, it is possible to consider the relativistic corrections as a perturbation and determine the corrections to the levels (16.49) and, as a consequence, the effect on the Zeeman lines (***Paschen–Back effect***). We will not discuss now this problem because we will find the result as a particular case in the framework of the discussion of the intermediate field.

The case we are now going to discuss is – from the physical point of view – more interesting, inasmuch as more easily realizable: it is the case of the

Weak Field

A magnetic field is considered a weak field when it produces, on the levels of an atom, effects small with respect to the distances among the fine-structure levels. In other words, when the matrix elements of H' given by (16.47) among the states belonging to a given fine-structure multiplet are small with respect to the energy differences among the levels belonging to the multiplet. For example, in the case of the doublet $3\,^2P_{\frac{1}{2}}$, $3\,^2P_{\frac{3}{2}}$ of Na, with a separation of

2×10^{-3} eV, a field B of 10^4 gauss can still be considered small, because it produces effects of the order of

$$\mu_B B \simeq 10^{-4} \, \text{eV} \, .$$

In this case it is therefore legitimate to consider H' as a perturbation and only consider the matrix elements of H' among the eigenstates of H_{at} with the same energy, i.e. among the states

$$| E_J \, J \, J'_z \rangle \, , \qquad\qquad -J \le J'_z \le +J \qquad\qquad (16.50)$$

where E_J are the eigenvalues of H_{at}; the states (16.50) are – for the time being – the *exact* eigenstates of H_{at}.

Since H' commutes with J_z, the off-diagonal matrix elements of H' among the $2J+1$ states (16.50) are vanishing; therefore the diagonal elements directly give the first-order corrections to the energy levels. Since the states (16.50) are eigenstates of J_z, we are left with calculating the mean values

$$\langle E_J \, J \, J'_z \, | \, S_z \, | \, E_J \, J \, J'_z \rangle \, . \qquad\qquad (16.51)$$

Thanks to (16.27) (Wigner–Eckart theorem) the matrix elements between states *with the same* J, exactly as the mean values (16.51), equal the matrix elements of

$$\frac{\vec{S} \cdot \vec{J}}{\vec{J}^{\,2}} \, J_z \, . \qquad\qquad (16.52)$$

It follows that

$$\langle E_J \, J \, J'_z \, | \, S_z \, | \, E_J \, J \, J'_z \rangle = \gamma_J \, J'_z \qquad\qquad (16.53)$$

where

$$\gamma_J = \frac{1}{\hbar^2 \, J(J+1)} \, \langle E_J \, J \, J'_z \, | \, \vec{S} \cdot \vec{J} \, | \, E_J \, J \, J'_z \rangle \, . \qquad\qquad (16.54)$$

only depends on J and the considered level, but not on J'_z inasmuch as $\vec{S} \cdot \vec{J}$ is a scalar operator, i.e. it commutes with all the components of $\vec{J}$.

If we define the **Landé factor** g_J as

$$g_J = 1 + \gamma_J \qquad\qquad (16.55)$$

one has that the energy levels of the atom to the first order in the magnetic field are

$$E_J + \mu_B B \, g_J \, J'_z \, . \qquad\qquad (16.56)$$

The magnetic field removes the degeneracy on J'_z: each level corresponding to a given value of J splits up into $2J+1$ equally spaced Zeeman sublevels ($\Delta E = \mu_B B \, g_J$) but, contrary to the case discussed in Sect. 13.7, the distances among adjacent Zeeman sublevels depend on the level by which they are originated.

In the case discussed in Sect. 13.7 each line of the spectrum was split by the magnetic field into three components, due to the selection rules $\Delta M =$

± 1, 0 and to the fact that the Zeeman separation between adjacent levels was independent of the originating level (Fig. 13.1). Now, instead, with the levels given by (16.56), although the selection rules $\Delta J'_z = \pm 1$, 0 still apply, thanks to the fact that the separations among the Zeeman sublevels do depend on the originating level, each of the spectral lines will split up, due to the magnetic field, into a number of components that will depend on the considered case and, in general, will not be not equidistant from each other: this corresponds to what is observed in most cases (*anomalous Zeeman effect*).

In order to determine the number of components into which a given line of the spectrum splits up and the distances among these lines, the value of the Landé factor in (16.56) is needed for the levels involved in the transition: this can be achieved if we replace the exact eigenstates (16.50) of $H_{\rm at}$ with the approximate eigenstates (16.26): this is legitimate if the first-order treatment of the relativistic corrections (spin-orbit interaction etc.) is reliable. Indeed, by exploiting the identity

$$\vec{S} \cdot \vec{J} = \frac{1}{2}(\vec{J}^2 + \vec{S}^2 - \vec{L}^2)$$

and since the states (16.26) $|E_{LSJ}\, L\, S\, J\, J'_z\rangle$ are eigenstates of $\vec{J}^2$, $\vec{L}^2$, $\vec{S}^2$, (16.54) yields the Landé factor (16.54) (that in the present approximation we denote by g_{LSJ}):

$$g_{LSJ} = 1 + \frac{J(J+1) + S(S+1) - L(L+1)}{2\,J(J+1)} \tag{16.57}$$

and, correspondingly, the energies of the Zeeman levels are

$$E_{LSJ} + \mu_{\rm B} B\, g_{LSJ}\, J'_z \,. \tag{16.58}$$

In the present case the dependence of the Landé factor on the quantum numbers $L\,S\,J$ of the considered level is explicitly known: we are therefore able to determine the number of components into which each line of the spectrum splits up and the distances among such lines.

Let us consider, for example, the transitions among the levels of the doublet $3\,^2P_{\frac{1}{2}}$ and $3\,^2P_{\frac{3}{2}}$ and those the lowest term $3\,^2S_{\frac{1}{2}}$ of Na. The values of the Landé factor (16.57) respectively are:

$$^2P_{\frac{1}{2}} : \ g_{1\,\frac{1}{2}\,\frac{1}{2}} = \frac{2}{3} \,; \quad ^2P_{\frac{3}{2}} : \ g_{1\,\frac{1}{2}\,\frac{1}{2}} = \frac{4}{3} \,; \quad ^2S_{\frac{1}{2}} : \ g_{0\,\frac{1}{2}\,\frac{1}{2}} = 2 \,. \tag{16.59}$$

Due to the fact that the Landé factors of the levels $^2P_{\frac{1}{2}}$ and $^2P_{\frac{3}{2}}$ are different from that of the ground level $^2S_{\frac{1}{2}}$, all the transitions permitted by the selection rules $\Delta J'_z = \pm 1$, 0 give rise to lines with different frequencies: from the $^2P_{\frac{1}{2}}$ to the $^2S_{\frac{1}{2}}$ there occur four transitions, two of them being π lines ($\Delta J'_z = 0$), the other two being instead σ lines ($\Delta J'_z = \pm 1$) (Fig. 16.3a).

From the $^2P_{\frac{3}{2}}$ to the $^2S_{\frac{1}{2}}$ the permitted transitions are six, four of them being σ and two being π (Fig. 16.3b). In Figs. 16.3a and 16.3b we also report, in scale, the distances of the several lines from the line in absence of the field (dashed line). So, in the case of weak field, between the $3p$ and the $3s$ levels of Na there occur, in total, ten lines: if we increase the intensity of the magnetic field up to the point that the fine-structure separation between the $^2P_{\frac{1}{2}}$ and $^2P_{\frac{3}{2}}$ becomes negligible with respect

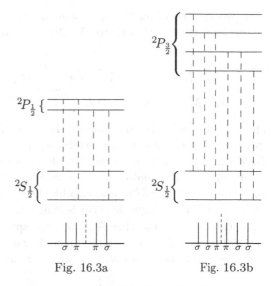

Fig. 16.3a   Fig. 16.3b

to the Zeeman separation (the case of very strong field), these ten lines merge together in such a way as to reproduce the normal Zeeman effect (three equidistant lines).

But the latter is not the only case in which the normal Zeeman effect occurs: from (16.57) it turns out that for all the levels with $S = 0$, and therefore $J = L$, one has

$$g_{L\,0\,J} = 1$$

so that there occurs the normal Zeeman effect in all the transitions between levels with $S = 0$, as e.g. the singlet levels of He (parahelium).

We finally discuss the case of:

Intermediate Field

In this case the Zeeman separations are comparable with the fine-structure ones. The problem is solved by means of perturbation theory for quasi-degenerate levels we have discussed in Sect. 12.4: both the relativistic terms (the spin-orbit interaction and the others) and the interaction (16.47) with the external magnetic field are taken together as the perturbation. The unperturbed Hamiltonian is (16.8) inclusive of only the Coulombic interactions and the unperturbed eigenvectors are given by (16.12), with $E^0 = E_{LS}$. We must therefore diagonalize matrices of dimensions $(2L + 1) \times (2S + 1)$: however, since the perturbation commutes with J_z, all the matrix elements between states with different values of J_z' are vanishing, so any matrix is block-diagonalizable, one block for each of the $2J^{\max} + 1 = 2(L + S) + 1$ possible values of J_z in the states (16.12), and the dimension of each of them equals the number of states with the same J_z'. The dimensions of these matrices are, as a consequence: 1 in correspondence with $J_z' = \pm(L + S)$, 2 with $J_z' = \pm(L + S - 1)$, $\cdots$ and the maximum dimension equals the number of terms the fine-structure multiplet consists of.

If we only consider the spin-orbit interaction from among the relativistic corrections, then – thanks to (16.29) – the matrices one has to diagonalize are:

$$\langle E_{LS}\ L\ M'\ S\ S'_z\ |\ \Lambda \vec{L}\cdot\vec{S} + \frac{e\,B}{2m\,c}\,(L_z + 2S_z)\ |\ E_{LS}\ L\ M''\ S\ S''_z\rangle$$

$$M',\ M'' = -L,\cdots,+L\ ;\qquad\qquad S',\ S'' = -S,\cdots,+S\ .\quad (16.60)$$

The problem is easily solvable in the case of the $2p$ levels of Li (which is one of the few cases in which the fine-structure separation is comparable with the Zeeman separations for reasonable values of the magnetic field) inasmuch as one has to diagonalize a 6×6 matrix that splits up into two blocks of dimension 1 ($j'_z = \pm3/2$) and two blocks ($j'_z = \pm1/2$) of dimension 2. In the basis (16.12) the magnetic term is diagonal, the states with $j'_z = \pm3/2$ are eigenstates of $\vec{J}^2$ and therefore also the spin-orbit term $\Lambda \vec{L}\cdot\vec{s}$ is such: due to (16.30), the value of the latter in such states is $A_{LS}\,\hbar^2/2$; as a consequence the energies of the states with $j'_z = \pm3/2$ are

$$E^{(j'_z = \pm\frac{3}{2})} = E^0 + \frac{1}{2}\,A_{LS}\,\hbar^2 \pm 2\mu_{\mathrm{B}}B = E_{\frac{3}{2}} \pm 2\mu_{\mathrm{B}}B \qquad (16.61)$$

where $E_{\frac{3}{2}} = E^0 + \frac{1}{2}\,A_{LS}\,\hbar^2$ is the energy of the level $P_{\frac{3}{2}}$ in absence of the magnetic field: in this case the result coincides with that of weak field.

The matrix representing the spin-orbit term relative to the states with $j'_z = 1/2$ (the notation is $|\,m\,s'_z\,\rangle$) is

$$\begin{pmatrix} \langle 1\ -\frac{1}{2}\,|\,A_{LS}\,\vec{L}\cdot\vec{s}\,|\,1\ -\frac{1}{2}\rangle & \langle 1\ -\frac{1}{2}\,|\,A_{LS}\,\vec{L}\cdot\vec{s}\,|\,0\ +\frac{1}{2}\rangle \\ \langle 0\ +\frac{1}{2}\,|\,A_{LS}\,\vec{L}\cdot\vec{s}\,|\,1\ -\frac{1}{2}\rangle & \langle 0\ +\frac{1}{2}\,|\,A_{LS}\,\vec{L}\cdot\vec{s}\,|\,0\ +\frac{1}{2}\rangle \end{pmatrix}.\qquad (16.62)$$

In order to calculate the elements of the matrix (16.62) in general one should know the Clebsch–Gordan coefficients (see (15.22)), in such a way as to express the vectors $|\,m\,s'_z\,\rangle$ in terms of the eigenvectors of $\vec{J}^2$ (and therefore of $\vec{L}\cdot\vec{s}$). In this particular case this is not needed: the diagonal terms are immediately calculated because, thanks to the selection rules $\Delta M = \pm1$ for L_x and L_y, and $\Delta s'_z = \pm1$ for s_x and s_y, only the term $L_z\,s_z$ from $\vec{L}\cdot\vec{s}$ contributes, so (16.62) has the form

$$\begin{pmatrix} -\frac{1}{2}A_{LS}\,\hbar^2 & \alpha \\ \alpha^* & 0 \end{pmatrix}.$$

Furthermore, since the eigenvectors of (16.62) are the eigenvectors of $\vec{J}^2$ with $j = \frac{1}{2}$ and $j = \frac{3}{2}$, thanks again to (16.30) we know that the eigenvalues of (16.62) are $-A_{LS}\,\hbar^2$ ($j = \frac{1}{2}$) and $A_{LS}\,\hbar^2/2$ ($j = \frac{3}{2}$), therefore $|\alpha|^2 = A_{LS}^2\,\hbar^4/2$. In conclusion, the complete matrix of the perturbation relative to the states with $j'_z = 1/2$ is

$$\begin{pmatrix} -\frac{1}{2}A_{LS}\,\hbar^2 & \alpha \\ \alpha^* & \mu_{\mathrm{B}}B \end{pmatrix} \qquad (16.63)$$

whose eigenvalues are

$$\delta E_{\pm}^{(j_z'=\frac{1}{2})} = \frac{1}{2}\left(\mu_B B - \frac{1}{2}A_{LS}\,\hbar^2 \pm \sqrt{\frac{9}{4}A_{LS}^2\,\hbar^4 + A_{LS}\,\hbar^2\,\mu_B B + \mu_B^2\,B^2}\right).$$

Since the energies of the levels $P_{\frac{1}{2}}$ and $P_{\frac{3}{2}}$ respectively are $E^0 - A_{LS}\,\hbar^2$ and $E^0 + \frac{1}{2}A_{LS}\,\hbar^2$ and calling $\delta E = \frac{3}{2}A_{LS}\,\hbar^2$ their separation (that in Li is about $0.42 \times 10^{-4}\,\mathrm{eV}$), one has

$$E_+^{(j_z'=\frac{1}{2})} = E_{\frac{3}{2}} + \frac{1}{2}\left(\mu_B B + \sqrt{(\delta E)^2 + \frac{2}{3}\delta E\,\mu_B B + \mu_B^2 B^2} - \delta E\right) \quad (16.64)$$

$$E_-^{(j_z'=\frac{1}{2})} = E_{\frac{1}{2}} + \frac{1}{2}\left(\mu_B B - \sqrt{(\delta E)^2 + \frac{2}{3}\delta E\,\mu_B B + \mu_B^2 B^2} + \delta E\right). \quad (16.65)$$

Equations (16.64) and (16.65) have been written in such a way as to emphasize the originating level: indeed, for $B \to 0$, the terms in parentheses are vanishing.

As for the states with $j_z' = -1/2$, it is sufficient to make the replacement $\mu_B B \to -\mu_B B$ in (16.63), whence

$$E_+^{(j_z'=-\frac{1}{2})} = E_{\frac{3}{2}} + \frac{1}{2}\left(-\mu_B B + \sqrt{(\delta E)^2 - \frac{2}{3}\delta E\,\mu_B B + \mu_B^2 B^2} - \delta E\right) \quad (16.66)$$

$$E_-^{(j_z'=-\frac{1}{2})} = E_{\frac{1}{2}} + \frac{1}{2}\left(-\mu_B B - \sqrt{(\delta E)^2 - \frac{2}{3}\delta E\,\mu_B B + \mu_B^2 B^2} + \delta E\right). \quad (16.67)$$

The weak field result – expressed by (16.58) with the Landé factors given by (16.59) – is recovered by expanding (16.64) ÷ (16.67) to the first order in $\mu_B B/\delta E$. The case of strong field – corresponding to taking into account the first-order effect of the spin-orbit interaction on the energy levels (16.49) – is obtained by expanding to the first order in $\delta E/\mu_B B$: in the latter case one has

$$
\begin{aligned}
E^{(j_z'=+\frac{3}{2})} &\to E^{(1,+\frac{1}{2})} &=& \quad E_{\frac{3}{2}} + 2\mu_B B \\[4pt]
E^{(j_z'=+\frac{1}{2})} &\to E^{(0,+\frac{1}{2})} &=& \quad \tfrac{1}{3}\left(2E_{\frac{3}{2}} + E_{\frac{1}{2}}\right) + \mu_B B \\[4pt]
E^{(j_z'=-\frac{1}{2})} &\to E^{(-1,+\frac{1}{2})} &=& \quad \tfrac{1}{3}\left(E_{\frac{3}{2}} + 2E_{\frac{1}{2}}\right) \\[4pt]
E^{(j_z'=+\frac{1}{2})} &\to E^{(1,-\frac{1}{2})} &=& \quad \tfrac{1}{3}\left(E_{\frac{3}{2}} + 2E_{\frac{1}{2}}\right) \\[4pt]
E^{(j_z'=-\frac{1}{2})} &\to E^{(0,-\frac{1}{2})} &=& \quad \tfrac{1}{3}\left(2E_{\frac{3}{2}} + E_{\frac{1}{2}}\right) - \mu_B B \\[4pt]
E^{(j_z'=-\frac{3}{2})} &\to E^{(-1,-\frac{1}{2})} &=& \quad E_{\frac{3}{2}} - 2\mu_B B\,.
\end{aligned}
\quad (16.68)
$$

In Fig. 16.4 (in the next page) the energy levels given by (16.61) and by (16.64) ÷ (16.67) are reported as functions of the ratio $\mu_B B/\delta E$: for high

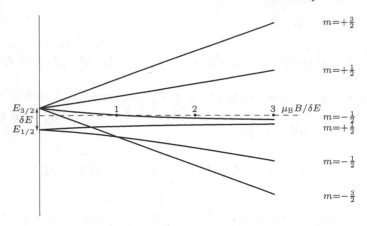

Fig. 16.4

values of the field ($\mu_{\mathrm{B}} B \gtrsim 2 \, \delta E$ they are in the same order (top-down) as the levels given by (16.68).

In Fig. 16.4 it is also interesting to note that the two lines with $m = +1/2$ do not cross each other, and the same is true for those with $m = -1/2$. This is in compliance with the **no-crossing theorem** by E.P. Wigner and J. von Neumann according to which *levels with the same symmetry do not cross each other*: in this case the symmetry is the rotation around the direction of the magnetic field, and 'the same symmetry' means 'the same m'.

We look for necessary conditions in order that, given the Hamiltonian $H(\lambda) \equiv H_0 + \lambda V$, two eigenvalues $E_1(\lambda)$ and $E_2(\lambda)$ collapse for some (real) value $\bar{\lambda}$ of the parameter λ: $E_1(\bar{\lambda}) = E_2(\bar{\lambda}) \equiv \bar{E}$. If this is the case, then for $\lambda_0 \equiv \bar{\lambda} - \delta\lambda$ the eigenvalues of $H(\lambda_0)$ are 'quasi–degenerate' and we can apply the perturbation theory for quasi-degenerate levels we have discussed in Sect. 12.4: therefore the matrix

$$H_{ij}(\bar{\lambda}) = \langle \, E_i(\lambda_0) \mid H(\bar{\lambda}) \mid E_j(\lambda_0) \, \rangle, \quad i,j = 1,2$$

must be a multiple of the identity, i.e.:

$$H_{11}(\bar{\lambda}) = H_{22}(\bar{\lambda}) = \bar{E}, \qquad\qquad H_{12}(\bar{\lambda}) = 0. \qquad (16.69)$$

Equations (16.69) are three *real* equations for the unknown $\bar{\lambda}$; the condition $H_{12}(\bar{\lambda}) = 0$ is satisfied if the states corresponding to the eigenvalues $E_1(\lambda)$ and $E_2(\lambda)$ possess different symmetry properties, that is if they are eigenvectors with different eigenvalues of an operator which for any λ commutes with the Hamiltonian $H(\lambda)$: different values of m, in the case we are considering. In this case the occurrence of level crossing depends on whether the first of (16.69) has a real solution; otherwise, except for special cases, level crossing cannot occur. This explains the statement that level crossing can occur only for states with different symmetry, while states of equal symmetry repel each other, as illustrated in Fig. 16.4 where we can also see that the $m = -3/2$ line crosses both a $m = +1/2$ and a $m = -1/2$ line.

Chapter 17

Elementary Theory of Scattering

17.1 Introduction

In a scattering experiment a beam of particles is fired against a target and the angular distribution of the scattered particles is observed. In physics almost every measurement process involves a scattering experiment, even if the beam does not consist of particles, but is rather a wave, e.g. sound or electromagnetic waves (the two terms "diffusion" and "scattering" are interchangeable, they reflect the wave-particle dualism of quantum mechanics: the former emphasizes the wave-like aspect of the phenomenon, the latter the particle-like one). Even disregarding the high-energy experiments with particle accelerators, for instance all the experiments performed in order to investigate the structure of the atoms (energy levels, degeneracy etc.), as those discussed in Chapt. 13, are scattering experiments: electromagnetic radiation is sent to the atom and from the observation of the scattered radiation the properties of the "scatterer" are deduced. The same occurs if the target consists of molecules rather then atoms.

This fundamental subject involves a multitude of situations: from the simplest one of the scattering of a particle by a fixed and structureless target, to the more complicated ones in which new particles are produced, or where the target has an internal structure that is changed by the particle undergoing the scattering process; not to mention multiple scattering with interference between the scattered waves, as in the Bragg and Davisson–Germer experiments.

Most of the image-based diagnostic techniques in medical physics (RX, Computerized Tomography, NMR, Ecography, etc.), are examples of scattering processes and the list could go on endlessly.

From the historical point of view, the first compendium of scattering theory is that by J.W.S. Rayleigh "The Theory of Sound" (1880) where all the relevant aspects of diffusion are discussed. Among the first scattering experiments with particles, the most known and important for their impact on the knowledge of the atomic structure, are the already mentioned Geiger–Marsden

© Springer International Publishing Switzerland 2016
L.E. Picasso, *Lectures in Quantum Mechanics*, UNITEXT for Physics,
DOI 10.1007/978-3-319-22632-3_17

ones in 1909, performed under the direction of Ernest Rutherford (see Sect. 1.1).

The present chapter is meant as an introduction to the theory of scattering, therefore we will concentrate only on the simplest case of the scattering between two structureless particles, with the aim at giving the basic building blocks of the subject.

17.2 The Cross Section

Suppose that the target consists of n independent scattering centers and the beam of a flux of $\mathcal{N}$ incident particles per unit area and unit time. $\mathcal{N}$ is called the *luminosity* of the beam. We assume that the luminosity is small enough so that there is no interference between the particles of the beam and that the scattering centers are sufficiently far apart so that any collision process involves only one of them.

The number $\Delta n(\theta, \phi)$ of particles that, as the result of the interaction between the incident particles and the scatterers, emerge per unit time in a small solid angle $\Delta\Omega$ around the direction identified by the polar angles θ and ϕ with respect to the direction of the incident particles, is proportional to n, $\mathcal{N}$ and $\Delta\Omega$:

$$\Delta n(\theta, \phi) = \sigma(\theta, \phi) \mathcal{N} n \Delta\Omega \tag{17.1}$$

the proportionality constant $\sigma(\theta, \phi)$ is the *differential cross section*. The *total cross section* is

$$\sigma_{\text{tot}} = \int \sigma(\theta, \phi) \, d\Omega . \tag{17.2}$$

The differential cross section $\sigma(\theta, \phi)$ has the dimension of an area: $\sigma(\theta, \phi) \Delta\Omega$ equals the portion of area of the incident beam (orthogonal to the direction of the beam) that is crossed by the same number of particles that are scattered within the solid angle $\Delta\Omega$ by a single scatterer.

The cross sections are measured in *barn* (b): $1\,\text{b} = 10^{-24}\text{cm}^2$ or, more frequently, in multiples of the $\text{mb} = 10^{-3}\,\text{barn}$.

Since the concept of cross section is a classical one, we can illustrate it in classical terms: suppose that the target consists of n rigid objects of cross-sectional area A. For instance, if the scatterers are rigid spheres of radius a, their cross-sectional area is πa^2. In the unit time $\mathcal{N} \times (n \times A)$ projectiles hit the scatterers and each of them is scattered, therefore σ_{tot} is just A. Thus, in particular, if the scatterers are rigid spheres, the classical total cross section is πa^2, independent on the energy of the projectiles. It is therefore surprising that the quantum mechanical cross section varies from 4 times the classical one at low energies to $2\pi a^2$ at high energies. But the surprise is not justified, because in the classical case one has to do with projectiles that are particles while in the quantum case one has to do with waves and, as in optics, diffraction plays an essential role: notice that low energies means $\lambda \gg a$ and that $4\pi a^2$ just equals the surface of the entire sphere; even at high energies ($\lambda \ll a$) the shadow of a sphere, because of diffraction, is not the geometrical one.

If the particles of the target are not fixed, as for instance in an experiment with colliding beams, and have mass m_t while the particles of the beam (those that will be revealed by the detectors) have mass m_b then, as discussed in Sect. 11.2, the problem reduces to that of a particle of mass $m = m_b m_t/(m_b + m_t)$ subject to the interaction potential $V(|\vec{r}_b - \vec{r}_t|)$ between the two particles that, for simplicity, we have assumed spinless. Of course, one must then translate the results from the center of mass frame to the laboratory frame where, for instance, the target particle is initially at rest.

Before discussing how the cross section can be calculated, we consider a one-dimensional scattering problem since it involves many issues that we will encounter in the three-dimensional case.

17.3 One-Dimensional Case

In Sect. 8.3 we discussed the one-dimensional problem of a particle sent against a potential barrier: this is the prototype of a scattering process and we carried on the discussion in terms of the stationary states of the system. We noticed that this was a choice of convenience because the physical problem of sending particles against a "barrier" and then looking for the reflected and the transmitted ones is not stationary.

The physical problem should be discussed in terms of an initial state represented by a wave packet very distant from the "barrier" and moving towards it and then studying its time evolution that, as we foresee, will give rise to both a reflected packet and a transmitted one. Therefore this is a time-evolution problem and, as such, it is rather complicated and its discussion requires several approximations.

Suppose, as in Sect. 8.3, that the potential $V(x)$, now not necessarily a "barrier", vanishes for both $x < 0$ and $x > a$. The initial state is given in the Schrödinger representation by a wavefunction $\psi(x, t = 0)$ whose support is confined in a finite region $x < 0$ very far from the potential ("scatterer"), and whose mean velocity $v_0 \equiv p_0/m$ is positive. Then

$$\psi(x, t = 0) = \int_{-\infty}^{+\infty} \alpha(k')\, e^{i k' x}\, dk' \tag{17.3}$$

with $\alpha(k')$ concentrated around $k' = k_0 = p_0/\hbar$.

In order to determine the time evolution of the state we must express $\psi(x, t = 0)$ in terms of the eigenfunctions of the Hamiltonian: let us suppose, for simplicity, that there are no bound states. In this case we know from Sect. 8.3 that in the regions $x < 0$ and $x > a$, where the potential vanishes, the eigenfunctions of the Hamiltonian (that were indeed called "scattering states") are given, with slightly modified notations with respect to those of Sect. 8.3, by

$$\phi_k(x) = \begin{cases} e^{i k x} + B(k)\, e^{-i k x} & x \leq 0 \\ C(k)\, e^{i k x} & x \geq a \end{cases} \qquad k > 0 \tag{17.4}$$

for particles coming from the left, and

$$\phi_k(x) = \begin{cases} C(k)\,\mathrm{e}^{\mathrm{i}\,k\,x} & x \le 0 \\ \mathrm{e}^{\mathrm{i}\,k\,x} + B(k)\,\mathrm{e}^{-\mathrm{i}\,k\,x} & x \ge a \end{cases} \qquad k < 0 \qquad (17.5)$$

for particles coming from the right. The expression of $\phi_k(x)$ for $0 \le x \le a$ depends on the potential $V(x)$. Hence

$$\psi(x,t) = \int_{-\infty}^{+\infty} \alpha(k')\,\phi_{k'}(x)\,\mathrm{e}^{-\mathrm{i}E't/\hbar}\mathrm{d}k', \qquad E' = \frac{\hbar^2 k'^2}{2m}. \qquad (17.6)$$

In (17.6) we have written $\alpha(k')$ as in (17.3), instead of a different function $\tilde{\alpha}(k')$: we show that actually $\tilde{\alpha}(k') = \alpha(k')$. To this end we must make the following approximation: in (17.3) $\big($and consequently also in (17.6)$\big)$ we limit the integration only to the positive values of k', i.e.

$$\psi(x,\,t=0) \simeq \int_0^{+\infty} \alpha(k')\,\mathrm{e}^{\mathrm{i}\,k'\,x}\,\mathrm{d}k'. \qquad (17.7)$$

The legitimacy of this approximation is due to the fact that $\alpha(k')$ is concentrated around $k' = k_0 = p_0/\hbar$ and since k_0 is positive, $\alpha(k')$ is negligible (although not zero) for negative values of k'. We then must show that for $t = 0$ $\psi(x,0)$ given by (17.6) (approximately) coincides with $\psi(x,0)$ given by (17.7), that is

$$\int_0^{\infty} \alpha(k')B(k')\,\mathrm{e}^{-\mathrm{i}\,k'\,x}\,\mathrm{d}k' = 0, \qquad x \le 0$$

$$\int_0^{\infty} \alpha(k')C(k')\,\mathrm{e}^{+\mathrm{i}\,k'\,x}\,\mathrm{d}k' = 0, \qquad x \ge a. \qquad (17.8)$$

A further approximation is necessary, still justified by the fact that $\alpha(k')$ is significantly different from 0 only near k_0: we replace $B(k')$ with $B(k_0)$ and $C(k')$ with $C(k_0)$. In this case the first integral in (17.8) equals $B(k_0)\,\psi(-x,0)$ which vanishes since $\psi(-x,0) \ne 0$ only for $x > 0$. A similar argument holds for the second integral in (17.8).

We now come back to (17.6) to determine the time evolution of the wave packet. In addition to the already mentioned approximations we expand E' around $k' = k_0$ and keep only the linear terms:

$$E' \equiv \frac{\hbar^2 k'^2}{2m} = \frac{\hbar^2 k_0^2}{2m} + \hbar^2\frac{k_0}{m}(k' - k_0) + \cdots = -\frac{\hbar^2 k_0^2}{2m} + \hbar v_0\,k' + \cdots \quad (17.9)$$

where the omitted term is $\hbar^2(k'-k_0)^2/2m$ and is responsible for the spreading of the packet.

Because of (17.4), (17.6) gives rise to three terms: the first one is

$$\mathrm{e}^{+\mathrm{i}E_0t/\hbar}\int_0^{\infty} \alpha(k')\,\mathrm{e}^{\mathrm{i}\,k'(x-v_0t)}\,\mathrm{d}k' \qquad x < 0 \qquad (17.10)$$

that, apart from the lower limit of integration and the irrelevant phase factor $e^{+iE_0t/\hbar}$, is $\psi(x - v_0t, 0)$, i.e. the initial packet moving toward the scattering center with velocity v_0. Since it has a finite extension, i.e. $\psi(x, 0) = 0$ for, say $x < -L$, for $t > L/v_0$ vanishes: $x - v_0t < -L$ for any $x < 0$ (remember that (17.10) concerns the region $x < 0$).

Always in the region $x < 0$ we have a second term

$$e^{+iE_0t/\hbar} B(k_0) \int_0^\infty \alpha(k') e^{-ik'(x+v_0t)} \, dk' \qquad x < 0 \qquad (17.11)$$

that equals $B(k_0)\psi(-x - v_0t, 0)$ and is zero for $t = 0$ $(-x > 0)$. For sufficiently large values of t the argument $-x - v_0t$ becomes negative and correspondingly $B(k_0)\psi(-x - v_0t, 0)$ becomes different from zero. The physical meaning is clear: it is a wave packet specular to the initial one, reduced in intensity by the factor $|B(k_0)|^2$ $\left(|B(k_0)|^2 = 1 - |C(k_0)|^2 \right.$, see (8.26)) travelling away from the scatterer along the negative x-axis: as expected it is the reflected wave packet.

Finally, in the region $x > 0$ we have (omitting the phase factor $e^{+iE_0t/\hbar}$)

$$C(k_0) \int_0^\infty \alpha(k') e^{ik'(x-v_0t)} \, dk' = C(k_0) \, \psi(x - v_0t, 0) \qquad (17.12)$$

that, for sufficiently large t, corresponds to the transmitted wave packet travelling with velocity v_0 past the scatterer.

We now briefly discuss the conditions of validity of the above results, both to illustrate their physical meaning and to ascertain their mutual compatibility.

The momentum-space wavefunction $\alpha(k)$ of the initial state has been supposed to be concentrated in a neighborhood of $k = k_0$, with $\Delta k \ll k_0$. This condition is equivalent to $\Delta x \gg \lambda_0$, where Δx (approximately) gives the extension of the wavefunction in coordinate space. Therefore the initial state practically consists of a plane wave of finite extension, i.e. – in the spirit of the discussion of Sect. 6.9 – we can refer to it as "an approximate eigenvector of p". For example, with reference to the neutron interferometry experiment discussed in Sect. 3.3, the neutrons – in that case of very low energy – had a wavelength $\lambda_0 \simeq 1\,\text{Å}$, while the longitudinal dimension Δx was estimated to be of the order of $10^{-2}\,\text{cm}$, therefore $\Delta k/k_0 \simeq \lambda_0/\Delta x \simeq 10^{-6}$.

Since we neglected the spreading of the wave packet during the time evolution, we must ascertain that in the whole scattering process, from the initial state to the detection of the scattered particle, the wave packet does not spread appreciably. Just to give an idea of the approximation involved, we take the expression of the spreading of a free wave packet, as given by (9.34). If T is the duration of the process, we must require that $\Delta x(T) \simeq \Delta x(0)$, i.e. $T \lesssim m\,\Delta x/\Delta p = \Delta x/\Delta v$, namely the distance covered by the packet is $D \lesssim \Delta x \times (k_0/\Delta k)$.

Moreover, when the wave packet reaches the detector it must have completely gone beyond the region where the potential is nonvanishing: this requires that $T \gg L/v_0$. Therefore T must satisfy the following inequalities

$$\frac{L}{v_0} \ll T \lesssim \frac{\Delta x}{\Delta v} \quad \text{or, equivalently} \quad L \ll D \lesssim \Delta x \times \frac{k_0}{\Delta k}$$

which is possible as long as $\Delta x/L \gg \Delta k/k_0$.

17.4 Three-Dimensional Case

Except for the case of a wave impinging onto a large slab of optical material, in all real situations scattering occurs in three dimensions. The discussion of the previous section was aimed at showing how a scattering problem should be faced as a non-stationary problem. It also underlines the essential role of the stationary states, whose determination in the one-dimensional case is a prerequisite for the determination of the reflection and transmission coefficients and, in the three-dimensional case, as we will see, of the scattering amplitude i.e. of how the incident particles are 'scattered' by a given target.

Thinking in terms of wave packets, a scattering process consists of an initial (three-dimensional) wave packet travelling towards the region where the potential is different from zero, whence it gets diffused in all directions. Actually, it is not necessary to assume that the potential vanishes outside a finite region, it is sufficient that it decreases at infinity faster than $1/r$ (therefore the Coulomb potential is excluded and requires a special treatment). With this picture in mind we look for stationary (improper) solutions of the Schrödinger equation consisting of a plain wave with given momentum, representing the incoming particle, plus a wave 'exiting' in all directions from the scattering center, i.e. the diffused wave:

$$\psi_E(\vec{r}) = e^{i\vec{k}\cdot\vec{r}} + \chi_E(r,\theta,\phi) \tag{17.13}$$

(θ and ϕ being the polar angles with respect to $\vec{k}$).

We are interested in the asymptotic form of this eigenfunction of the Hamiltonian since the detection process occurs very far from the scattering center. Since asymptotically the particle is free, $\psi_E(\vec{r})$ must (asymptotically) satisfy the Schrödinger equation for a free particle: the first term in (17.13) is already a solution, hence $\big($see (10.38) and (11.6)$\big)$

$$\left(-\frac{\hbar^2}{2m}\frac{1}{r}\frac{\partial^2}{\partial r^2}r + \frac{\vec{L}^2}{2mr^2} \right)\chi_E(r,\theta,\phi) = E\,\chi_E(r,\theta,\phi) \tag{17.14}$$

with $\vec{L}^2$ given by (10.38).

In the asymptotic region the term $\vec{L}^2/2mr^2$, the only one containing the angles, must be omitted because of the r^2 in the denominator, therefore (17.14) reduces to

$$-\frac{1}{r}\frac{\partial^2}{\partial r^2}r\,\chi_E(r,\theta,\phi) = \vec{k}^2\,\chi_E(r,\theta,\phi), \qquad \vec{k}^2 = 2m\,E/\hbar^2 \tag{17.15}$$

whose solution is of the form

$$\chi_E(r,\theta,\phi) = f_{\vec{k}}(\theta,\phi)\, g(r)$$

where $f_{\vec{k}}(\theta,\phi)$ is undetermined, while the reduced radial function $r\,g(r)$ satisfies the second order equation:

$$\frac{\partial^2}{\partial r^2}\left(r\,g(r)\right) = -\vec{k}^2\left(r\,g(r)\right).$$

Hence

$$g(r) = \alpha\,\frac{e^{ik\,r}}{r} + \beta\,\frac{e^{-ik\,r}}{r}. \tag{17.16}$$

Once multiplied by $a(\vec{k})\,e^{-i\,Et/\hbar}$ and integrated over $\vec{k}$ to form a wave packet, only the first term in (17.16) gives rise to a wave moving away from the scatterer. Thus the asymptotic solution we are after is of the form

$$\psi_E(r,\theta,\phi) \overset{r\to\infty}{\Longrightarrow} e^{i\vec{k}\cdot\vec{r}} + f_{\vec{k}}(\theta,\phi)\,\frac{e^{ik\,r}}{r} \tag{17.17}$$

where $f_{\vec{k}}(\theta,\phi)$, the **scattering amplitude**, is the unknown quantity of the problem and, obviously, depends on the potential $V(\vec{r})$.

In order to determine $f_{\vec{k}}(\theta,\phi)$ let us go back to the (complete) Schrödinger equation:

$$\left(-\frac{\hbar^2}{2m}\Delta + V(\vec{r})\right)\psi_E(\vec{r}) = E\,\psi_E(\vec{r}) \tag{17.18}$$

with $\psi_E(\vec{r})$ given by (17.13). Then (17.18) can be rewritten as

$$\left(\vec{k}^2 + \Delta\right)\chi = \frac{2m}{\hbar^2}V(\vec{r})\,(e^{i\vec{k}\cdot\vec{r}} + \chi) \equiv U(\vec{r})\,(e^{i\vec{k}\cdot\vec{r}} + \chi). \tag{17.19}$$

Equation (17.19) is a form of the **Lippmann–Schwinger equation** and can be solved iteratively by considering the potential V as a perturbation. At the first order, i.e. neglecting χ in the right-hand side of (17.19), we have:

$$\left(\vec{k}^2 + \Delta\right)\chi = U(\vec{r})\,e^{i\vec{k}\cdot\vec{r}}. \tag{17.20}$$

Equation (17.20) can be solved the following way: consider the function

$$G_k(\vec{r}) \equiv -\frac{1}{4\pi}\,\frac{e^{ik\,r}}{r} \tag{17.21}$$

which is a solution of the equation:

$$\left(\vec{k}^2 + \Delta\right)G_k(\vec{r}) = \delta(\vec{r}) \tag{17.22}$$

with $\delta(\vec{r})$ the Dirac's delta function in three dimensions (see (6.41)). [To prove (17.22) it is sufficient to note that for $r > 0$ $G_k(\vec{r})$ is a solution of (17.22) (see (17.16)); moreover, if we integrate both members of (17.22) over a sphere of radius ϵ, for the right-hand side we get 1 whereas for the left-hand side, making use of Green's theorem, we get

$$-\frac{1}{4\pi}\,\vec{k}^{\,2}\int d\Omega \int_0^\epsilon \frac{e^{ikr}}{r}\,r^2 dr - \frac{1}{4\pi}\int \frac{\partial}{\partial r}\frac{e^{ikr}}{r}\Big|_{r=\epsilon} \epsilon^2\, d\Omega\,.$$

The first term goes to 0 for $\epsilon \to 0$, the second is

$$-\frac{1}{4\pi}\int \left(-\frac{e^{ikr}}{r^2}+ik\frac{e^{ikr}}{r}\right)\Big|_{r=\epsilon}\epsilon^2\,d\Omega \;\to\; 1 \quad \text{for}\quad \epsilon \to 0$$

(alternatively one could use of $\Delta\,(1/r) = -4\pi\,\delta(\vec{r})$, which is nothing but Gauss' theorem of electrostatic).]

The solution of (17.20) is then:

$$\chi(\vec{r}) = \int G_k(\vec{r}-\vec{r}')\,U(\vec{r}')\,e^{i\vec{k}\cdot\vec{r}'}\,d\vec{r}'$$

$$= -\frac{1}{4\pi}\int \frac{e^{ik|\vec{r}-\vec{r}'|}}{|\vec{r}-\vec{r}'|}\,U(\vec{r}')\,e^{i\vec{k}\cdot\vec{r}'}\,d\vec{r}' \tag{17.23}$$

since

$$(\vec{k}^{\,2}+\Delta)\int G_k(\vec{r}-\vec{r}')\,U(\vec{r}')\,e^{i\vec{k}\cdot\vec{r}'}\,d\vec{r}' =$$
$$\int (\vec{k}^{\,2}+\Delta)G_k(\vec{r}-\vec{r}')\,U(\vec{r}')\,e^{i\vec{k}\cdot\vec{r}'}\,d\vec{r}' =$$
$$\int \delta(\vec{r}-\vec{r}')\,U(\vec{r}')\,e^{i\vec{k}\cdot\vec{r}'}\,d\vec{r}' = U(\vec{r})\,e^{i\vec{k}\cdot\vec{r}}\,.$$

The function $G_k(\vec{r}-\vec{r}')$ is the **Green function** of (17.20) such that $\chi(\vec{r})$ is an outgoing wave.

Equation (17.23) is the first-order solution of the Lippmann–Schwinger equation which is sufficient for our purposes. Anyway, if the second order is needed, it is sufficient to insert $\chi(\vec{r})$ given by (17.23) into the right-hand side of (17.19) and proceed as above (the expression of the second-order term will be given in the last section).

To get the scattering amplitude we need the asymptotic behaviour of $\chi(\vec{r})$. This is obtained by inserting the expansion $\big(|\vec{r}| \gg |\vec{r}'|\big)$

$$|\vec{r}-\vec{r}'| = r - \frac{\vec{r}\cdot\vec{r}'}{r} + O(1/r)$$

into the Green function G_k. Keeping only the lowest order term we find $(\hat{r} \equiv \vec{r}/r)$

$$-\frac{1}{4\pi}\frac{e^{k|\vec{r}-\vec{r}'|}}{|\vec{r}-\vec{r}'|} \;\xrightarrow{r\to\infty}\; -\frac{1}{4\pi}\frac{e^{ik(r-\hat{r}\cdot\vec{r}')}}{r}$$

which yields for $\chi(\vec{r})$ $(\vec{k}_f \equiv k\hat{r},\;\; \vec{k}_i \equiv \vec{k})$

$$\chi(\vec{r}) \xrightarrow{r\to\infty} -\frac{1}{4\pi}\frac{e^{ikr}}{r}\int e^{-i\vec{k}_f\cdot\vec{r}'}U(\vec{r}')e^{i\vec{k}_i\cdot\vec{r}'}\,d\vec{r}' = f_{\vec{k}}(\theta,\phi)\frac{e^{ikr}}{r} \tag{17.24}$$

with

$$f_{\vec{k}}(\theta, \phi) = -\frac{1}{4\pi} \frac{2m}{\hbar^2} \int e^{-i\vec{k}_{f}\cdot\vec{r}'} V(\vec{r}') e^{i\vec{k}_{i}\cdot\vec{r}'} d\vec{r}', \qquad V = \frac{\hbar^2}{2m} U \qquad (17.25)$$

that depends on the angles θ and ϕ only through $\vec{k}_{f}$.

The above expression for $f_{\vec{k}}(\theta, \phi)$ is called the **Born approximation** and is the first term of the Born series for the scattering amplitude.

Since $f_{\vec{k}}(\theta, \phi)$ is the first term of the perturbative expansion in which V is considered as a perturbation, it is necessary to investigate the condition of its validity. This is done by considering that in (17.19) we have neglected the term $V\chi$ with respect to $(\hbar^2 k^2/2m)\chi$, therefore we can presume that the Born approximation is more reliable at high energies. This hand-waving argument is confirmed by the study of the convergence of the Born series.

Notice that within the Born approximation the cross section $\sigma(\theta, \phi)$ does not depend on the sign of the potential: $V(\vec{r})$ and $-V(\vec{r})$ give rise to the same cross section. This result is clearly a drawback of the Born approximation.

Having determined the asymptotic form of the stationary scattering states, it is now possible to carry on the discussion in terms of wave packets. Except for the obvious different notations due to the dimensionality of the space and the fact that in the present case the space is not separated in two disjoint regions by the potential, the discussion is similar to that of Sect. 17.3 above: initially we have the wave packet

$$\psi_0(\vec{r}; t = 0) = \int \alpha(\vec{k}') e^{i\vec{k}'\cdot\vec{r}} d\vec{k}' \qquad (17.26)$$

with $\alpha(\vec{k}')$ concentrated around $\vec{k}' = (k_0, 0, 0)$ with $k_0 > 0$; it has finite extension not only in the x-direction $\left(\text{say } -|x_1| \leq x \leq -|x_2|\right)$, but also in the transverse directions, since it must probe all the region where the potential is significantly different from zero. For large t, i.e. after the interaction with the potential, two wave packets are present in the asymptotic region: the first is the incoming one that, with the same approximation expressed by (17.9), goes on with the same intensity as before the interaction:

$$\psi_0(\vec{r}; t) \simeq \psi_0(x - v_0 t, y, z; t = 0) \qquad (17.27)$$

(this point deserves a particular discussion and we shall came back to it in the next section), the second is the scattered wave:

$$\psi_{\text{sc}}(r, \theta, \phi; t) \simeq \int \alpha(\vec{k}') f_{\vec{k}'}(\theta, \phi) \frac{1}{r} e^{i k'(r - v_0 t)} d\vec{k}' \qquad (17.28)$$

and, as we did after (17.8), we assume that $f_{\vec{k}'}(\theta, \phi)$ is almost constant where $\alpha(\vec{k}')$ is significantly different from zero and replace it with $f_{\vec{k}_0}(\theta, \phi)$. Therefore (17.28) takes the form

$$\psi_{\text{sc}}(r, \theta, \phi; t) \simeq f_{\vec{k}_0}(\theta, \phi) \frac{1}{r} \int \alpha(\vec{k}') e^{i k'(r - v_0 t)} d\vec{k}'. \qquad (17.29)$$

By comparison with (17.27) in which the argument $x - v_0 t$ must be negative and less then $-|x_2|$, it is clear that $\psi_{\rm sc}$ is zero until $v_0 t > |x_2|$ and reaches the asymptotic region for $v_0 t \gg |x_1|$.

We can now go on calculating the cross section. The probability current density associated with the initial packet is

$$\vec{J}_0 = -{\rm i}\,\frac{\hbar}{2m}\left(\psi_0^*\,\nabla\,\psi_0 - \psi_0\,\nabla\,\psi_0^*\right) \simeq |\psi_0|^2 \times \frac{\vec{k}_0}{m} = \left(|\psi_0|^2 v_0,\,0,\,0\right) \qquad (17.30)$$

(the $\simeq$ sign is due to having taken $\vec{k}'$ outside the integrals). The current density associated with the scattered packet in the asymptotic region, keeping only the leading terms in $1/r$, has only the radial component:

$$J_r = \frac{|f_{\vec{k}_0}(\theta,\phi)|^2 |\psi_0|^2 v_0}{r^2} \qquad (17.31)$$

since the angular components are proportional to $1/r^3$. If we have a beam of N particles all prepared in the same way, then the luminosity $\mathcal{N}$ is given by $N J_0$ and the number of particles scattered per unit time and per scatterer within a given solid angle $\Delta\Omega$ (not in the forward direction, where also the incoming packet is present) is $N J_r\, r^2 \Delta\Omega$. Therefore the differential cross section is

$$\sigma(\theta,\phi) = N J_r\, r^2 \Delta\Omega / N J_0 \Delta\Omega = J_r\, r^2 / J_0 = |f_{\vec{k}_0}(\theta,\phi)|^2 \quad (\theta \neq 0) \qquad (17.32)$$

and the total cross section

$$\sigma_{\rm tot} = \int |f_{\vec{k}_0}(\theta,\phi)|^2\,{\rm d}\Omega\,.$$

Another way of writing $f_{\vec{k}_0}(\theta,\phi)$ given by (17.25) is the following: if by $\widetilde{V}$ we denote the Fourier transform of the potential $V(\vec{r})$ and by $\vec{q}_{\rm fi}$ the **momentum transfer**, i.e. $\vec{q}_{\rm fi} \equiv \hbar\,(\vec{k}_{\rm f} - \vec{k}_{\rm i})$, then:

$$f_{\vec{k}_0}(\theta,\phi) = -4\pi^2 m\hbar\,\langle\,\vec{p}_{\rm f}\mid V\mid \vec{p}_{\rm i}\,\rangle = -\frac{\sqrt{2\pi}\,m}{\hbar^2}\,\widetilde{V}(\vec{q}_{\rm fi}/\hbar) \qquad (17.33)$$

with $|\vec{q}_{\rm fi}|$ ranging from 0 to $2\hbar\,k_0$. Therefore, by means of scattering experiments, information on the Fourier transform of the potential is obtained and therefore indirectly on $V(\vec{r})$ itself. In particular, from the properties of the Fourier transform, to probe the structure of $V(\vec{r})$ at small distances high momentum transfers are needed: it is precisely for this reason that the experiments by Geiger and Marsden (see Sect. 1.1) allowed Rutherford to reject Thomson's model of the atom in favour of the one in which the positive charge is concentrated in a nucleus; today, to investigate the structure of the "elementary" particles, higher and higher energies are required.

17.5 The Optical Theorem

We have remarked that, contrary to the one-dimensional case where the transmitted wave has an intensity lower than that of the incoming one (in all cases in which a reflected wave is present), in the three-dimensional case the incoming wave packet goes on, after interacting with the potential, with the same intensity as before the interaction.

This fact seems to violate the conservation of probability, i.e. the unitarity of the time-evolution operator: indeed, initially there is the normalized incoming wave packet; then, for large t, besides the translated incoming wave packet, there also is the scattered wave with, obviously, a nonvanishing norm. Differently, but equivalently, consider the states given by (17.13): since these states, albeit improper, are stationary, the flux of the probability current density (8.28) through a large sphere should be zero: this is true for the plane wave, but there also is a nonvanishing contribution from the scattered wave.

This is only an apparent contradiction since the norm of $\psi_0 + \chi$ is not the sum of the norms of ψ_0 and of χ, but there is also the interference term; the same is true for the flux.

The optical theorem that we are going to prove is nothing else but the consequence of imposing the vanishing of the flux of the current density associated with the stationary states given by (17.13) through a sphere of radius r in the limit $r \to \infty$. The plane wave gives rise to a constant current density (equal to $\hbar k/m$ in the x-direction), therefore its contribution vanishes; the current density associated with the scattered wave (17.24) is radial and is given by

$$J_r = \frac{\hbar k}{m} \frac{1}{r^2} |f_{\vec{k}}(\theta, \phi)|^2$$

therefore the flux through the sphere equals $(\hbar k/m)\,\sigma_{\mathrm{tot}}$. Finally, there are the interference terms:

$$-\frac{i\hbar k}{2m} (\psi_0^* \nabla \chi + \chi^* \nabla \psi_0 - \text{c.c.})$$

and since we are interested in the flux through a sphere, only the radial component is relevant. Thus the following integrals should be calculated:

$$\mathrm{i}k\, r^2 \int \left(\mathrm{e}^{-\mathrm{i}kr\cos\theta} f_{\vec{k}}(\theta, \phi) \frac{\mathrm{e}^{\mathrm{i}kr}}{r} + f_{\vec{k}}^*(\theta, \phi) \frac{\mathrm{e}^{-\mathrm{i}kr}}{r} \cos\theta\, \mathrm{e}^{\mathrm{i}kr\cos\theta} \right) \sin\theta\, \mathrm{d}\theta\, \mathrm{d}\phi - \text{c.c.}$$

where in the expression

$$\frac{\partial}{\partial r} \chi(r) = \mathrm{i}k\, f_{\vec{k}}(\theta, \phi) \frac{\mathrm{e}^{\mathrm{i}kr}}{r} - f_{\vec{k}}(\theta, \phi) \frac{\mathrm{e}^{\mathrm{i}kr}}{r^2}$$

the second term has been ignored since negligible with respect to the first.

We only give a hint for the calculations that are lengthy although not difficult: after some partial integrations with respect to the θ-variable, the r-dependence disappears and eventually we end up with

$$4\pi \mathrm{i}\, f_{\vec{k}}(0) - 4\pi \mathrm{i}\, f_{\vec{k}}^{*}(0) = -8\pi\, \Im m f_{\vec{k}}(0)$$

(when $\theta = 0$ the azimuthal variable ϕ is irrelevant) and, after restoring all the omitted coefficients, we conclude that the flux through the sphere is zero if $k\,\sigma_{\mathrm{tot}} - 4\pi\, \Im m f_{\vec{k}}(0) = 0$, i.e.

$$\sigma_{\mathrm{tot}} = \frac{4\pi}{k}\, \Im m f_{\vec{k}}(0) \,. \tag{17.34}$$

This is the **optical theorem** which assures the probability conservation: the interference between the plane wave and the wave scattered in the forward direction results in a decrease of the intensity of the "transmitted" wave that compensates the probability of scattering in all directions different from the forward one. It is a general theorem of wave propagation whose importance was already emphasized in the work of Rayleigh on the theory of sound.

In order to get a better insight in the meaning of the above result, let us go back to the one-dimensional case discussed in Sect. 17.3. To mimic the three-dimensional case we rewrite (17.4) in the following way:

$$\phi_k(x) = \mathrm{e}^{\mathrm{i}kx} + \chi(x)\,; \qquad \chi(x) = \begin{cases} \mathrm{i}\tau\, \mathrm{e}^{\mathrm{i}kx} & x \geq a \\ \rho\, \mathrm{e}^{-\mathrm{i}kx} & x \leq 0 \end{cases} \tag{17.35}$$

(τ and ρ are respectively for "transmitted" and "reflected" amplitude, the "i" just for convenience). The conservation of probability is expressed $\big($see (8.26)$\big)$ by $|\rho|^2 + |1 + \mathrm{i}\tau|^2 = 1$, that is

$$|\rho|^2 + |\tau|^2 = 2\, \Im m\, \tau \tag{17.36}$$

which is the analogous of (17.34), with $|\rho|^2 + |\tau|^2$ in place of the total cross section. From (17.35) and (17.36) we learn that it is *all* the forward scattered wave $\mathrm{i}\tau\, \mathrm{e}^{\mathrm{i}kx}$ that interferes with the plane wave $\mathrm{e}^{\mathrm{i}kx}$ to guarantee the probability conservation, but only its imaginary part is related to the reduction of the intensity of the wave in the forward direction ($x > a$ in this case), exactly as in optics the imaginary part of the refraction index is related to the absorbtion.

Before ending this section we notice that in the Born approximation (17.33) the optical theorem is not satisfied since the forward amplitude $f_{\vec{k}}(0)$ is real.

17.6 Central Potential

If scattering occurs between two particles without internal structure (spin, electric-dipole moment, etc.), then the potential is spherically symmetric, i.e. it depends only on the distance between the two particles: $V = V(r)$. In this case the scattering amplitude $f_{\vec{k}}(\theta,\phi)$ does not depend on the azimuthal angle ϕ, the polar axis being parallel to $\vec{k}$. Indeed, the solutions of the Schrödinger equation asymptotically of the form (17.17) are uniquely determined by (the potential and) the momentum $\vec{k}$ of the incoming particle, therefore are invariant under rotations around the polar axis: $\vec{k}$ and $V(r)$ are unchanged. As a consequence $f_{\vec{k}}(\theta, \phi) = f_{\vec{k}}(\theta, \phi + \alpha)$. In this case the Born approximation gives:

$$f_{\vec{k}}(\theta) = -\frac{m}{2\pi\hbar^2} \int e^{-i\vec{k}_{\mathrm{fi}}\cdot\vec{r}'} V(r')\, \mathrm{d}\vec{r}' = -\frac{2m}{\hbar^2 \kappa} \int \sin(\kappa r')\, V(r')\, r'\mathrm{d}r' \quad (17.37)$$

with $\kappa = k_{\mathrm{fi}} \equiv |\vec{k}_{\mathrm{fi}}| = 2\,k\,\sin(\theta/2)$ (the integration over the angles has been performed by provisionally taking $\vec{k}_{\mathrm{fi}}$ as polar axis) and, as expected, does not even depend on the direction of $\vec{k}_{\mathrm{i}}$.

Moreover, when the potential is spherically symmetric, all the components of the angular momentum $\vec{L}$ commute with the Hamiltonian. In this case it is convenient to express the plane wave in (17.13) as a superposition of functions of given angular momentum. The polar axis being parallel to $\vec{k}$, only the spherical harmonics with $m = 0$ contribute to the expansion and consequently, in accordance with (11.8) and (10.41), we have

$$e^{i\vec{k}\cdot\vec{r}} = \sum_{l=0}^{\infty} i^l \sqrt{4\pi(2l+1)}\, j_l(kr)\, Y_{l,0}(\theta)\,. \quad (17.38)$$

The factor $i^l\sqrt{4\pi(2l+1)}$ is there for convenience, the radial functions $j_l(kr)$ are the **Bessel functions**: $j_l(kr)Y_{l,0}(\theta)$ are the simultaneous eigenfunctions of H_0, $\vec{L}^2$, L_z, where H_0 is the free Hamiltonian. Equation (17.38) is known as the **partial wave** expansion of the plane wave.

The potential $V(r)$ will then give rise to the scattered wave $\chi(\vec{r})$ but, because of the angular momentum conservation, all the components of the sum in (17.38) are scattered independently of one another and consequently the scattering amplitude is given by a sum analogous to (17.38):

$$f_k(\theta) = \sum_{l=0}^{\infty} \sqrt{4\pi(2l+1)}\, \gamma_l(k) Y_{l,0}(\theta) \quad (17.39)$$

where γ_l accounts for the contribution of the partial wave in (17.38) with the same angular momentum l.

From (17.39) and (10.41) the total cross section is then the sum of the cross sections σ_l relative to the single partial waves:

$$\sigma_{\mathrm{tot}}(k) = \sum_{l=0}^{\infty} \sigma_l(k) = \sum_{l=0}^{\infty} 4\pi\,(2l+1)|\gamma_l(k)|^2\,. \quad (17.40)$$

By the optical theorem (17.34) the coefficients $\gamma_l(k)$ can be written as

$$\gamma_l(k) = \frac{1}{k}\, e^{i\,\delta_l(k)}\, \sin\delta_l(k)\,. \quad (17.41)$$

Indeed, since $Y_{l,0}(0) = \sqrt{(2l+1)/4\pi}$, $|\gamma_l|^2 = \Im m\, \gamma_l/k$. If we put $\tilde{\gamma}_l = k\,\gamma_l$ and $\tilde{\gamma}_l = a + ib$, we get $a^2 + b^2 = b > 0$ and, letting $a^2/b = \cos^2 \delta_l$, $b = \sin^2 \delta_l$, we have $a = \pm \sin \delta_l \cos \delta_l$. Then $\tilde{\gamma}_l(k) = \pm e^{\pm i\,\delta_l(k)} \sin \delta_l(k)$. The two solutions are equivalent since they are obtained one from the other by $\delta_l \to -\delta_l$ then, taking the one with the $+$ sign, (17.41) is obtained. Hence (17.39) and (17.40) can be written as

$$f_k(\theta) = \frac{1}{k} \sum_{l=0}^{\infty} \sqrt{4\pi(2l+1)}\; e^{i\,\delta_l(k)} \sin \delta_l(k)\, Y_{l,0}(\theta) \qquad (17.42)$$

$$\sigma_{\text{tot}}(k) = \sum_{l=0}^{\infty} \sigma_l = \sum_{l=0}^{\infty} \frac{4\pi}{k^2}(2l+1) \sin^2 \delta_l(k)\,. \qquad (17.43)$$

The δ_l's are called **phase shifts** and from (17.43) we get the **unitarity bound**

$$\sigma_l(k) \le \frac{4\pi}{k^2}(2l+1) \qquad (17.44)$$

("unitarity" since it is a consequence of the optical theorem, i.e. of the flux conservation). When the unitarity bound is saturated: $\sigma_l(k) = (4\pi/k^2)(2l+1)$, the scattering is said to be resonant.

The importance of the partial wave expansion can be appreciated by considering the behaviour of the functions $j_l(kr)$ near the origin: after (11.34) we noticed that the behaviour near the origin is the same for all radial functions solutions of (11.9), provided the potential does not diverge at the origin faster than the centrifugal one, hence $j_l(kr) \simeq (kr)^l$. Therefore, at low energies, i.e. low k, only the first few angular momenta contribute to the scattering amplitude: the higher the angular momentum the lesser is the Bessel function near the origin, where usually the potential is greater. This is the same argument we used in Sect. 14.5 when discussing the alkali atoms. For this reason the partial wave expansion is particularly useful at low energies, while at high energies the Born approximation is more suited.

In order to illustrate explicitly how (17.39) follows from (17.38), we consider the case of the s-wave (i.e. $l = 0$) scattering by a potential $V(r)$ whose support is $0 < r < a$.

As we know (Sect. 11.1), the problem for the reduced radial function $u_0(r) \equiv u_{l=0}(r) \equiv r\, R_0(r)$ reduces to the one-dimensional problem of a particle constrained in the region $r > 0$: since the centrifugal potential is absent ($l = 0$), the equation for $u_0(r)$ is

$$-\frac{\mathrm{d}^2 u_0(r)}{\mathrm{d}r^2} + U(r)\, u_0(r) = k^2 u_0(r)\,, \qquad u_0(0) = 0\,, \quad U(r) = \frac{2m}{\hbar^2} V(r)\,.$$

The solution is unique up to a factor and is given by

$$u_0(r) = \begin{cases} A\, \varphi(r) & \text{for } r < a\,, \quad \varphi(0) = 0 \\ \sin\left(kr + \delta(k)\right) & \text{for } r > a \end{cases} \qquad (17.45)$$

where A and $\delta(k)$ are to be determined by imposing the continuity conditions at $r = a$:

$$\begin{cases} A\,\varphi(a) = \sin\left(ka + \delta(k)\right) \\ A\,\varphi'(a) = k\,\cos\left(ka + \delta(k)\right). \end{cases}$$

Then

$$\tan\left(ka + \delta(k)\right) = \frac{k\,\varphi(a)}{\varphi'(a)}. \qquad (17.46)$$

and for the radial function $R_0(r)$ for $r > a$ we can write

$$R_0(r) = \frac{C}{2i}\left(e^{+i\delta(k)}\,\frac{e^{ikr}}{kr} - e^{-i\delta(k)}\,\frac{e^{-ikr}}{kr}\right), \qquad r > a \qquad (17.47)$$

with C a constant to be determined in such a way that the difference between $R_0(r)$ and the free solution (i.e. the one without the potential)

$$j_0(kr) = \frac{\sin kr}{kr} = \frac{1}{2i}\left(\frac{e^{ikr}}{kr} - \frac{e^{-ikr}}{kr}\right)$$

is, for $r > a$, an outgoing wave (the scattered wave), as in (17.13). In this way we get

$$C = e^{i\delta(k)}, \qquad R_0(r) - j_0(kr) = \frac{1}{2ik}\left(e^{2i\delta(k)} - 1\right)\frac{e^{ikr}}{r}$$

and for $\chi_{l=0}$ (see (17.13), (17.24) and (17.39))

$$\chi_{l=0}(r) = \frac{1}{2ik}\left(e^{2i\delta(k)} - 1\right)\frac{e^{ikr}}{r}$$

hence

$$f_{l=0} = \frac{1}{2ik}\left(e^{2i\delta(k)} - 1\right) = \frac{1}{k}\,e^{i\delta(k)}\sin\delta(k).$$

As a consequence $\delta(k)$ in (17.45), the phase difference between the asymptotic wavefunction with the potential and the free wavefunction, is just the phase shift $\delta_0(k)$ of (17.41), which is determined by (17.46).

Since the s-wave scattering is particularly relevant at low energies, we take into consideration the $k \to 0$ limit of the phase shift $\delta_0(k)$, still considering a potential different from zero in the region $0 < r < a$.

From (17.46) $\tan\left(ka + \delta_0(k)\right)$ is of order $O(k)$ so the same is for $\delta_0(k)$, therefore the limit of $\delta_0(k)/k$ exists and

$$a_s \equiv -\lim_{k \to 0}\frac{\delta_0(k)}{k} \qquad (17.48)$$

is known as the **scattering length** (the reason of the minus sign will be clarified in the next section). In terms of a_s:

$$\lim_{k \to 0}\sigma_0(k) = \lim_{k \to 0}\frac{4\pi}{k^2}\sin^2\delta_0(k) = 4\pi\,a_s^2. \qquad (17.49)$$

The scattering length a_s has a simple geometrical meaning: consider the tangent to the curve $y = \sin\left(kx + \delta_0(k)\right)$ at $x = a$; its equation is

$$y = \sin\left(ka + \delta_0(k)\right) + k(x - a)\cos\left(ka + \delta_0(k)\right).$$

This straight line cuts the x-axis at $x_0(k) = \left(ka - \tan\left(ka + \delta_0(k)\right)\right)/k$ and

$$\lim_{k \to 0} x_0(k) = -\lim_{k \to 0} \frac{\delta_0(k)}{k} = a_s.$$

17.7 Applications

The first application concerns the scattering from a hard sphere:

$$V(r) = \begin{cases} \infty & \text{for } r < a \\ 0 & \text{for } r > a. \end{cases}$$

This example is important since, as already mentioned in Sect. 17.2, it exhibits a striking disagreement with the classical result.

We will consider only the low energy case: $ka \ll 1$ so that, in accordance with the discussion of the previous section, we take into consideration only s-wave scattering. In this case $u_0(r)$ is zero for $r \leq a$ and is given by $\sin k(r-a)$ for $r \geq a$, then $\delta_0 = -ka$, $a_s = a$ and

$$\sigma_0(k) = \frac{4\pi}{k^2}\sin^2 ka \simeq 4\pi a^2 \qquad (17.50)$$

since $ka \ll 1$.

In the light of (17.50) the meaning of (17.49) is that at low energies, that is at large wavelength, any potential is equivalent to a hard sphere of radius equal to the scattering length a_s (this is the reason of the minus sign in the definition (17.48)).

The high energy limit is not so easy to calculate, neither can we make use of the Born approximation since it diverges.

In the above discussion there was no approximation, apart from the restriction to the s-wave scattering. In the next example, instead, we will make use of the Born approximation: we apply (17.37) to the case of the (three-dimensional) square-well potential

$$V(r) = \begin{cases} V_0 & r < a \\ 0 & r > a \end{cases} \qquad (17.51)$$

(we do not specify the sign of V_0 since the cross section in the Born approximation is independent of it). From (17.37) we get

$$f_k(\theta) = -\frac{2\,mV_0}{\hbar^2\kappa}\int_0^a \sin(\kappa r')\,r'\mathrm{d}r' = -\frac{2\,mV_0}{\hbar^2\,\kappa^3}\left(\sin\kappa a - \kappa a\cos\kappa a\right) \qquad (17.52)$$

where $\kappa = 2\,k\,\sin(\theta/2)$.

From the graph in Fig. 17.1 of the func-
tion $\eta(\kappa a) \equiv (\sin \kappa a - \kappa a \cos \kappa a)/(\kappa a)^3$
it is clear that the scattering occurs almost
completely within the region $\kappa a < 4$, i.e.
$ka \sin(\theta/2) < 2$, which means that at
high energies ($ka \gg 1$) the scattering is
concentrated within the diffraction peak
$\theta \lesssim 4 \times \lambda/a$.

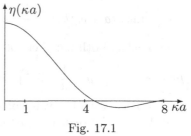

Fig. 17.1

In Fig. 17.2a we report the differential cross section:

$$|f_k(\theta)|^2 = \left(\frac{2\,mV_0\,a^3}{\hbar^2}\right)^2 \frac{\left[\sin\big(2ka\sin(\theta/2)\big) - 2ka\sin(\theta/2)\cos\big(2ka\sin(\theta/2)\big)\right]^2}{\big(2ka\sin(\theta/2)\big)^6}$$

for three values of ka: $ka = 0.5, 1, 2$. The common value of the three curves
at $\theta = 0$ is $(1/9)\,(2\,mV_0\,a^3/\hbar^2)^2$. In accordance with the discussion in Sect.
17.6 we notice that the lower the value of ka, the more the differential cross
section is independent of θ: for $ka \to 0$ $f(\theta)$, and consequently $\sigma(\theta)$, become
constant, i.e. only the s-wave is scattered.

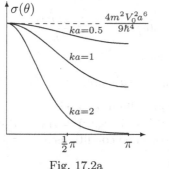
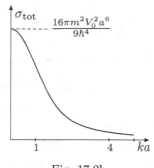

Fig. 17.2a   Fig. 17.2b

In Fig. 17.2b we report the total cross section:

$$\sigma_{\text{tot}}(k) = \int |f_k(\theta)|^2 \, d\Omega$$

(the integral can easily be calculated taking $x = 2ka\sin(\theta/2)$ as variable of
integration). At high energies, i.e. $ka \gg 1$,

$$\sigma_{\text{tot}}(k) \simeq \frac{\sigma_{\text{tot}}(k = 0)}{16(ka)^2} \qquad ka \gg 1.$$

Let us now calculate the $l = 0$ total cross section σ_0 to first order in the
potential and in the limit $ka \to 0$ using the method of the partial waves.
If $V_0 > 0$, the function $\varphi(r)$ in (17.45) is proportional to $\sinh \kappa r$ and from
(17.46) we get

$$\tan\left(ka + \delta_0(k)\right) = \frac{k}{\kappa}\tanh\kappa a, \qquad \kappa = \sqrt{\frac{2\,mV_0}{\hbar^2} - k^2}. \tag{17.53}$$

By expanding both members of (17.53) to first order in k and V_0 we find

$$ka + \delta_0 \simeq \frac{k}{\kappa}\left(\kappa a - \frac{1}{3}(\kappa a)^3\right) \;\Rightarrow\; \delta_0 \simeq -\frac{k}{3}\frac{2\,mV_0 a^3}{\hbar^2} \;\Rightarrow\; \sigma_0 = \frac{16\,\pi m^2 V_0^2 a^6}{9\hbar^4}$$

in accordance with the value of σ_{tot} at $k = 0$ reported in Fig. 17.2b.

If $V_0 < 0$ the (exact) equation determining δ_0 is

$$\tan\left(ka + \delta_0(k)\right) = \frac{k}{\kappa}\tan\kappa a, \qquad \kappa = \sqrt{\frac{2\,m|V_0|}{\hbar^2} + k^2}$$

different, as it must be, from (17.53) relative to the case $V_0 > 0$, but within the same approximations only the sign of δ_0 changes, not σ_0. Notice, however, the connection between the sign of the phase shift and that of the potential.

In the last application we determine the scattering length for the potential

$$V(r) = v_0\,\delta(r - a) \tag{17.54}$$

and compare the result with that obtained by means of the Born approximation. The effect of the Dirac δ-potential is to give rise to a discontinuity in the first derivative of the reduced radial function $u_0(r)$ at the point $r = a$, given by

$$\Delta u_0'(a) = \frac{2m\,v_0}{\hbar^2}u_0(a) = g_0\,u_0(a), \qquad g_0 \equiv \frac{2m\,v_0}{\hbar^2}$$

$\Big[$this can be proved by integrating the equation

$$u_0''(r) + k^2 u_0(r) = g_0\,\delta(r - a)\,u_0(r) \tag{17.55}$$

between $a - \epsilon$ and $a + \epsilon$ and then taking the limit $\epsilon \to 0\Big]$.

At $k \approx 0$ the solution of (17.55) for $r < a$ is $u_0(r) = A\sin kr \approx A\,kr$, while for $r > a$ it is $\sin(kr + \delta_0) \approx kr + \delta_0$. The conditions at $r = a$ are

$$\begin{cases} A\,ka = ka + \delta_0 \\ A\,k \;= k - g_0\times(ka + \delta_0) \end{cases}$$

then

$$\delta_0 = -\frac{g_0\,a^2}{1 + g_0 a}\,k, \qquad a_s = -\lim_{k\to 0}\frac{\delta_0}{k} = \frac{g_0 a^2}{1 + g_0 a} \tag{17.56}$$

and from (17.42) we get

$$f_k = \frac{1}{k}\,e^{i\,\delta_0(k)}\sin\delta_0(k)$$

therefore also

$$a_s = -\lim_{k\to 0} f_k\,.$$

In the Born approximation, from (17.37) we have

$$f_{\vec{k}}(\theta, \phi) = -\frac{2m\,v_0 a}{\hbar^2}\frac{\sin\kappa a}{\kappa} \overset{k\to 0}{=} -g_0 a^2 = -a_s^{\text{Born}}$$

which coincides with a_s given in (17.56) only to the first order in g_0, i.e. v_0.

17.8 Collision of Identical Particles

When the collision occurs between two identical particles it is no longer possible to distinguish the beam-particle from the target-particle. Indeed, the *state* (*not* the state-vector) does not change if the two particles are interchanged.

It is easier to discuss the problem in the center-of-mass frame: this choice corresponds to many real situations where head-on collisions occur between two beams of identical particles with the same energy, as the proton-proton collisions within the Intersecting Storage Rings (ISR) that operated at CERN some decades ago, or within the Large Hadron Collider (LHC) still operating these days. In these cases the center-of-mass and the laboratory frames coincide.

To begin with we discuss the effect of the symmetry or antisymmetry of the space part of the wavefunction on the scattering amplitude. The energy eigenfunctions $\Psi(\vec{r}_1, \vec{r}_2)$ can be written as the product of the wavefunction $\Phi(\vec{r}_1 + \vec{r}_2)$ of the center-of-mass, and of $\psi(\vec{r})$, $\vec{r} = \vec{r}_1 - \vec{r}_2$, the wavefunction in the center-of-mass frame, i.e. of the relative motion: since $\Phi(\vec{r}_1 + \vec{r}_2)$ is symmetric with respect to the exchange of $\vec{r}_1$ with $\vec{r}_2$, it is only $\psi(\vec{r})$ that must be symmetrized or antisymmetrized. Therefore instead of (17.17) we must write

$$\psi_E(r, \theta, \phi) \overset{r \to \infty}{\longrightarrow} \left(e^{i\vec{k}\cdot\vec{r}} \pm e^{-i\vec{k}\cdot\vec{r}} \right) + \left[f_{\vec{k}}(\theta, \phi) \pm f_{\vec{k}}(\pi - \theta, \phi + \pi) \right] \frac{e^{ikr}}{r} \quad (17.57)$$

since $\vec{r} \to -\vec{r}$ is equivalent to $r \to r$, $\theta \to \pi - \theta$, $\phi \to \phi + \pi$.

The differential cross section is then given by

$$
\begin{aligned}
\sigma(\theta, \phi) &= \left| f_{\vec{k}}(\theta, \phi) \pm f_{\vec{k}}(\pi - \theta, \phi + \pi) \right|^2 \\
&= \left| f_{\vec{k}}(\theta, \phi) \right|^2 + \left| f_{\vec{k}}(\pi - \theta, \phi + \pi) \right|^2 \\
&\quad \pm 2\Re e \left[f_{\vec{k}}(\theta, \phi) f_{\vec{k}}^*(\pi - \theta, \phi + \pi) \right].
\end{aligned}
\quad (17.58)
$$

Only the last term in (17.58) is the consequence of the symmetrization demanded by the indistinguishability of the two particles: if the two particles where distinguishable in principle, but not by the detectors, as for instance two protons (spin 1/2) in two orthogonal spin states, or as in classical physics where identical particles are always distinguishable, then the interference term in (17.58) would be absent and the detector at the angles θ, ϕ would respond both when particle 1 is scattered in the direction θ, ϕ, and also when particle 1 is scattered in the opposite direction $\pi - \theta$, $\phi + \pi$, since in that case it is particle 2 that hits the detector. In this case:

$$\sigma^{\text{dist}}(\theta, \phi) = \left| f_{\vec{k}}(\theta, \phi) \right|^2 + \left| f_{\vec{k}}(\pi - \theta, \phi + \pi) \right|^2. \quad (17.59)$$

If the interaction between the two particles is spin-independent, then $V = V(r)$ and the scattering amplitude does not depend on ϕ, moreover $\sigma(\theta)$ is symmetrical about $\theta = \pi/2$ in the center-of-mass frame.

Since the above discussion does not take the spin into account, it is correct as it stands only for spinless particles (and in this case the + sign must be taken in (17.57) and (17.58) whenever the $\pm$ sign appears).

Suppose now that the spin of the two colliding particles is $1/2$. If, for instance, they are in the *same* spin-state (triplet state), then the space part of the wavefunction must be antisymmetrical and the interference term in (17.58) must be taken with the $-$ sign. If instead the spin state of any pair of colliding particles is the singlet state, then the interference term must be taken with the + sign.

Usually the two beams are unpolarized, i.e. they are a uniform statistical mixture of spin states. Therefore, on the average, every four collisions 3 are in the triplet state and 1 in the singlet state and the interference term in (17.58) must be taken 3 times with the $-$ sign and only once with the + sign. This means that every four collisions the interference term contributes only $3 - 1 = 2$ times, with the $-$ sign. Then, for unpolarized beams of spin $1/2$ particles, (17.58) reads:

$$\sigma^{\text{unpol}}(\theta, \phi) = \left| f_{\vec{k}}(\theta, \phi) \right|^2 + \left| f_{\vec{k}}(\pi - \theta, \phi + \pi) \right|^2$$
$$- \frac{1}{2} \times 2\Re e \left[f_{\vec{k}}(\theta, \phi) \, f_{\vec{k}}^*(\pi - \theta, \phi + \pi) \right]. \qquad (17.60)$$

For spin 1 particles (or atoms) there are $3 \times 3 = 9$ spin states, 6 symmetric ($S = 2, 0$) and 3 antisymmetric ($S = 1$) and in this case the space part of the wavefunction has the same symmetry as the spin part, so every 9 collisions the interference term contributes $6 - 3 = 3$ times, with the + sign. Hence the factor in front of the interference term in this case is $+3/9 = 1/3$. In general, for particles of spin s the above factor is $(-1)^{2s}/(2s + 1)$.

17.9 The Reciprocity Theorem and the Detailed Balance

Let us now introduce a different notation for the scattering amplitude: instead of $f_{\vec{k}}(\theta, \phi)$ we write $f(\vec{k}_{\text{f}}, \vec{k}_{\text{i}})$. The two notations, with $\vec{k} = \vec{k}_{\text{i}}$, clearly are equivalent.

Reciprocity theorem:

$$f(\vec{k}_{\text{f}}, \vec{k}_{\text{i}}) = f(-\vec{k}_{\text{i}}, -\vec{k}_{\text{f}}). \qquad (17.61)$$

This theorem is a consequence of the unitarity of the time evolution and of the fact that if the transition $\vec{k}_{\text{i}} \to \vec{k}_{\text{f}}$ is possible, then also the *time reversed* transition $-\vec{k}_{\text{f}} \to -\vec{k}_{\text{i}}$ is possible: this is true if the potential in the Schrödinger representation is real, therefore it fails in the presence of a magnetic interaction as $\vec{L} \cdot \vec{B}$, since the angular momentum in the Schrödinger representation is imaginary (actually, also in classical physics the motion of a charged particle in a magnetic field cannot be inverted).

The reciprocity theorem is the generalization to three dimensions of the result reported in Sect. 8.3 that the transmission and reflection coefficients of the scattering from a barrier are independent of the direction of the incoming particle.

We do not give the proof of (17.61) that explicitly makes use of the invariance under *time reversal*, but we show that (17.61) is satisfied by the terms of the Born series. The first term (17.25) clearly is invariant under $\vec{k}_{\rm f}, \vec{k}_{\rm i} \to -\vec{k}_{\rm i}, -\vec{k}_{\rm f}$. The second term is $\left(G_k \text{ is the Green's function (17.21)} \right)$

$$-\frac{1}{4\pi} \left(\frac{2m}{\hbar^2} \right)^2 \iint e^{-i\vec{k}_{\rm f}\cdot\vec{r}''} V(\vec{r}'') \, G_k(|\vec{r}'' - \vec{r}'|) V(\vec{r}') \, e^{i\vec{k}_{\rm i}\cdot\vec{r}'} \, {\rm d}\vec{r}' \, {\rm d}\vec{r}''$$

and satisfies (17.61) thanks to $G_k(|\vec{r}'' - \vec{r}'|) = G_k(|\vec{r}' - \vec{r}''|)$. Similarly for the higher orders that symbolically write as $\langle \vec{k}_{\rm f} \mid V G_k V G_k V \cdots V \mid \vec{k}_{\rm i} \rangle$, apart from factors.

If the potential is invariant under space inversion then $f(\vec{k}_{\rm f}, \vec{k}_{\rm i}) = f(-\vec{k}_{\rm f}, -\vec{k}_{\rm i})$.

Combining the above result with the reciprocity theorem we get the *detailed balance:*

$$f(\vec{k}_{\rm f}, \vec{k}_{\rm i}) = f(\vec{k}_{\rm i}, \vec{k}_{\rm f}). \tag{17.62}$$

Both the reciprocity theorem and the detailed balance find important applications in statistical physics: kinetic theory of gases, chemical kinetics, ... and are tied to names such as J. C. Maxwell and L. Boltzmann.

Chapter 18

The Paradoxes of Quantum Mechanics

18.1 Introduction

We have already mentioned on many occasions that the probabilistic interpretation of quantum mechanics, the so called Göttingen (Born) – Copenhagen (Bohr) interpretation, was not accepted by the entire community of physicists and scientists as outstanding as Einstein, Schrödinger and de Broglie were among its opponents.

Let us immediately say that nobody ever challenged the results of quantum mechanics: those we have discussed in the last chapters only are a small part of the several applications that were developed in the thirties of the twentieth century and that range from molecular physics to solid state physics, nuclear physics, ... ; it is needless to cite all the fields in which important results have been obtained: it appears that the obtaining of new results in any field of physics seems to be limited only by the development of calculative techniques, both analytic and computational.

The subject we want to address is another one: is quantum mechanics – or, better, the paradigms it proposes – *the* (ultimate) theory of *all* natural phenomena, or rather – this was Einstein's firm conviction – is it only a provisional and incomplete version of what will be the *true* theory?

The crucial problem that divided, and still divides, the community of physicists is the probabilistic nature of quantum mechanics that, in general, denies the possibility of predicting with certainty the behaviour of the single system: indeed Einstein compared quantum mechanics to thermodynamics, that is an incomplete theory inasmuch as it exclusively determines the behaviour of the ensemble, giving up the description of the single atom.

The problem remained for many years (until the works by J. Bell in the sixties) a problem exclusively philosophical in nature, and it is known that on such kind of problems it is not possible to find a solution on which all agree. The problem whether the theory should be causal, or not, links with the philosophical concept of the existence, or not, of an objective reality that does not depend on the observations (for instance: does the electron, at any given time, possess well determined position and velocity, independently of what we

© Springer International Publishing Switzerland 2016

L.E. Picasso, *Lectures in Quantum Mechanics*, UNITEXT for Physics,

DOI 10.1007/978-3-319-22632-3_18

can measure?) and it is clear that realism and positivism are philosophical positions that not only influence the political belief, but also are influenced by it.

Much has been said and written on such problems and we have no pretension to give in a few pages a precise idea about the set of problems and the reasons of the ones and of the others: we will limit ourselves to present some of the so called "paradoxes of quantum mechanics", i.e. the tip of the iceberg of the discussions among the great physicists from the thirties of the past century on.

In order to clear the discussion of any possible misunderstanding, we recall that "paradox" does not mean "contradiction" (if it were so, the inconsistency of quantum mechanics should already have been shown: indeed this point seems to be beyond question), but rather "contrary to common sense".

18.2 The de Broglie Box

An electron is in a stationary state within a (ideal) box endowed with perfectly reflecting walls. By taking all the precautions the case requires, the box is split into two equal boxes by means of a sliding diaphragm, then the two boxes – let us say the right and the left one – are taken far apart from each other: one is taken to Rome, the other to Paris.

According to quantum mechanics the electron is in a state $|A\rangle$ that is a superposition of the state $|r\rangle$, in which it is in the right box, and of the state $|l\rangle$, in which it is in the left box:

$$|A\rangle = |r\rangle + |l\rangle .$$ (18.1)

Then the probability of finding the electron in each box is $1/2$.

Suppose now that the right box (that in Rome) is opened and the electron found. Problem: was the electron there before the opening of the box (and then was the left box already empty), or the presence of the electron in Rome and its absence in Paris only is a consequence of the measuring process, that consists in opening the box and establishing (e.g. by means of a Heisenberg microscope) whether the electron is inside it? The ones maintain that the above is a non-scientific problem, the others – i.e. the supporters of a *realistic* theory – stick to the first hypothesis and then conclude that quantum mechanics is not a complete theory, inasmuch as the description given by (18.1) does not represent reality: it does not say in which box the electron is. In addition they reply to the supporters of the non-scientific character of the problem that exactly this kind of statements precludes the search for a "complete" theory.

The paradoxical, i.e. "contrary to common sense", aspects of the above example are essentially two: the superposition principle applied to 'too orthogonal' states (the electron in Rome, the electron in Paris), and the effect of the measuring process that produces the instantaneous collapse of the wavefunction (or "reduction of the wave packet"), initially nonvanishing *both* in Rome and in Paris, and then nonvanishing *either* in Rome *or* in Paris. Many classical paradoxes are variations on such themes and they differentiate from one another according to how much they put the accent on either one of the two aspects of the problem. In conclusion, it is worth analyzing a little more deeply the paradox of the de Broglie box.

First of all, we ask ourselves if necessarily, in accordance with the principles of quantum mechanics, should one claim that the electron is in the state (18.1), i.e. in a *coherent* superposition of the states $|r\rangle$ and $|l\rangle$: if we want to talk about superposition, we must be able to distinguish by means of experiments – i.e measurements – if the state of the electron is either $|r\rangle+|l\rangle$, or $|r\rangle-|l\rangle$, $|r\rangle+i|l\rangle$, $\cdots$, and this is possible only if we measure observables that have nonvanishing matrix elements between $|r\rangle$ and $|l\rangle$.

But: do such observables exist? Certainly it is possible to write many $f(q,p)$ suitable for the case at hand, and therefore, perhaps, *in principle*, such observables exist. We have said 'perhaps' because, if such observables existed, it should be possible, by means of a couple of measurements, to let the electron pass from the box in Rome into that in Paris: indeed such an observable could have $|r\rangle\pm|l\rangle$ as eigenstates so that, with a nonvanishing probability, with two measurements (the first, such observable; the second, the opening of a box) the following transitions $|r\rangle\rightarrow|r\rangle\pm|l\rangle\rightarrow|l\rangle$ would be possible; and this seems to be in contradiction with the statement that the two boxes are impenetrable, i.e. with the very assumptions of the paradox. But, apart from this difficulty (it suffices to invent a more flexible paradox), there always is the problem that the two boxes are macroscopically separated; may we be satisfied with the statement that, *in principle* the observables we need do exist? Would it be too much of a scandal (from the philosophical point of view) to state that, among the paradigms of quantum mechanics, there also is that according to which the observables only are those we can either realize in at least one of the laboratories existing in the world, or that in any way we are able to construct? Let us provisionally accept this idea and explore its consequences.

If observables allowing for the observation of a possible phase factor in the superposition of the states $|r\rangle$ and $|l\rangle$ are not available, in any event, after the boxes have been separated, the electron is not in a pure state, but in the statistical mixture $\{|r\rangle,\frac{1}{2};|l\rangle,\frac{1}{2}\}$. At this point another step must be taken: we must state that in a statistical mixture any single system the ensemble is made of is in a well determined state, much like a coin that has been tossed and not yet observed: it is in a well determined state, but we only know which is the probability it is heads or tails. So, before opening the de Broglie boxes, the electron is either in Rome or in Paris: the conclusion is that maintained by the realists, but it is reached without denying the fundamental concepts of quantum mechanics, i.e. – if the proposed paradigms are accepted – from within quantum mechanics.

It should be evident that this is the standpoint of the present author, therefore it is only one among the many points of view and, as such, can be refused by the reader, also because if analyzed a little more, it exhibits aspects that are all but painless. The first objection that can be raised is that, even if observables with non vanishing elements between the states with the electron in Rome and the electron in Paris are not available today, the possibility that somebody will succeed in constructing them tomorrow cannot be excluded: so what we today call a statistical mixture tomorrow we will call a pure state.

The objection is legitimate, indeed mandatory, and the reply consistent with the above discussion is that what defines a physical system is the set of the observables, and the properties of the states do depend on which observables are at one's disposal: in other words, the primary entity is the set of observables (this also is the point of view of the formulation of quantum mechanics due to Haag and Kastler we have hinted at in the end of Sect. 4.8), and the attributes ("pure" or "not pure") of a state are not intrinsic, but depend on the observables that define the system, so they may change along with the technological progress that provides us with new instruments, i.e. that allows us to enrich the algebra of observables: clearly, according to this point of view the concept of state looses quite a lot of its ontological meaning (in the sense of an entity endowed with properties independent of the contingent situation), and maybe today not many are willing to accept this fact.

Another big problem of quantum mechanics is that of the measurement process: we will talk about it in Sect. 18.4 and we will analyze it in the light of the ideas exposed above: we now prefer to illustrate another classical paradox.

18.3 Schrödinger's Cat

The problems raised by the paradox of de Broglie box are even more dramatized by the paradox proposed by Schrödinger, inasmuch as involving a poor cat unaware of what may happen to him.

In a big box there are a cat, an ampoule of cyanide, and a device that, in some instant, emits a photon; the photon impinges on a semi-transparent mirror and, if it crosses it nothing happens while, if is reflected by it, it triggers an amplification process whose final step is the breaking of the ampoule of cyanide.

Which is the destiny of the cat after the emission of the photon?

According to quantum mechanics the state of the (maximally schematized) system is a superposition of the two states | alive cat, unbroken ampoule ⟩ and | dead cat, broken ampoule ⟩ so the cat is neither dead nor alive. But, when we open the box, the cat is found either dead or alive: it is therefore the observation, i.e. the measurement, that produces the collapse of the wavefunction that, in the present case, may have dramatic consequences for the unlucky cat. Obviously for the realists the cat in the box is either dead or alive, independently of the fact that its destiny be ascertained by opening the box: so quantum mechanics is not a complete theory because it is not able to predict the fate of the cat.

It is clear that the present paradox proposes again the same problems as de Broglie's: superposition of states, here macroscopic, there macroscopically separated, and the crucial, but casual, role of the measurement process.

An aspect of Schrödinger's paradox that usually is not emphasized is that indeed quantum mechanics does not allow one to talk about the state of the cat after the emission of the photon: the system is a composite one (we have schematized the system as cat + ampoule) and the state is an "entangled" one (see Sect. 15.1):

$$|A\rangle = |\heartsuit, \uparrow\rangle + |\heartsuit, \downarrow\rangle \qquad (18.2)$$

($\uparrow$ and $\downarrow$ respectively stand for 'unbroken ampoule' and 'broken ampoule'), and in this case the cat is not in a pure state: it is not correct to state that the cat (or another living being, as in a less bloody paradox, that of Wigner's friend) is in a state that is superposition of dead cat and alive cat: even if the state of the *system* were given by (18.2), i.e. if the system were in a pure state (but we put forward the warnings we have expressed about (18.1)), the single subsystem is described by a statistical mixture.

Let us consider the entire system, not only the cat: to the present day there exist no observables with matrix elements between $|\heartsuit, \uparrow\rangle$ and $|\heartsuit, \downarrow\rangle$ therefore, if (just for the sake of discussion) the point of view expressed in the previous section is accepted, one must conclude (much as for the de Broglie electron) that after the emission of the photon the cat is either dead or alive, and therefore, in the absence of observables able to emphasize the quantum nature of the system expressed by the superposition principle, the system is classical (as the tossed coin). Tomorrow, when the observables that today do not exist will become available, we will be able, by means of two measurements, to resurrect dead cats and fix broken ampoules of cyanide: therefore tomorrow the system cat+ampoule will be a quantum system with pure states of the type (18.2).

18.4 What Is a Measurement?

The section title is a too embarrassing one: also in this case blue streaks, thousand of written pages ... .

Essentially the problem is the following: a measurement is the interaction between the system (an electron, a photon, ... : let us call it the "microsystem") and an instrument that is a macroscopical object. An instrument that measures the observable ξ on the microsystem should (according to quantum mechanics) work more or less in the following way: in the beginning, before the measurement, the microsystem and the instrument are separated and the 'grand-system' consisting of microsystem + measuring instrument is the factorized state

$$|A, \Xi_0\rangle \equiv |A\rangle|\Xi_0\rangle, \qquad |A\rangle = \sum_i a_i|\xi_i\rangle \qquad (18.3)$$

where $|\Xi_0\rangle$ is the initial state of the instrument, $|A\rangle$ that of the microsystem and $|\xi_i\rangle$ the eigenstates of the observable ξ measured by the instrument; in the end the grand-system should be in the statistical mixture

$$\{ |\,\xi_i\,,\,\varXi_i\,\rangle\,,\,|a_i|^2 \} \tag{18.4}$$

where $|\,\varXi_i\,\rangle$ are the possible states of the instrument after the interaction (the pointer of the instrument either on ξ_1, or ξ_2 etc. At this point the measurement has taken place both in the case we realize the result by observing it and in the contrary case.

We have said the grand-system 'should be' ... instead of 'is' ... because, if it is true that quantum mechanics applies to all systems, either microscopical or not (if it were not so, it would be possible to show that quantum mechanics is not a consistent theory), then it applies also to the grand-system consisting of microsystem + measuring instrument. The grand-system has a Hamiltonian and, therefore, a unitary time evolution operator; therefore the state after the interaction is well determined, in blatant contradiction with the postulate according to which the result of the measurement is only statistically determined: in other words, a unitary operator cannot map pure states into statistical mixtures.

Now, if it is true that the supporters of the Copenhagen interpretation can ignore the de Broglie and Schrödinger paradoxes and maintain that if they go against the common sense, too bad! (i.e. all the worse for the common sense), they can hardly ignore the problem of measurement, because it appears that in this case we are facing a contradiction that is internal to the theory.

Let us imagine to analyze a measurement in its different phases: for example, a photon arrives at a photo-multiplier, interacts with an atom and an electron is emitted – and this process certainly is described by quantum mechanics – then the electron ionizes some atoms and some electrons are emitted ... the amplification process carries on until the display showing the final result is activated; it is clear that, in order to save the postulate of quantum mechanics concerning the measurement process, at a certain point the above chain of events must stop being determined by a unitary operator and must become probabilistic, i.e. the state of the grand-system, a pure state until that moment, must become a statistical mixture. But in which point of the chain does this happen? Until we are dealing with interactions among electrons and atoms or even solid state physics phenomena (in the electronics of the display), it cannot be maintained that the time evolution is not governed by a Hamiltonian – no matter how complicated, but in any event a Hamiltonian. So it has even been proposed (Wigner) that the act of taking conscience of the result is the point of the chain in which the causal time evolution breaks down: the human brain does not obey the law of unitary time evolution!

Another possible way out consists in saying that the grand-system interacts with the external environment, then we are dealing with a super-grand-system and the state of this super-system is an entangled (pure) state of the grand-system and of the external environment: in this case the grand-system is not in a pure state (as the cat in the case of the cat+ampoule system), but in a statistical mixture.

If the solutions consisting in attributing the collapse of the wavefunction either to the human intervention or to the interaction with the external environment (or similar ones) are not considered satisfying, then the problem is still open, at least in the framework of the "orthodox" (i.e. the Copenhagen) interpretation of quantum mechanics.

If we accept the paradigms proposed in Sect. 18.2, we must say the time evolution of the state of the grand-system is formally determined by a unitary time evolution operator, but since the state evolves towards an entangled state of microsystem and instrument, when one arrives at a point in which there no longer exist observables with nonvanishing matrix elements between the states involved in the superposition, then such a state is a statistical mixture: it is not important to know exactly in which point this happens or, more realistically, if this happens gradually: the important thing is that in the end we will *formally* have the state

$$\sum_i a_i \,|\,\xi_i \,,\, \Xi_i\,\rangle = U(t)\,|\,A\,,\, \Xi_0\,\rangle$$

but actually it is the statistical mixture (18.4).

18.5 The Einstein–Podolsky–Rosen Paradox

Several versions of the Einstein–Podolsky–Rosen (EPR) paradox exist: we will start with the original formulation published in 1935, even if it is not very elegant, inasmuch using improper states.

In the work of the above Authors there is no room for folklore (boxes, cats, nice friends, ...); more than a paradox, it is a theorem: if you accept some premises (that we will enunciate in the sequel) then we show that quantum mechanics is not a complete theory; if you are convinced that quantum mechanics is a complete theory, then necessarily you must give up the above premises.

The premises made by the authors are the following.

Principle of reality: *there exists something as "the real state" of a physical system, that exists objectively and independently of whatever observation or measurement and that can be, in principle, be described by the means of physics.*

As a matter of fact, the above principle is not explicitly enunciated by the Authors in the 1935 paper, however it is at the basis of Einstein thinking, as he expressed it on several occasions.

The next two points are instead the literal quotation of what has been written by the authors.

Whatever the meaning assigned to the term complete, the following requirement for a complete theory seems to be a necessary one: every element of the physical reality must have a counterpart in the physical theory. We shall call this the **condition of completeness.**

. . . . . .

A comprehensive definition of reality is[, however,] unnecessary for our purpose. We shall be satisfied with the following criterion, which we regard as reasonable. If, without in any way disturbing a system, we can predict with certainty (i.e., with probability equal to unity) the value of a physical quantity, then there exists an element of physical reality corresponding lo this physical quantity. (Criterion for the physical reality)

At this point the line of reasoning of EPR is the following: we show you that even if we cannot simultaneously (and precisely) measure q and p, they are elements of the physical reality, therefore, within a complete theory, both their values must be part of the description of the system. Given that this is not so in quantum mechanics, the latter is not a complete theory.

Demonstration (by contradiction): let us consider the system consisting of the two particles 1 and 2 on a straight line. Let us assume that the value of $q_1 - q_2$ is known (for example $x_1 - x_2 = 10\,\text{km}$). If we accept quantum mechanics (i.e. if quantum mechanics is a complete theory), then neither p_1 nor p_2 can be known, instead $P = p_1 + p_2$ can be known ($q_1 - q_2$ and $p_1 + p_2$ are compatible observables) and let (for example) $P' = 0$ (P' is the value of P).

If p_1 is measured (let p_1' be the result), then it is possible to predict the value of p_2 ($= -p_1'$) without interacting with particle 2. Then, by the reality criterion, we conclude that p_2 is an element of the physical reality of particle 2 and therefore (definition of complete theory) it must have a counterpart in the theory.

If instead q_1 is measured, then it is possible to predict the value of q_2, still without interacting with particle 2. So we conclude that also the value x_2 of q_2 is an element of the physical reality of particle 2 and must have a counterpart in the theory.

So x_2 and p_2', even if cannot be simultaneously known, both are elements of the physical reality of particle 2, hence both must simultaneously be contained in the description of the system, i.e. in its wavefunction – which is false, therefore the quantum-mechanical description of the physical reality given by the wavefunction is not complete.

The Authors conclude by saying that one could object that their criterion of reality is not enough restrictive:

Indeed, one would not arrive at our conclusion if one insisted that two or more physical quantities can be regarded as simultaneous elements of reality only when they can be simultaneously measured or predicted. On this point of view, since either one or the other, but not both simultaneously, of the quantities p_2 and q_2 can be predicted, they are not simultaneously real. This makes the reality of p_2 and q_2 depend upon the process of measurement carried out on the first system, which does not disturb the second system in any way. No reasonable definition of reality could be expected to permit this.

In the above passage the Authors implicitly make use of the following

Einstein locality principle: if at the moment of the measurement [on the first system] the two systems do not interact any longer, no real change can take place in the second system.

As we have said in the beginning of the section, there exist several versions of the EPR paradox: we conclude by presenting the Bohm–Aharonov version (1957).

At a certain instant t_0 a particle decays into two spin $1/2$ particles not interacting with each other. Let us assume that the system consisting of the two particles is in a singlet spin state: such an assumption is fully compatible with quantum mechanics; indeed, for example, the decaying particle has zero angular momentum and, owing to nature of the interaction responsible for the decay, the two particles are produced in a state with vanishing orbital angular momentum: therefore, due to the conservation of total angular momentum, also their spin must be vanishing.

At time $t_1 > t_0$ the observable s_{1z} is measured and suppose the value $+1/2$ (in units of $\hbar$) is found. Since $S_z' = 0$ and s_{1z} commutes with S_z, it follows that $s_{2z}' = -1/2$.

Then, by the reality criterion and the locality principle, $s_{2z}' = -1/2$ even at time t_0, i.e. before the measurement on particle 1 was performed. Moreover, since $S_z' = 0$ also for particle 1 the conclusion that, at time t_0, $s_{1z}' = +1/2$ applies. Therefore, before the measurement took place, the state on which we have made the measurement was

$$|A\rangle = |s_{1z}' = +1/2, s_{2z}' = -1/2\rangle$$

and the measurement made at time t_1 is, as in classical physics, a mere 'verification'. If instead the measurement of s_{1z} at time t_1 provides the result $-1/2$, then (always thanks to the reality criterion and the locality principle) we conclude that this particular system was, already at time t_0, in the state

$$|B\rangle = |s_{1z}' = -1/2, s_{2z}' = +1/2\rangle.$$

So we arrive at a contradiction: the system (before the measurement) is not in a singlet spin state, but in a statistical mixture of the states $|A\rangle$ and $|B\rangle$. Furthermore: angular momentum is not conserved: both the state $|A\rangle$ and the state $|B\rangle$ are superpositions of triplet and singlet spin states, therefore a measurement of the total spin of the two particles may also yield $S = 1$ as a result.

In conclusion: reality criterion, locality principle and (completeness of) quantum mechanics cannot coexist.

18.6 Bell's Theorem

The last statement of the previous section strongly suggests the following way out: hopefully it might be possible to complete quantum mechanics in such a way that all of its predictions (which are of statistical nature) are correct and, at the same time, locality is preserved.

By "completing quantum mechanics" we mean what was the firm belief of Einstein: the state of a system is not fully described by the wavefunction, but additional variables (the so called 'hidden variables' already mentioned in Sect. 3.6) are necessary to make the theory deterministic. Thus, for instance, when photons all in a given polarization state $e_{\vartheta\varphi}$ are sent on a birefringent crystal, some of them will emerge in the extraordinary ray and the other in the ordinary ray, depending on the value of the hidden variable ε that can be different for the various photons. Therefore there must exist a function $A(\varepsilon)$ which gives the result of every single measurement: for instance $A = +1$ in correspondence of those values of ε which determine that the photon will emerge in the extraordinary ray, $A = 0$ in the other cases.

The value of the variable ε pertaining to any single photon is presently unknown but, in the 'final' theory of the type envisaged by Einstein, the hidden variable would have a dynamical significance and laws of motion. Presently, in the "not yet complete quantum mechanics", in the birefringent crystal experiments with a large number of photons, there will be a distribution of the values ε with (unknown) density $\rho(\varepsilon)$: if $\rho(\varepsilon)$ is normalized to 1, the mean value of the results will be

$$\int \rho(\varepsilon)A(\varepsilon)\,d\varepsilon$$

and $\rho(\varepsilon)$ should be such that the (statistical) result given by quantum mechanics ($\cos^2\vartheta$, i.e. Malus' law) is reproduced. In the present case, as well as in other 'simple' cases, it is possible to find a density-function ρ with the required property (J.Bell 1964) (and this would account for the achievements of quantum mechanics). The same statement is not necessarily true in situations, as the one of the Bohm–Aharonov version of the EPR paradox, where there are correlations between distant particles, and where the problem of locality shows up.

In the hidden variables framework, locality means that the results of measurements on system (particle) 1 are given by a function $A_1(\varepsilon_1)$ and those on the (distant) system 2 by $A_2(\varepsilon_2)$, with the function A_1 depending *only* on the state of system 1 and on the measured observable, but *not* on those (state and observable) relative to system 2, and viceversa. So, for instance,

$$\int \rho(\varepsilon_1, \varepsilon_2)A_1(\varepsilon_1)A_2(\varepsilon_2)\,d\varepsilon_1 d\varepsilon_2$$

for a suitable density function $\rho(\varepsilon_1, \varepsilon_2)$ should reproduce the mean value of the product of the results of the two observables measured on the two systems, as given by quantum mechanics.

It is at this point that ***Bell's theorem*** intervenes:

No physical theory of local hidden variables can ever reproduce all of the predictions of quantum mechanics.

The proof, although quite simple, will not be given here. With reference to the above discussion, the relevant point in Bell's theorem is that *not always* it is possible to find a density function ρ that reproduces the result of quantum mechanics, if locality is required.

At this point it is clear that the hope of completing quantum mechanics by means of *local* additional variables in such a way that *all* its predictions are reproduced, is ruled out.

Bell's theorem provides some inequalities involving correlation functions, called **Bell inequalities**, to be satisfied by any local theory and that are violated by quantum mechanics: therefore the great merit of Bell's theorem is that now the question whether quantum mechanics or a deterministic local theory is the correct one, has been brought back to the experimental domain.

However, as we have already mentioned, the conflict between quantum mechanics and the local hidden-variables approach shows up in situations, of the kind of the Bohm–Aharonov version of the EPR paradox, that are very delicate from the experimental point of view. For this reason many experiments have been performed, at first with contradictory results, until those by A. Aspect et al. in the years 1980–82. In such experiments pairs of photons in an entangled polarization state, rather than spin 1/2 particles (as in the Bohm–Aharonov version of EPR paradox), are measured in order to test the so called Clauser–Horn–Shimony–Holt inequalities, a generalization of the Bell inequalities. These experiments, held to date as the more reliable ones by the scientific community, show – with 'almost absolute' certainty – the validity of quantum mechanics as opposed to any theory of local (hidden) variables.

Clearly, it is still open the possibility of completing quantum mechanics in a deterministic way by means of non-local (hidden) variables; in any event, both with quantum mechanics in its present form and in its potential completion, locality has to be given up: we report here, verbatim, the conclusions of Bell's paper

In a theory in which parameters are added to quantum mechanics to determine the results of individual measurements, without changing the statistical predictions, there must be a mechanism whereby the setting of one measuring device can influence the reading of another instrument, however remote. Moreover, the signal involved must propagate instantaneously, so that a theory could not be Lorentz invariant.

Needless to say, this situation has attracted the interest of the philosophers: we do not delve into this discussion, however, just for its provocative content, we quote the J. Bell proposal (reported by P. Davies): the universe is superdeterministic. This means that (also) our thoughts, actions, decisions are predetermined, and this applies also to which experiments will be performed, and to their results. Hence:

"there is no need for a faster-than-light signal to tell particle A what measurement has been carried out on particle B, because the universe, including particle A, already knows what that measurement, and its outcome, will be."

Appendix

Physical Constants

Electronvolt	eV	1.6×10^{-12} erg
Speed of light	c	3×10^{10} cm/s
Elementary charge	e	4.8×10^{-10} esu $= 1.6 \times 10^{-19}$ C
Electron mass	m_e	0.91×10^{-27} g $= 0.51\,\mathrm{MeV}/c^2$
Hydrogen mass	m_H	1.7×10^{-24} g $= 939\,\mathrm{MeV}/c^2$
Planck constant	h	6.6×10^{-27} erg s $= 4.1 \times 10^{-15}$ eV s
Reduced Planck constant	$\hbar = \dfrac{h}{2\pi}$	1.05×10^{-27} erg s $= 0.66 \times 10^{-15}$ eV s
Boltzmann constant	k_B	$1.38 \times 10^{-16}\,\mathrm{erg/K} \simeq \dfrac{1}{12000}\,\mathrm{eV/K}$
Avogadro constant	N_A	$6.03 \times 10^{23}\,\mathrm{mol}^{-1}$
Fine structure constant	$\alpha = \dfrac{e^2}{\hbar c}$	$7.297 \times 10^{-3} \simeq \dfrac{1}{137}$
Bohr radius	$a_B = \dfrac{\hbar^2}{m_e e^2}$	$0.53\,\text{Å} = 0.53 \times 10^{-8}$ cm
Bohr magneton	$\mu_B = \dfrac{e\hbar}{2m_e c}$	$0.93 \times 10^{-20}\,\mathrm{erg/G} = 5.8 \times 10^{-9}\,\mathrm{eV/G}$
Rydberg constant	$R_\infty = \dfrac{e^2}{2a_B hc}$	$109737\,\mathrm{cm}^{-1}$
Compton wavelength	$\lambda_c = \dfrac{h}{m_e c}$	$0.024\,\text{Å}$
Classical electron radius	$r_e = \dfrac{e^2}{m_e c^2}$	2.8×10^{-13} cm
Atomic unit of energy	$\dfrac{e^2}{a_B} = \alpha^2 m_e c^2$	27.2 eV
A useful mnemonic rule	$h\,c$	$12400\,\mathrm{eV\,Å}$

© Springer International Publishing Switzerland 2016
L.E. Picasso, *Lectures in Quantum Mechanics*, UNITEXT for Physics,
DOI 10.1007/978-3-319-22632-3

Appendix

Physical Constants

Index

© Springer International Publishing Switzerland 2016   351
L.E. Picasso, *Lectures in Quantum Mechanics*, UNITEXT for Physics,
DOI 10.1007/978-3-319-22632-3

Printed in the United States
By Bookmasters

Printed in the United States
By Bookmasters